SIMPLE ADAPTIVE STRATEGIES

From Regret-Matching to
Uncoupled Dynamics

World Scientific Series in Economic Theory
(ISSN: 2251-2071)

Series Editor: Eric Maskin *(Harvard University, USA)*

Published

Forthcoming

World Scientific Series in Economic Theory – Vol. 4

SIMPLE ADAPTIVE STRATEGIES

From Regret-Matching to Uncoupled Dynamics

Sergiu Hart

The Hebrew University of Jerusalem, Israel

Andreu Mas-Colell

Universitat Pompeu Fabra, Barcelona

with the collaboration of

Yakov Babichenko
Amotz Cahn
Yishay Mansour
David Schmeidler

 **World Scientific**

NEW JERSEY · LONDON · SINGAPORE · BEIJING · SHANGHAI · HONG KONG · TAIPEI · CHENNAI

Published by

World Scientific Publishing Co. Pte. Ltd.

5 Toh Tuck Link, Singapore 596224

USA office: 27 Warren Street, Suite 401-402, Hackensack, NJ 07601

UK office: 57 Shelton Street, Covent Garden, London WC2H 9HE

Library of Congress Cataloging-in-Publication Data

Hart, Sergiu.
  Simple adaptive strategies : from regret-matching to uncoupled dynamics / by Sergiu Hart
(The Hebrew University of Jerusalem, Israel) & Andreu Mas-Colell (Universitat Pompeu Fabra,
Spain).
     p. cm -- (World Scientific series in economic theory, ISSN 2251-2071 ; v. 4)
     Includes bibliographical references and index.
     ISBN 978-9814390699
     1. Games of strategy (Mathematics) 2. Heuristic algorithms. I. Mas-Colell, Andreu. II. Title.
  QA270.H37 2012
  519.3--dc23

                                                                    2012023203

British Library Cataloguing-in-Publication Data
A catalogue record for this book is available from the British Library.

In-house Editor: Alisha Nguyen

Typeset by Stallion Press
Email: enquiries@stallionpress.com

Printed in Singapore by World Scientific Printers.

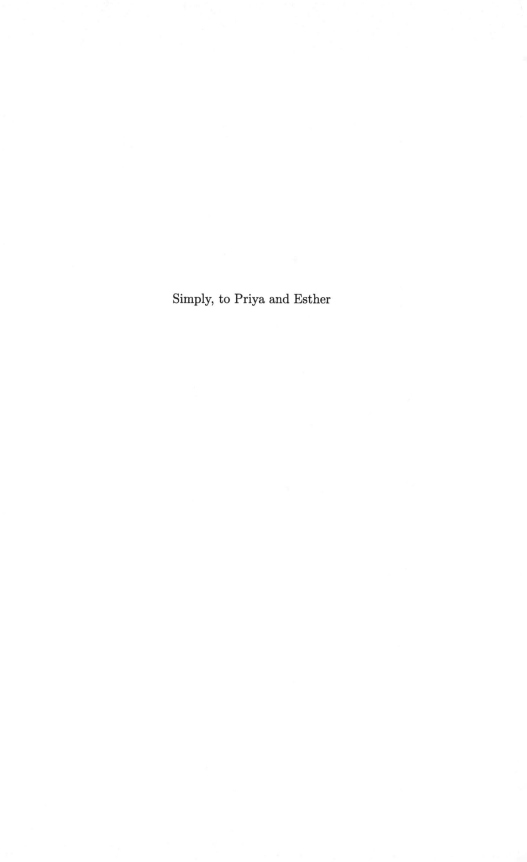

Simply, to Priya and Esther

FOREWORD

Nash equilibrium—a configuration of strategies from which no player has the incentive to deviate unilaterally—is the central solution concept in non-cooperative game theory. That is, in both the theoretical and applied literatures, it constitutes the leading prediction of how players make their strategic choices. Left unstated in its definition, however, is how a Nash equilibrium might actually come about.

One possibility is that equilibrium arises through introspection: a player expects others to play their Nash equilibrium strategies, and so he finds it payoff-maximizing to do the same. But this still doesn't explain why he would anticipate that other players will use equilibrium strategies, i.e., why he would have "equilibrium expectations." As the literature on rationalizability makes clear, even common knowledge of players' rationality and uniqueness of equilibrium do not guarantee such expectations.

However, it is conceivable that through repeated experience of the same game, a player might *learn* to expect others to play equilibrium strategies. This is the issue that Sergiu Hart and Andreu Mas-Colell have been exploring in their important work on adaptive dynamics. Interestingly, they conclude that learning—at least, learning of a reasonably simple kind—will *not*, in general, lead to Nash equilibrium expectations (unless players know one another's payoff functions or there is some other way in which players are "coupled"). However, a form of adaptive learning, called *regret matching*—wherein a player is apt to use a strategy that would have done well in previous play—will lead in the long run to *correlated* equilibrium: a Nash equilibrium in which players' randomizations need not be independent of one another.

I am delighted that Sergiu and Andreu have agreed to collect twelve of their most significant papers in this area—together with a new and detailed introduction—in this volume.

Eric Maskin

Editor-in-Chief, *World Scientific Series in Economic Theory*

CONTENTS

Chapter 3.   A General Class of Adaptive Strategies   47
Sergiu Hart and Andreu Mas-Colell

Chapter 4.   A Reinforcement Procedure Leading to Correlated Equilibrium   77
Sergiu Hart and Andreu Mas-Colell

Chapter 5.   Regret-Based Continuous-Time Dynamics   99
Sergiu Hart and Andreu Mas-Colell

Chapter 6.   General Procedures Leading to Correlated Equilibria   125

Amotz Cahn

Chapter 10.   How Long to Equilibrium?
The Communication Complexity of Uncoupled
Equilibrium Procedures   215
Sergiu Hart and Yishay Mansour

Part IV Dynamics and Equilibria   251

INTRODUCTION

Sergiu Hart* and Andreu Mas-Colell[†]

The general framework of this volume is that of *game theory*, with multiple participants ("players") who interact repeatedly over time. The players may be people, corporations, nations, computers—even genes. See the *Handbook of Game Theory* (Aumann and Hart 1994, 2002, 2004; Young and Zamir 2012).

Many of the standard concepts of game theory are *static* by their very nature. This is certainly true of the central concept of strategic equilibrium, both in its classical form of *Nash equilibrium* (Nash 1950) and in its extended form of *correlated equilibrium* (Aumann 1974). Indeed, an equilibrium situation is such that, once the players happen to find themselves in it, no player has an incentive to move away from it. However, these equilibrium concepts say nothing about how such situations are reached, i.e., about their *dynamic* basis.

Yet, it is of utmost importance—theoretically as well as in applications—to study dynamic processes, and relate them to appropriate static solutions. This is so for at least two reasons. On the one hand, it is of interest in its own right to analyze processes that to some degree reflect observed patterns of behavior (we could call them "natural" dynamics). On the other hand, the significance of an equilibrium solution has to depend on how easy it is to reach: the justification of a concept that turns out to be dynamically hard to reach is shaky.

In this volume we study the connections between dynamics and equilibria. Our goal is to characterize interesting classes of natural dynamics

---

*Center for the Study of Rationality, Department of Economics, and Institute of Mathematics, The Hebrew University of Jerusalem, Feldman Building, Givat Ram, 91904 Jerusalem, Israel. *E-mail*: hart@huji.ac.il; *URL*: http://www.ma.huji.ac.il/hart.

[†]Department of Economics and Business, Universitat Pompeu Fabra, and Barcelona Graduate School of Economics, Ramon Trias Fargas 25-27, 08005 Barcelona, Spain. *E-mail*: andreu.mas-colell@upf.edu.

for which convergence to Nash or correlated equilibria can be guaranteed, and classes for which it cannot (i.e., where "impossibility" results hold).

Simple Adaptive Strategies

A *strategy* in a long-term interaction provides instructions on what to do after every possible history of play. It specifies what to do in the first period, how to react in the second period to what happened in the first period, and so on. Many (in fact, most) of these strategies are extremely complex and therefore hardly practical. In this volume our interest is restricted to strategies that are *simple*—and therefore easy to implement—and *adaptive*—and therefore related to real behavioral traits. Note that we are not adding an "optimality" requirement, and so we are placing ourselves in a bounded rationality framework. In this context we are interested in studying what can be reached, and what cannot be reached, by players who use simple adaptive strategies.

Regret Matching

Our research has identified a salient class of strategies with some remarkable properties, the leading element of which is the *regret matching* strategy. On the one hand, these strategies are based on natural measures of regret; roughly speaking, the regret is the possible increase in the payoff that one would have received had one used a different action in the past. These regret measures are used to determine one's propensity to switch to a different action: the greater the regret for an action, the greater the propensity to switch to it. As such, these are very similar to strategies obtained in the behavioral literature. They certainly qualify as instances of simple and adaptive strategies. On the other hand, these strategies, despite their deceptive simplicity, lead in the long run to behavior that is similar to that obtained from fully rational considerations, namely, correlated equilibria (Aumann 1987).

Uncoupled Dynamics

One may well ask whether it is possible to use variants of regret-based strategies to reach Nash equilibria. A long history of failure in the search for dynamic procedures guaranteeing convergence to Nash equilibria should indicate that the answer to this question is bound to be negative. And indeed it is. Yet, the reason for the failure turns out to be quite instructive. It is not due to some aspect of the adaptive property, but rather to one aspect of the simplicity property: namely, that in the strategies under consideration players do not make use of the payoffs of the other players. While the actions of the other players may be observable, their objectives and reasons for playing those actions are not known (and at best can only be inferred). This simple informational restriction—which we call *uncoupledness*—is the key property that makes Nash equilibria very hard if not impossible to reach.

Dynamics and Equilibria

At this point the connections between static equilibrium concepts and dynamics can be summarized as follows: Nash equilibrium is a "dynamically hard" concept, whereas correlated equilibrium is a "dynamically easy" concept. Indeed, regret matching and the general class of regret-based strategies—which are simple, adaptive, and natural—lead to correlated equilibria, whereas any strategies that are uncoupled cannot reasonably lead to Nash equilibria in all games.

We now provide a more detailed overview of this volume.

PART I: Correlated Equilibria

A Nash equilibrium is a situation where "each player's strategy is optimal against those of the others" (Nash 1950). In other words, it is a combination of strategies of all players such that no player can gain by unilaterally changing his strategy.

A correlated equilibrium (Aumann 1974) is a Nash equilibrium where players receive payoff-irrelevant information ("signals") before playing the game. But since this information does not affect the game itself, does it matter? Yes, it does: it may well be used by the players when making their strategic choices.

When the signals are (stochastically) independent across the players, the resulting correlated equilibrium is clearly just a standard (mixed-strategy) Nash equilibrium: the signals can only be used as a private randomizing device. But when the signals are correlated, new equilibria emerge. In case the information is public (i.e., the same signal is sent to every player, and therefore it is commonly observed and commonly known), then, after each realization of the signal, the resulting play has to be a Nash equilibrium of the original game (with no signals). Thus, what we get over-all is a probabilistic mixture (i.e., a convex combination) of Nash equilibria of the original game, with weights that are precisely the probabilities of the various possible signals; call this a *publicly correlated equilibrium*. For example, let NE1 and NE2 be two distinct Nash equilibria of the original game, and assume that 60% of the days are sunny and 40% are cloudy, and this is observed by all players before they play the game. Then there is a (publicly) correlated equilibrium consisting of playing NE1 on sunny days and NE2 on rainy days, which means that NE1 is played with probability 60% and NE2 with probability 40% (in general, this behavior cannot be a Nash equilibrium of the original game, since a Nash equilibrium requires the players' choices to be independent).

Even more interesting correlated equilibria can occur when the signals are neither independent nor fully correlated. New phenomena emerge then. For example, one may get correlated equilibria whose payoffs Pareto domi-nate the payoffs of all Nash equilibria of the original no-signals game (e.g., in the so-called Chicken game; see Section 6 in Chapter 11).

Existence of Correlated Equilibria

Consider now a finite game; that is, there are finitely many players, each having finitely many strategies. Does a correlated equilibrium necessarily exist? The answer is clearly "yes," since Nash equilibria exist (Nash 1950), and every Nash equilibrium is also a correlated equilibrium.

While this provides a correct mathematical proof for the existence of correlated equilibria, it has a drawback. The set of correlated equilibria is characterized by a set of linear inequalities (the "incentive compatibility" constraints of the players), and is therefore a convex polytope in a finite-dimensional Euclidean space. To prove nonemptiness of such a set using fixed-point theorems (the use of a fixed-point theorem, such as that of Brouwer or Kakutani, is needed to prove the existence of Nash equilibria) looks in such a context like "overkill." One would like to prove this result more elementarily, specifically, in a framework of linear algebra and linear inequalities.

Such an elementary proof is provided by Sergiu Hart and David Schmeidler, "Existence of Correlated Equilibria," *Mathematics of Operations Research* 14 (1989), 18–25, which is **Chapter 1** of this volume.[1] It rests on an appeal to the minimax theorem (which is essentially equivalent

---

[1] Another elementary proof is provided by Nau and McCardle (1990).

to the duality theorem of linear programming). To wit: given a finite n-person game, one constructs an auxiliary two-person zero-sum game, with the auxiliary row player, call it ROW, choosing an n-tuple of strategies in the original game, and the auxiliary column player, call it COL, choosing one of the incentive constraints for a correlated equilibrium. It is seen that an optimal (mixed) strategy for ROW is then precisely a correlated equilibrium in the original game; its existence then follows by applying (twice) the minimax theorem of two-person zero-sum games.

While providing the "right" proof is often a worthwhile formal exercise, in this case it had the added advantage that it eventually led to the research included in this volume. Our basic, first idea was to use the above auxiliary game and apply to it the simple dynamic of "fictitious play" (Brown 1951)—which, for two-person zero-sum games, converges and leads to optimal strategies (Robinson 1951)—in order to obtain a dynamic that leads to correlated equilibria of the original n-person game (which, as seen above, are precisely the optimal strategies of ROW). However, the resulting dynamic was unwieldy, as it required all the n players to move in a coordinated way. It took many modifications and transformations until we succeeded in "uncoupling" the players (cf. Part III) and getting the simple regret matching strategies (see Part II).

PART II: Regret Matching

We start by considering our basic simple adaptive strategy, "regret matching," introduced by Sergiu Hart and Andreu Mas-Colell, "A Simple Adaptive Procedure Leading to Correlated Equilibrium," *Econometrica* 68 (2000), 1127–1150, which is **Chapter 2** of this volume. Consider a game that is repeatedly played over time; regret matching is defined by the following rule:

> *Regret Matching*: Switch next period to a different action with a probability that is proportional to the regret for that action, where regret is defined as the increase in the payoff had such a change always been made in the past.

That is, consider a player deciding on his action (it is convenient to use the term "action" for the choice in the underlying one-shot game, and "strategy" for the choice in the repeated game) at a certain time period. Let U be the average payoff the player has obtained up to now, and let j be the action that he played in the previous period. For each alternative

action k different from j, let $V(k)$ be the average payoff the player would
have obtained had he played k instead of j every time in the past that
he actually played j. The *regret* $R(k)$ for action k is then defined as the
amount, if any, by which $V(k)$ exceeds the actual payoff U; i.e., $R(k) =$
$V(k) - U$ if $V(k) \geq U$, and $R(k) = 0$ otherwise. Regret matching stipulates
that each action k different from the previous period's action j is played
with a probability that is proportional to its regret $R(k)$, and, with the
remaining probability, the same action j as in the last period is played
again. (Formally, the probability of playing action k equals $cR(k)$ for every k
different from j, and it equals the remaining probability $1 - \sum_{k \neq j} cR(k)$ for
$k = j$; here, c is an appropriate fixed constant; see footnote 5 in Chapter 2.)
We could thus say that the player contemplates, first, whether to continue
to play next period the same action j as in the previous period, or to switch
to a different action k. If the latter, he looks at what would have happened
to his average payoff had he always replaced j by k in the past (since one is
looking at the long-run average payoff, it makes sense to consider replacing
j by k not just in the previous period, but also in all the other periods
in the past when j was played; after all, the effect of one period becomes
negligible as the number of periods increases). The player compares what
he got, U, to what he would have gotten, $V(k)$. If the alternative payoff is
no higher, i.e., if $V(k) \leq U$, then he has no regret for k (the regret $R(k)$ for
k equals 0) and he does not switch to action k. If the alternative payoff is
higher, i.e., if $V(k) > U$, then the regret for k is positive ($R(k)$ equals the
increase $V(k) - U$) and the player switches to action k with a probability
that is proportional to this regret.

The main result of Chapter 2 (Hart and Mas-Colell 2000) is:

Regret Matching Theorem: If each player plays a regret matching strategy, then the joint distribution of play converges to the set
of correlated equilibria of the underlying game.

The term "joint distribution of play" (also known as the "empirical distribution" or "sample distribution") refers to the relative frequency with
which each combination of actions of all players has been played (more
on this below). The Regret Matching Theorem says that, for almost every
history of play, the joint distribution of play converges to the set of correlated equilibria of the underlying (one-shot) game. This means that,
from some time on, the joint distribution is close (i.e., within ε) to a correlated equilibrium, or, equivalently, that it is a correlated approximate

(i.e., an ε-) equilibrium. The convergence here is to the set of correlated equilibria, not necessarily to a point in that set. Observe that it is the empirical distributions that become essentially correlated equilibria—not the actual play. What our result implies is that the long-run statistics of play of "regret-matchers," and that of fully rational players (who play a correlated equilibrium each period; see Aumann 1987), become essentially indistinguishable.

Clearly, regret matching, as well as its generalizations below, has the character of an adaptive strategy; in fact, it embodies commonly used rules of behavior. For instance, if all the regrets are zero ("there is no regret"), then a regret matching player will continue to play the same action of the previous period (as in the common saying, "never change a winning team"). When some regrets are positive, actions may change—with probabilities that are proportional to the regrets: the higher the payoff would have been from switching to another action in the past, the higher the tendency is to switch to that action now. Again, this seems to fit standard behavior (recall ads such as "Had you invested in A rather than B, you would have gained X more by now. So switch to A now!," the sense of urgency being related to the magnitude of X). In the learning, experimental, and behavioral literature there are various models that bear a likeness to regret matching; see Bush and Mosteller (1955), Roth and Erev (1995), Erev and Roth (1998), Camerer and Ho (1998, 1999), Selten, Abbink, and Cox (2005), Kahneman (2011, Ch. 32), and others; probably the closest are the Erev–Roth models. Regret measures also feature in the recent neuroeconomics literature on decision-making; see Camille et al. (2004) and Coricelli, Dolan, and Sirigu (2007). Also, incorporating regret measures into the utility function has been used to provide alternative theories of decision-making under uncertainty; see Bell (1982) and Loomes and Sugden (1982).

Another interesting aspect captured by regret matching has to do with the "sluggishness" of decision-making. It has been observed that people tend to have too much "inertia" in their decisions: they stick to their current action for a disproportionately long time (as in the "status quo bias"; see Samuelson and Zeckhauser 1988 and Moshinsky 2003; see also the subsection "Regret" in Chapter 32 in Kahneman 2011). Regret matching has built-in inertia: the probability of not switching (i.e., of repeating the previous period's action) is always strictly positive, and in fact regret matching generates behavior where the same action is played over and over again for long time intervals.

Regret matching is not just adaptive but also very simple (we could say unsophisticated!). Players neither develop beliefs nor reply optimally.

At each time period there are "propensities" of play, which are adjusted over time in a simple and natural way. The computation that determines these propensities involves at each period a straightforward additive updating of the appropriate regret.

There are other dynamics leading to correlated equilibria: "calibrated learning" (i.e., best-reply to calibrated forecasts: Foster and Vohra 1997), and "conditional smooth fictitious play eigenvector strategy" (Fudenberg and Levine 1999). However, these dynamics require the players to make at each period a complex mathematical computation (specifically: compute an eigenvector), which, in our view, makes them neither simple nor easily interpretable in adaptive behavioral terms.

Now where does the "correlation" in the Regret Matching Theorem come from? The answer is, of course, that it arises from the commonly observed history of play. Indeed, each player's action is determined by his regrets, which are in turn determined by the history. Thus, the exogenous signals in the definition of correlated equilibria are now *endogenously* generated—by the regret matching strategies themselves.[2]

It is a standard hypothesis that the players observe the history of play ("standard monitoring"), which determines the joint distribution of play. Therefore, having players determine their actions based on the joint distribution of play (rather than having each player consider only his own frequencies of play, i.e., the corresponding marginal distribution) does not go beyond the commonly used monitoring framework: it is information that the players possess anyway. In fact—and this is a significant behavioral observation—people react to the joint distribution, not only to the marginals: people are very much aware of coincidences, signals, communications, and so on (even to the point of overdoing it and interpreting random phenomena and spurious correlations as meaningful). It should be emphasized that while at each stage the players randomize independently of one another, this does not imply that the joint distribution of play should be independent across players (i.e., the product of its marginal distributions), nor should it become so in the long run.

To summarize: reasonable models of play can—and should—take into account the joint distribution of play.

---

[2] While the history of play is commonly observed, it does *not* follow that one obtains *publicly* correlated equilibria; the reason is that each player plays according to his own regrets. These regrets are correlated (since they are based on the common history), but in general they are far from being fully correlated.

Generalized Regret-Based Strategies

The regret matching strategy appears to be very specific: the play probabilities are directly proportional to the regrets. It is natural to enquire whether this delicate construct is necessary for the result of the Regret Matching Theorem. What would happen were the probabilities proportional to, say, the squares of the regrets? In another direction, we could also ask for the connections between regret matching and other dynamics leading to correlated equilibria, particularly variants of conditional smooth fictitious play (e.g., Fudenberg and Levine 1999a).

This leads us to consider a large class of adaptive heuristics that are based on regrets. Specifically, instead of the switching probability being proportional to the regret, i.e., equal to $cR(k)$, we now allow this switching probability to be given by a general function $f(R(k))$ of the regret $R(k)$, provided that f is sign-preserving (i.e., $f(x) > 0$ for $x > 0$ and $f(0) = 0$) and regular (which here means Lipschitz continuous). Call the resulting strategies *generalized regret matching* strategies, or *regret-based* strategies.

The following result is based on Sections 3.2 and 5.1 of Sergiu Hart and Andreu Mas-Colell, "A General Class of Adaptive Strategies," *Journal of Economic Theory* 98 (2001), 26–54, which is **Chapter 3** of this volume, and proved as Theorem 4.1 in[3] Amotz Cahn, "General Procedures Leading to Correlated Equilibria," *International Journal of Game Theory* 33 (2004), 21–40, which is **Chapter 6** of this volume:

Generalized Regret Matching Theorem. If each player plays some generalized regret matching strategy then the joint distribution of play converges to the set of correlated equilibria of the underlying game.

In fact, the full class of generalized regret matching strategies (for which the above theorem holds) is even larger; see Section 3.2 in Chapter 3 and Section 4 in Chapter 4.

As a special case, consider the family of functions $f(x) = cx^r$, where $r \geq 1$ and $c > 0$ is an appropriate constant. At one extreme, when $r = 1$, we have regret matching. At the other extreme, the limit as r goes to infinity results in the probability of switching being equally divided among those actions $k \neq j$ with maximal regret (i.e., those k with $R(k) = \max_{\ell \neq j} R(\ell)$).

---

[3]Based on the master's thesis of Amotz Cahn, written under the supervision of Sergiu Hart.

This yields a variant of fictitious play, which, however, no longer belongs to the admissible class of generalized regret strategies (it is not continuous), and, indeed, the result of the Generalized Regret Matching Theorem above does not hold for it (see Section 4 in Chapter 3). Of course, the result does hold for any finite $r \geq 1$, which for very large r leads to "smooth conditional fictitious play"; see Section 4.5 in Chapter 6.

The case of the unknown game ("complete uncoupledness")

Consider now the apparently hopeless situation where each player knows initially only his own set of actions, and is informed, after each period of play, of his realized payoff. He does not know what game he is playing, that is, how many players there are and what their actions and payoffs are. Moreover, he does not even know his own payoff function—but only the payoffs he did actually receive every period. This is essentially a standard stimulus-response setup, which we called "the case of the unknown game" (also known in the literature as "payoff-based" and "radically uncoupled"; the current terminology refers to this case as "complete uncoupledness"— see Babichenko 2011).

While at each period the player knows his realized average payoff U, he cannot know his alternative payoffs $V(k)$ and his regrets $R(k)$: he knows neither what the other players did, nor what his payoff would have been had he played an alternative action k instead. Yet, we can still define a *proxy regret* measure, by using the payoffs he got when he did actually play k. In Sergiu Hart and Andreu Mas-Colell, "A Reinforcement Procedure Leading to Correlated Equilibrium," in *Economic Essays*, edited by Gerard Debreu, Wilhelm Neuefeind, and Walter Trockel, Springer (2001), 181–200, which is **Chapter 4** of this volume, we show exactly how to carry this out (it involves certain adjustments and also a small degree of experimentation). The surprising result is that convergence to correlated approximate equilibria is obtained also for proxy-regret matching strategies.

Unconditional Regrets and Hannan Consistency

The regret for action k has been defined relative to the action j of the previous period. Consider instead a rougher measure, namely, the increase in the average payoff, if any, were one to replace *all* past plays, and not just the j-plays, by k. This yields the *unconditional regret* for action k, denoted $R^{\mathrm{u}}(k)$ (see Section 4 (c) in Chapter 2; these regrets are also known as "external regret," with "internal regrets" for the original regrets). The resulting *unconditional-regret matching* strategy prescribes play probabilities at each period that are directly proportional to the vector of unconditional regrets;

i.e., the probability of playing k is $R^{\mathrm{U}}(k)/\sum_\ell R^{\mathrm{U}}(\ell)$, the unconditional regret of k divided by the sum of the unconditional regrets for all actions ℓ (unlike regret matching, here we do not use a constant proportionality factor c, but simply normalize the vector of unconditional regrets to get a probability vector).

A strategy of a player is said to be *Hannan-consistent* if it guarantees, for any strategies of the other players, that all the unconditional regrets vanish in the limit with probability one. We have:

Unconditional Regret Theorem. Unconditional regret matching is Hannan-consistent. Moreover, if all players play unconditional regret matching, then the joint distribution of play converges to the Hannan set of the stage game.

This is Theorem B in Chapter 2 (Hart and Mas-Colell 2000). It is proved via Blackwell's Approachability Theorem (Blackwell 1956a, 1956b). The *Hannan set*[4] (see Hannan 1957, Moulin and Vial 1978, Hart and Mas-Colell 2003a [Chapter 5]) is defined as the set of joint distributions of play where no player can gain unilaterally by playing the same constant action at all periods, irrespective of any signal he gets. Note the contrast with correlated equilibrium, where the no-gain test would be for changes conditional on the signal. The set of correlated equilibria is contained in the Hannan set (and the two sets coincide when every player has at most two strategies); moreover, the Hannan distributions that are independent across players are precisely the Nash equilibria of the game.

Hannan-consistent strategies have been constructed by, among others, Hannan (1957), Blackwell (1956b), Foster and Vohra (1993, 1998), Fudenberg and Levine (1995), and Freund and Schapire (1999) (many of these strategies are smoothed-out variants of fictitious play, which, by itself, is not Hannan-consistent). In comparison, our unconditional regret matching appears to be the simplest.

Continuous Time

The regret-based dynamics up to this point have been discrete-time dynamics: the time periods were $t = 1, 2, \ldots$. It is natural to study also continuous-time models, where the time t is a continuous variable, and the changes

---

[4]Also known as the set of "coarse correlated equilibria." The Hannan set refers to the one-shot game (it is a set of joint distributions), whereas Hannan-consistency refers to the repeated game (it is a long-run property of a strategy).

in the players' actions are governed by appropriate differential equations. In Sergiu Hart and Andreu Mas-Colell, "Regret-Based Continuous-Time Dynamics," *Games and Economic Behavior* 45 (2003), which is **Chapter 5** of this volume, we show that the results carry over to this framework. In fact, some of the proofs become simpler.

Summary and Applications

For an extensive survey, with precise pointers to the relevant papers, of the results on regret matching and general regret-based strategies, the reader may consult Chapter 11 in Part IV.

Interestingly, regret matching is also successfully considered nowadays in various areas of practical application. The fact that game-theoretic adaptive procedures, such as the various regret matching strategies in this volume, are both simple and decentralized (i.e., they can be carried out "locally" without the need to communicate with a "central authority") makes them appealing and useful in many setups where one needs to make efficient use of limited resources. Some examples: "cognitive radio," which refers to wireless transceivers that change their parameters dynamically in response to their environment (Maskery, Krishnamurthy, and Zhao 2009; Wang, Wu, and Liu 2010); traffic, congestion, and Voronoi diagrams (Arslan, Marden, and Shamma 2007; Kalam, Gani, and Seneviratne 2008); sensor networks (Krishnamurthy, Maskery, and Yin 2008); neural networks (Marchiori and Warglien 2008); statistical analysis of large datasets in medical diagnosis (Gambin et al. 2009). In many of these cases regret matching (the extremely simple unconditional regret matching, or the somewhat more sophisticated conditional regret matching) turn out to yield quite efficient results. It would be interesting to understand what exactly lies behind this apparent efficiency, as it does not follow from the general results presented in this volume (while there are correlated equilibria that are "better" than, say, Nash equilibria, in general there are also "worse" correlated equilibria).

PART III: Uncoupled Dynamics

As explained so far, regret matching and its generalizations fit the equilibrium concept of correlated equilibrium. What about Nash equilibrium? Nash equilibria belong to the set of correlated equilibria but, unless this latter set happens to be a singleton, or there is a pure strict Nash equilibrium, regret matching will not generate dynamics converging to Nash equilibrium. This, we hasten to add, came as no surprise, since dynamic procedures ensuring convergence to Nash equilibria—such as fictitious play—have only been obtained for quite restricted classes of games (such as two-person zero-sum games, potential games, 2×2 games). Intuitively speaking, regret matching seemed much too simple and much too adaptive to hope for a convergence result to Nash equilibria for general classes of games. But can this statement be made precise? This is the question that motivated the line of work of Part III, which we summarize now.

All the dynamic processes we have considered until now share the following characteristic, which is part of the simplicity, rather than the adaptiveness, requirement: what a player does at any moment of time does not depend on the payoff or the utility functions of the other players. It is this property, which we called *uncoupledness*, that we came to view as key and focused on in our research. It is related to the notions of "privacy-preserving" in mechanism design, "decentralized" procedures in economics, and "distributed" computations in computer science.

Uncoupledness for Deterministic Dynamics

In Sergiu Hart and Andreu Mas-Colell, "Uncoupled Dynamics Do Not Lead to Nash Equilibrium," *American Economic Review* 93 (2003), 1830–1836, which is **Chapter 7** of this volume, we defined the uncoupledness notion and established its relevance to the issue of determining Nash convergent mechanisms by means of a simple model of deterministic dynamics in continuous time. Essentially, we exhibited a class of games where any such dynamics (expressed by means of differential equations) on the state space of (continuous) action combinations, which is uncoupled, will necessarily fail to converge to (the unique) Nash equilibrium for some games.

Uncoupledness for Stochastic Dynamics

The above could not be the end of the story, however. One needed to consider the standard framework of discrete time and stochastic dynamics (which is also the context in which regret matching was originally formulated). Now, it is not difficult to see that deterministic and stochastic variants of exhaustive search ("keep looking until you find a Nash equilibrium") can be embedded into an uncoupled framework. However, exhaustive search is hardly an attractive adjustment mechanism. Thus the question was which additional considerations are important and natural (and in particular rule out exhaustive search and recover the impossibility result). These turned out to be, on the one hand, "finite recall," and on the other hand, the "speed of convergence."

An additional stimulus for our research at this point was the work of Foster and Young (2003, 2006), Germano and Lugosi (2007), Kakade and

Foster (2004), Young (2004, 2009). They formulated stochastic, adaptive mechanisms (that include some form of experimentation) that yield trajectories that are most of the time close to Nash equilibria. It was therefore important to clarify the relationships between these two lines of research and to find the demarcation line, so to speak, between possibility and impossibility.

In Sergiu Hart and Andreu Mas-Colell, "Stochastic Uncoupled Dynamics and Nash Equilibrium," *Games and Economic Behavior* 57 (2006), 286–303, which is **Chapter 8** of this volume, we explore the implications of recall restrictions—i.e., restrictions on how many past periods of play players remember—in a stochastic dynamic context. These restrictions certainly matter. We show, for example, that for the case where a pure strategy Nash equilibrium exists, convergence cannot be assured in the case of one-period recall. Thus, not only "best-replying to the last period" cannot lead to Nash equilibrium in general, but no uncoupled dynamic where players base their decisions only on the previous period's play can do so. But, if one increases the recall to two or more periods, then convergence to pure Nash equilibria can be guaranteed (by strategies reminiscent of exhaustive search). For the general case where all the Nash equilibria may be mixed, a more refined analysis is required, yet, again, if the recall is short, convergence is not assured, while if it sufficiently long (but still finite), then it is. However, no finite length of recall turns out to be sufficient to obtain the convergence to Nash equilibria of the period-by-period behavior probabilities; this can be obtained only within the broader context of finite memory (which leads us to the next chapter).

Uncoupledness for Finite Memory Dynamics

Recall limitations are just one way to capture the idea that the past can influence the future only through a finite number of parameters. More generally, one could appeal to the notion of finite memory, or, equivalently, strategies that can be implemented by finite automata. In the present context, this is explored in[5] Yakov Babichenko, "Uncoupled Automata and Pure Nash Equilibria," *International Journal of Game Theory* 39 (2010), 483–502, which is **Chapter 9** of this volume. The finite recall results turn out to generalize nicely: for example, to reach a pure Nash equilibrium

---

[5]Based on the master's thesis of Yakov Babichenko, written under the supervision of Sergiu Hart.

(assuming it exists), the number of states must be strictly larger than the number of actions.

How Long to Equilibrium?

Next, consider the important issue of the time it takes to reach Nash equilibria: can one estimate, or bound, the number of periods until an (approximate) equilibrium is reached? If that number of periods turns out to be extremely large, then there may be little use for such dynamics.

This question is addressed in Sergiu Hart and Yishay Mansour, "How Long to Equilibrium? The Communication Complexity of Uncoupled Equilibrium Procedures," *Games and Economic Behavior* 69 (2010), 107–126, which is **Chapter 10** of this volume.

The way one proceeds is by using a tool from theoretical computer science: communication complexity. It turns out that an uncoupled dynamic reaching an equilibrium is nothing but a so-called "distributed computational procedure." Informally, a distributed computational procedure consists of a number of agents, each one initially possessing some private information (the "inputs"), which through communication reach a situation where they all agree on a certain result (the "output"). The communication complexity is the minimal number of communication rounds that is needed to go from the private inputs to the common output.

An uncoupled dynamic reaching an (approximate) equilibrium does indeed fit this framework. The private inputs are the payoff functions (each player knows only his own), the communication phase consists of playing the game repeatedly, and the end result is the (approximate) equilibrium reached. The communication complexity of the uncoupled dynamics gives the minimum number of periods needed to reach the (approximate) equilibrium (this connection was made by Conitzer and Sandholm 2004 for two-person games; for communication complexity in general, see Yao 1979 and Kushilevitz and Nisan 1997).

In Chapter 10 it is shown that the number of periods needed to reach a Nash equilibrium—pure or mixed—can be exponential in the number of players n; when n is large, this becomes quickly unreasonably long. Since this is the time exhaustive search takes, the conclusion is that there are games where any uncoupled dynamic will take essentially as long as exhaustive search does to reach Nash equilibria. We emphasize that these exponential lower bounds apply to *any* uncoupled dynamic; additional requirements on the dynamics, such as finite recall and memory, or various incentives and rationality desiderata, can only increase the number of periods required.

At this point one may wonder whether we are perhaps asking for too much. After all, the description of the game is exponential in the number of players n; even a single player's payoff function is so (since the number of action combinations, for each one of which we need to specify a payoff, is exponential in n). It may thus seem unreasonable to expect a dynamic process to lead to equilibrium in a number of periods that is significantly shorter than the time it takes to describe the game. Yet, this is what happens for correlated equilibria: regret-based strategies do reach correlated (approximate) equilibria in a time that is polynomial in the number of players n (and one can also reach exact correlated equilibria in polynomial time; see Theorems 17 and 30 in Chapter 10). Perhaps surprisingly, this implies, in particular, that each player ends up looking only at a very small proportion of the entries in his payoff matrix. The technically savvy will note that, since correlated (approximate) equilibria are given by a linear (in n) number of inequalities, there always exist solutions whose support is polynomial in n; particular instances are the correlated approximate equilibria obtained in polynomial time by the regret-based procedures.

Indeed, impossibility results of the kind discussed here in Part III show why various past attempts to prove that certain dynamics always converge to equilibria were doomed to fail.

PART IV: Dynamics and Equilibria

Part IV ties together the various results of the previous chapters.

Adaptive Heuristics

The paper of Sergiu Hart, "Adaptive Heuristics," *Econometrica* 73 (2005), 1401–1430, which is **Chapter 11** in this volume, starts with a rough but hopefully useful classification of various dynamics into three classes: *learning dynamics, adaptive heuristics*, and *evolutionary dynamics*. Learning dynamics refers to situations where players start with certain Bayesian beliefs (on the game being played and on their opponents' strategies), which they update as the play progresses (e.g., see Kalai and Lehrer 1993 and the ensuing literature); evolutionary dynamics refers to situations where forces of selection and mutation change the populations' composition; and adaptive heuristics refers to simple rules of behavior that make a player move, roughly speaking, in seemingly payoff-improving directions. One can understand the distinctions in terms of the degree of rationality and cognitive optimization of the participants: high for learning dynamics, low for evolutionary dynamics, and in between for adaptive heuristics (see Section 2 in Chapter 11 for further details and references).

Regret matching and the generalized regret dynamics of Part II are simple rules of behavior, based on natural regret measures, and as such clearly qualify as adaptive heuristics. On the one hand, they serve by their very nature as a sort of bridge between rational and behavioral viewpoints. On the other hand, they establish a solid connection to correlated equilibria (rather than Nash equilibria). Thus simple adaptive behavior can lead in the long run to outcomes that embody full rationality (i.e, correlated equilibria; recall Aumann 1987).

Dynamics and Equilibria

More than sixty years after John Nash's Ph.D. thesis (1950), where the notion of strategic equilibrium—now known as "Nash equilibrium"—was introduced, we have learned that *there are no general, natural dynamics leading to Nash equilibrium*. This statement is proposed in the commentary of Sergiu Hart, "Nash Equilibrium and Dynamics," *Games and Economic Behavior* 71 (2011), 6–8, which is **Chapter 12** of this volume. "General" refers to dynamics that operate in all games, rather than only in some specific class of games (such as two-person zero-sum games, or two-person potential games, where such dynamics do exist). "Leading to Nash equilibrium" means that at some time the dynamic reaches a Nash equilibrium (or a neighborhood of a Nash equilibrium) and stays there from then on; therefore we do not include here the dynamics that spend most of the time (formally: $1 - \varepsilon$ of the time, for small $\varepsilon > 0$) near Nash equilibria, but never remain there. Finally, we take "natural" to mean simple and adaptive.

The papers in this volume, particularly in Part III, show that the lack of dynamics to which we pointed above is not a deficiency of the existing literature, but rather a result of the inherent difficulty and even impossibility of reaching Nash equilibria. Uncoupledness—which is nothing but a simplicity property concerning the amount of information players possess—severely restricts the possibility of dynamics to lead to Nash equilibria. In Chapter 8 it is shown that, with limited (small) recall, it is impossible for such dynamics to always reach Nash equilibria, and in Chapter 10 we see that uncoupledness by itself, without any further assumptions of simplicity or otherwise, already makes, in some cases, the number of periods needed to reach Nash equilibria unreasonably large.

In short, the evidence points to Nash equilibria as being a "dynamically hard" concept, whereas correlated equilibria are, in contrast, "dynamically easy."[6]

Directions of Research

The general program to which the research presented in this volume belongs may be viewed as a two-pronged approach. On the one hand, one tries to demarcate the border between those classes of dynamics where convergence to a certain equilibrium concept can be obtained and those where it cannot (i.e., where an "impossibility" result holds). On the other hand, one looks for natural and interesting dynamics—dynamics that are related to actual behavior and yield useful insights. While significant advances have been made in both approaches (such as the research presented in this volume, and the line of work initiated by Foster and Young and discussed in connection to Chapter 8 above), and these have increased our understanding of the connections between dynamics and equilibria, the general picture is still far from complete.

In particular, one needs to study the convergence and trajectories properties of the various dynamics, investigate various classes of dynamics, sharpen further the distinctions between the equilibrium concepts on dynamical grounds, and use all these insights in interesting applications.

We hope that bringing these papers together into one volume will facilitate further pursuit of this fascinating research.

Acknowledgments

The first author acknowledges partial financial support provided by a European Research Council Advanced Investigator grant.

We thank Alisha Nguyen at World Scientific Publishing for her efficient handling of this volume all along. We are most grateful to Mike Borns at the Center for the Study of Rationality in Jerusalem, who undertook the arduous task of editing the whole book; without his work and effort this project would not have been completed.

Last but not least our research over the years has benefited from the inspiration and support of many people. The reader can find the acknowledgments in each chapter. We wish to convey to all of them our heartfelt thanks.

---

[6]This is certainly related to—though probably not fully explained by—Nash equilibria being fixed points of nonlinear maps, whereas correlated equilibria are solutions of linear inequalities.

Technical Note

The papers in this volume have been reset. The numeration of theorems, lemmas, footnotes, etc. used in the originally published papers has been kept. Minor errors and misprints have been corrected, and bibliographic references brought up to date (i.e., we cite here the final published version). For convenience, to each reference to one of the papers that appear in this volume we have appended the chapter number in square brackets.

Jerusalem and Barcelona, 2012

References

Arslan, G., J. R. Marden, and J. S. Shamma (2007), "Autonomous Vehicle-Target Assignment: A Game Theoretical Formulation," *ASME Journal of Dynamic Systems, Measurement and Control* (special issue on "Analysis and Control of Multi-Agent Dynamic Systems"), 129, 84–596.

Aumann, R. J. (1974), "Subjectivity and Correlation in Randomized Strategies," *Journal of Mathematical Economics*, 1, 67–96.

———(1987), "Correlated Equilibrium as an Expression of Bayesian Rationality," *Econometrica*, 55, 1–18.

Aumann, R. J. and S. Hart (eds.) (1994, 2002, 2004), *Handbook of Game Theory, with Economic Applications*, Vols. I–III. Amsterdam: Elsevier.

Babichenko, Y. (2010), "Uncoupled Automata and Pure Nash Equilibria," *International Journal of Game Theory*, 39, 483–502. [Chapter 9]

Bell, D. E. (1982), "Regret in Decision Making under Uncertainty," *Operations Research*, 30, 961–981.

Blackwell, D. (1956a), "An Analog of the Minmax Theorem for Vector Payoffs," *Pacific Journal of Mathematics*, 6, 1–8.

———(1956b), "Controlled Random Walks," in *Proceedings of the International Congress of Mathematicians* 1954, Vol. III. Amsterdam: North-Holland, pp. 335–338.

Brown, G. W. (1951), "Iterative Solutions of Games by Fictitious Play," in *Activity Analysis of Production and Allocation*, ed. by T. C. Koopmans. New York: Wiley, pp. 374–376.

Bush, R. R. and F. Mosteller (1955), *Stochastic Models for Learning*. New York: Wiley.

Cahn, A. (2004), "General Procedures Leading to Correlated Equilibria," *International Journal of Game Theory*, 33, 21–40. [Chapter 6]

Camerer, C. and T.-H. Ho (1998), "Experience-Weighted Attraction Learning in Coordination Games: Probability Rules, Heterogeneity, and Time-Variation," *Journal of Mathematical Psychology*, 42, 305–326.

———(1999), "Experience-Weighted Attraction Learning in Normal Form Games," *Econometrica*, 67, 827–874.

Camille, N., G. Coricelli, J. Sallet, P. Pradat-Diehl, J.-R. Duhamel, and A. Sirigu (2004), "The Involvement of the Orbitofrontal Cortex in the Experience of Regret," *Science*, 304, 1167–1170.

Conitzer, V. and T. Sandholm (2004), "Communication Complexity as a Lower Bound for Learning in Games," in *Proceedings of the Twenty-First International Conference on Machine Learning*, ed. by C. E. Brody. New York: ACM, pp. 185–192.

Coricelli, G., R. J. Dolan, and A. Sirigu (2007), "Brain, Emotion and Decision Making: The Paradigmatic Example of Regret," *Trends in Cognitive Sciences*, 11, 258–265.

Erev, I. and A. E. Roth (1998), "Predicting How People Play Games: Reinforcement Learning in Experimental Games with Unique, Mixed Strategy Equilibria," *American Economic Review*, 88, 848–881.

Foster, D. P. and R. V. Vohra (1993), "A Randomization Rule for Selecting Forecasts," *Operations Research*, 41, 704–709.

_____ (1997), "Calibrated Learning and Correlated Equilibrium," *Games and Economic Behavior*, 21, 40–55.

_____ (1998), "Asymptotic Calibration," *Biometrika*, 85, 379–390.

Foster, D. P. and H. P. Young (2003), "Learning, Hypothesis Testing, and Nash Equilibrium," *Games and Economic Behavior*, 45, 73–96.

_____ (2006), "Regret Testing: Learning to Play Nash Equilibrium without Knowing You Have an Opponent," *Theoretical Economics*, 1, 341–367.

Freund, Y. and R. E. Shapire (1999), "Adaptive Game Playing Using Multiplicative Weights," *Games and Economic Behavior*, 29, 79–103.

Fudenberg, D. and D. Levine (1995), "Consistency and Cautious Fictitious Play," *Journal of Economic Dynamics and Control*, 19, 1065–1090.

_____ (1999), "Conditional Universal Consistency," *Games and Economic Behavior*, 29, 104–130.

Gambin, A., E. Szczurek, J. Dutkowski, M. Bakun, and M. Dadlez (2009), "Classification of Peptide Mass Fingerprint Data by Novel No-Regret Boosting Method," *Computers in Biology and Medicine*, 39, 460–473.

Germano, F. and G. Lugosi (2007), "Global Convergence of Foster and Young's Regret Testing," *Games and Economic Behavior*, 60, 135–154.

Hannan, J. (1957), "Approximation to Bayes Risk in Repeated Play," in *Contributions to the Theory of Games*, Vol. III , Annals of Mathematics Studies 39, ed. by M. Dresher, A. W. Tucker, and P. Wolfe. Princeton: Princeton University Press, pp. 97–139.

Hart, S. (2005), "Adaptive Heuristics," *Econometrica*, 73, 1401–1430. [Chapter 11]

_____ (2011), "Nash Equilibrium and Dynamics," *Games and Economic Behavior*, 71, 6–8. [Chapter 12]

Hart, S. and Y. Mansour (2010), "How Long to Equilibrium? The Communication Complexity of Uncoupled Equilibrium Procedures," *Games and Economic Behavior*, 69, 107–126. [Chapter 10]

Hart, S. and A. Mas-Colell (2000), "A Simple Adaptive Procedure Leading to Correlated Equilibrium," *Econometrica*, 68, 1127–1150. [Chapter 2]

———— (2001a), "A General Class of Adaptive Strategies," *Journal of Economic Theory*, 98, 26–54. [Chapter 3]

———— (2001b), "A Reinforcement Procedure Leading to Correlated Equilibrium," in *Economic Essays: A Festschrift for Werner Hildenbrand*, ed. by G. Debreu, W. Neuefeind, and W. Trockel. Berlin: Springer, pp. 181–200. [Chapter 4]

———— (2003a), "Regret-Based Continuous-Time Dynamics," *Games and Economic Behavior*, 45, 375–394. [Chapter 5]

———— (2003b), "Uncoupled Dynamics Do Not Lead to Nash Equilibrium," *American Economic Review*, 93, 1830–1836. [Chapter 7]

———— (2006), "Stochastic Uncoupled Dynamics and Nash Equilibrium," *Games and Economic Behavior*, 57, 286–303. [Chapter 8]

Hart, S. and D. Schmeidler (1989), "Existence of Correlated Equilibria," *Mathematics of Operations Research*, 14, 18–25. [Chapter 1]

Kahneman, D. (2011), *Thinking, Fast and Slow*. New York: Farrar, Straus and Giroux.

Kakade, S. and D. P. Foster (2004), "Deterministic Calibration and Nash Equilibrium," in *Learning Theory*, ed. by J. Shawe-Taylor and Y. Singer. Berlin: Springer, pp. 33–48.

Kalai, E. and E. Lehrer (1993), "Rational Learning Leads to Nash Equilibrium," *Econometrica*, 61, 1019–1045.

Kalam, S., M. Gani, and L. Seneviratne (2008), "Fully Non-Cooperative Optimal Placement of Mobile Vehicles," in *17th IEEE International Conference on Control Applications*, September 3–5, 2008, San Antonio, Texas, USA, pp. 1025–1030.

Krishnamurthy, V., M. Maskery, and G. Yin (2008), "Decentralized Adaptive Filtering Algorithms for Sensor Activation in an Unattended Ground Sensor Network," *IEEE Transactions on Signal Processing*, 56, 6086–6101.

Kushilevitz, E. and N. Nisan (1997), *Communication Complexity*. Cambridge: Cambridge University Press.

Loomes, G. and R. Sugden (1982), "Regret Theory: An Alternative Theory of Rational Choice under Uncertainty," *Economic Journal*, 92, 805–824.

Marchiori, D. and M. Warglien (2008), "Predicting Human Interactive Learning by Regret-Driven Neural Networks," *Science*, 319, 1111–1113.

Maskery, M., V. Krishnamurthy, and Q. Zhao (2009), "Decentralized Dynamic Spectrum Access for Cognitive Radios: Cooperative Design of a Non-Cooperative Game," *IEEE Transactions on Communications*, 57, 459–469.

Moshinsky, A. (2003), "The Status-Quo Bias in Policy Judgements," Ph.D. Thesis, The Hebrew University of Jerusalem (mimeo).

Moulin, H. and J. P. Vial (1978), "Strategically Zero-Sum Games: The Class of Games Whose Completely Mixed Equilibria Cannot Be Improved upon," *International Journal of Game Theory*, 7, 201–221.

Nash, J. F. (1950), "Equilibrium Points in n-Person Games," *Proceedings of the National Academy of Sciences U.S.A.*, 36, 48–49.

Nau, R. and K. F. McCardle (1990), "Coherent Behavior in Noncooperative Games," *Journal of Economic Theory*, 50, 424–444.

Robinson, J. (1951), "An Iterative Method of Solving a Game," *Annals of Mathematics*, 54, 296–301.

Roth, A. E. and I. Erev (1995), "Learning in Extensive-Form Games: Experimental Data and Simple Dynamic Models in the Intermediate Term," *Games and Economic Behavior*, 8, 164–212.

Samuelson, W. and R. Zeckhauser (1988), "Status Quo Bias in Decision Making," *Journal of Risk and Uncertainty*, 1, 7–59.

Selten, R., K. Abbink, and R. Cox (2005), "Learning Direction Theory and the Winner's Curse," *Experimental Economics*, 8, 5–20.

Wang, B., Y. Wu, and K. J. Ray Liu (2010), "Game Theory for Cognitive Radio Networks: An Overview," *Computer Networks*, 54, 2537–2561.

Yao, A. C.-C. (1979), "Some Complexity Questions Related to Distributive Computing," in *Proceedings of the Eleventh ACM Symposium on Theory of Computing*. New York: ACM, pp. 209–213.

Young, H. P. (2004), *Strategic Learning and Its Limits*. Oxford: Oxford University Press.

———— (2009), "Learning by Trial and Error," *Games and Economic Behavior*, 65, 626–643.

Young, H. P. and S. Zamir (eds.) (2012), *Handbook of Game Theory*, Vol. IV. Amsterdam: Elsevier.

Part I

Correlated Equilibria

Chapter 1

EXISTENCE OF CORRELATED EQUILIBRIA*

Sergiu Hart and David Schmeidler

An elementary proof, based on linear duality, is provided for the existence of correlated equilibria in finite games. The existence result is then extended to infinite games, including some that possess no Nash equilibria.

1. Introduction

The standard proof of existence of *correlated equilibria* (defined by Aumann 1974) in finite games consists of showing, first, that Nash equilibria are correlated, and second, that every game has at least one Nash equilibrium. The first argument is trivial; the second requires however the use of a fixed-point theorem.[1]

Originally published in *Mathematics of Operations Research*, 14 (1989), 18–25.

*David Schmeidler: Tel Aviv University, Israel.
Received May 5, 1986; revised May 19, 1987.
AMS 1980 subject classification. Primary: 90D13; Secondary: 90A14.
IAOR 1973 subject classification. Main: Games.
OR/MS Index 1978 subject classification. Primary: 231 Games.
Keywords: correlated equilibrium; infinite games; linear duality.
This paper supersedes our paper "Correlated Equilibria: An Elementary Proof of Existence," Tel Aviv University (version 1, October 1985; version 2, February 1986).
Hart's research was partially supported by grants from the National Science Foundation and the U.S.–Israel Binational Science Foundation. Part of this research was done while Schmeidler was visiting the Department of Economics at the University of Pennsylvania.
We would like to acknowledge useful discussions with Robert J. Aumann, Jean-François Mertens, Bezalel Peleg, and William Zame.

[1] Another proof of existence of correlated equilibria has been obtained by Reinhard Selten (private communication); it applies however only to the case where each player has two pure strategies.

Since the correlated equilibria form a convex set, defined by a set of explicit linear inequalities, it is reasonable to expect a simpler existence proof. We provide here (in Section 2) such an elementary proof, based on the standard Linear Duality Theorem. We actually chose to use an equivalent result, which seems more appropriate in this game-theoretic setup: the Minimax Theorem.[2]

We next consider infinite games, where the set of players and the sets of (pure) strategies are arbitrary. General results on the existence of correlated equilibria are obtained. Moreover, it is proved that a *countably additive* correlated equilibrium exists when the payoff functions are continuous and the strategy spaces are compact Hausdorff. All proofs are as "elementary" as possible: we use the result of the existence of correlated equilibria for finite games, together with a standard product compactness argument (e.g., Tychonoff's Theorem); this is equivalent to an infinite-dimensional separation theorem, but weaker than a fixed-point theorem (which is needed to show the existence of Nash equilibria). We study first, in Section 3, the simpler case where the sets of (pure) strategies are all finite; Section 4 then deals with the general case. We also provide several examples to illustrate the various difficulties.

2. Finite Games

A *finite game* (an "n-person game in normal or strategic form") is given as follows: Let $N = \{1, 2, \ldots, n\}$ be a finite set of *players*. For each $i \in N$, let S^i be a finite set of (pure) *strategies* of i. Let S be the set of n-tuples of strategies: $S = S^1 \times S^2 \times \cdots \times S^n$; an element of S is $s = (s^i)_{i \in N}$. For each $i \in N$ and $s \in S$ let

$$s^{-i} = (s^1, \ldots, s^{i-1}, s^{i+1}, \ldots, s^n)$$

denote the strategies played by everyone but i; thus $s^{-i} \in S^{-i} = \prod_{j \neq i} S^j$ and $s = (s^{-i}, s^i)$. Finally, for each $i \in N$, let[3] $h^i \colon S \to \mathbb{R}$ be the *payoff function* of player i; $h^i(s)$ is the payoff to i when the n-tuple of strategies s is played.

A *correlated equilibrium* (Aumann 1974, 1987) consists of a probability vector[4] $p = (p(s))_{s \in S}$ on S such that the following is satisfied for all $i \in N$

---

[2] Alternatively, we could have used the Theorem of the Alternative, or any other equivalent result.

[3] $\mathbb{R}$ denotes the real line.

[4] A probability vector is a vector whose coordinates are all nonnegative and sum up to 1.

and all $r^i, t^i \in S^i$:

$$\sum_{s^{-i} \in S^{-i}} p(s^{-i}, r^i)[h^i(s^{-i}, r^i) - h^i(s^{-i}, t^i)] \geq 0. \tag{1}$$

The interpretation is as follows: An n-tuple of strategies $r \in S$ is chosen at random (by a referee, say), according to the distribution p. Each player i is then told (only) his own coordinate r^i of r, and the original game is played. A correlated equilibrium results if the n-tuple of strategies in which each player i always plays the "recommended" r^i is a Nash equilibrium in this extended game (note that all players are assumed to know the distribution p). Condition (1) says that, whenever player i is told r^i, he will have no incentive to play t^i instead. (Note that one should have in (1) the conditional probability[5] $p(s^{-i}|r^i)$ instead of the joint probability $p(s^{-i}, r^i)$; multiplying by the marginal probability $p(r^i)$ then yields (1), which also holds trivially when $p(r^i) = 0$.)

Theorem 1. *Every finite game has a correlated equilibrium.*

Proof. Consider the following auxiliary two-person zero-sum game: Player I (the maximizer) chooses an n-tuple of strategies $s = (s^1, \ldots, s^n) \in S$; player II (the minimizer) chooses a triple (i, r^i, t^i), where $i \in N$ and $r^i, t^i \in S^i$. The payoff (from II to I) is: $h^i(s^{-i}, r^i) - h^i(s^{-i}, t^i)$ if $s^i = r^i$ and 0 otherwise. A correlated equilibrium in the original game is now easily seen to correspond to a strategy[6] of player I that guarantees a nonnegative payoff in this zero-sum game. By the Minimax Theorem, such a strategy exists if, for every *given* strategy of player II, there exists a strategy of player I yielding a nonnegative payoff. Let thus $y = (y^i(r^i, t^i))_{i \in N; r^i, t^i \in S^i}$ be a strategy of player II. We now need the following: □

Lemma. *Let $(a_{jk})_{j,k=1,\ldots,m}$ be nonnegative numbers. Then there exists a probability vector $x = (x_j)_{j=1,\ldots,m}$ such that, for any vector $u = (u_j)_{j=1,\ldots,m}$,*

$$\sum_{j=1}^{m} x_j \sum_{k=1}^{m} a_{jk}(u_j - u_k) = 0.$$

Proof. Denote the expression above by $\Phi(x, u)$. Since $\Phi(x, -u) = -\Phi(x, u)$, we need only to show that $\Phi(x, u) \geq 0$ for all u. Next, note

---

[5]We regard p as a probability distribution on the product space S; marginal and conditional probabilities are thus well defined.
[6]By "strategy" we mean, of course, "mixed strategy."

that it suffices to consider only probability vectors u, since one may add an arbitrary constant to all the coordinates of u (to make them nonnegative) and then multiply u by a positive scalar (to normalize it), without changing (the sign of) $\Phi(x, u)$. The function Φ is bilinear in x and u; therefore the Minimax Theorem applies to it: $\text{Max}_x \min_u \Phi(x, u) = \min_u \text{Max}_x \Phi(x, u)$, where x and u are both probability vectors. Given u, let j be such that $u_j = \text{Max}_k u_k$; for x the jth unit vector, we then have $\Phi(x, u) \geq 0$. This shows that the min Max is nonnegative, hence so is the Max min, implying the existence of a vector x as claimed.[7] $\square$

Proof of Theorem 1 (CONTINUED). For each i, apply the Lemma to $y^i(r^i, t^i))_{r^i, t^i \in S^i}$ (as (a_{jk})), to obtain a probability vector $(x^i(r^i))_{r^i \in S^i}$ such that, in particular,

$$\sum_{r^i} x^i(r^i) \sum_{t^i} y^i(r^i, t^i)[h^i(s^{-i}, r^i) - h^i(s^{-i}, t^i)] = 0 \qquad (2)$$

for every $s^{-i} \in S^{-i}$ (here, $u = (h^i(s^{-i}, r^i))_{r^i \in S^i}$). Define $x(r) = \prod_{j \in N} x^j(r^j)$ for each $r \in S$; x is clearly a strategy of player I in the auxiliary game. The payoff corresponding to the pair of strategies (x, y) (recall that y is the given strategy of player II) is

$$\sum_r \left[\prod_j x^j(r^j) \right] \left\{ \sum_i \sum_{t^i} y^i(r^i, t^i) \left[h^i(r^{-i}, r^i) - h^i(r^{-i}, t^i) \right] \right\}.$$

This may be rewritten as

$$\sum_i \sum_{r^{-i}} \left[\prod_{j \neq i} x^j(r^j) \right] \left\{ \sum_{r^i} x^i(r^i) \sum_{t^i} y^i(r^i, t^i) \left[h^i(r^{-i}, r^i) - h^i(r^{-i}, t^i) \right] \right\}$$

which equals zero by (2). This completes the proof. $\square$

Remark. An explicit formula for x in the Lemma is as follows: For each j in $M = \{1, \ldots, m\}$, let G_j be the set of all functions g from M into itself, such that, for every $k \neq j$, there is a positive integer r (depending on j) with $g^{(r)}(k) = j$ (we write $g^{(r)}$ for the composition of g with itself r times).

---

[7]See the Remark following the proof of Theorem 1, for an explicit formula for x.

Then

$$x_j = \sum_{g \in G_j} \prod_{k \neq j} a_{k,g(k)}$$

for all $j \in M$.

3. Infinitely Many Players

Consider now the case where the set of players may be infinite. We will assume in this section that the (pure) strategy sets of all players are finite, since the definitions, the results, the proofs—and the difficulties— are simpler and more transparent than in the general case (which is studied in Section 4).

A *game* (in normal or strategic form) consists of:[8] (i) a nonempty set of players N; (ii) for each $i \in N$, a nonempty set of (pure) strategies S^i; and (iii) for each $i \in N$, a bounded payoff function h^i from $S = \prod_{i \in N} S^i$ to $\mathbb{R}$.

In this section we assume that all the sets S^i are finite.

For each $i \in N$, denote $S^{-i} = \prod_{j \neq i} S^j$; thus, every $s \in S$ can be written as $s = (s^i, s^{-i})$ with $s^i \in S^i$ and $s^{-i} \in S^{-i}$. The sets S^i are endowed with the discrete topology,[9] and their product S with the resulting product topology (note that S is a compact Hausdorff space). Let Σ_0 denote the product σ-algebra on S: it is generated by the cylinders of the form $S^{-i} \times \{s^i\}$ for $i \in N$ and $s^i \in S^i$.

Let Σ be an algebra on S such that $\Sigma \supset \Sigma_0$ and all the payoff functions h^i are Σ-measurable. A *correlated equilibrium* (*with respect to Σ*) is a (finitely additive) probability measure p on the measurable space (S, Σ), such that the following[10] is satisfied for all $i \in N$ and all $r^i, t^i \in S^i$:

$$\int_{S^{-i} \times \{r^i\}} \left[h^i(s^{-i}, r^i) - h^i(s^{-i}, t^i) \right] dp(s) \geq 0. \tag{3}$$

(Note that, as in the finite case, we have multiplied by $p(r^i)$.)

---

[8]The difference between a "finite game" (as defined in Section 2) and a "game" is that the sets N and S^i are finite in the former and arbitrary in the latter. (Note that the functions h^i are always bounded in a finite game.)

[9]A standard reference for the concepts and results of topology, measure theory, linear spaces, etc., that are used in Sections 3 and 4 is Dunford and Schwartz (1958).

[10]The above conditions on Σ and the functions h^i guarantee that (3) is well defined.

Theorem 2. *Assume that, for each $i \in N$, the set S^i is finite.*

(i) *Let $\Sigma \supset \Sigma_0$ be an algebra on S and assume that, for each $i \in N$, the function h^i is bounded and Σ-measurable. Then there exists a correlated equilibrium with respect to Σ.*

(ii) *Assume that, for each $i \in N$, the function h^i is continuous. Then there exists a countably additive correlated equilibrium with respect to Σ_0.*

Remarks. 1. One may of course apply (i) to $\Sigma = 2^s$, in which case every function is Σ-measurable. Thus, if the payoff functions are bounded, then there always exists a (finitely additive) correlated equilibrium with respect to 2^s.

2. If h^i is a continuous function, then it is bounded (since S is compact) and Σ_0-measurable.

3. A real function on the product space S is continuous if and only if it is "almost finitely determined"; i.e., it is the uniform limit of functions depending only on finitely many coordinates (cf. Lemma 4.3 in Peleg 1969).

In the case that the payoff functions are continuous, Peleg (1969) has proved that there exists a Nash equilibrium, which is clearly also a countably additive correlated equilibrium with respect to Σ_0. However, his proof uses a fixed-point theorem, whereas the proof below is essentially an infinite-dimensional separation argument (or, equivalently, finite-dimensional separation together with compactness of an infinite product—which is the way our proof is presented).

Before proceeding to the proof of Theorem 2, it is instructive to consider two examples. The first one, due to Peleg (1969), is of a game that possesses no Nash equilibria (the payoff functions are not continuous); we exhibit a countably additive correlated equilibrium there (whose existence is *not* guaranteed by Theorem 2), and also a noncountably additive one, which shows that these equilibria (although they always exist) may well be quite "unreasonable." The second example, which is a slight modification of the first, possesses only noncountably additive correlated equilibria.

Example 1. Let N be the set of positive integers; for each $i \in N$ let $S^i = \{0, 1\}$ and $h^i(s) = s^i$ if $\Sigma_{i \in N} s^i < \infty$ and $h^i(s) = -s^i$ if $\Sigma_{i \in N} s^i = \infty$. Call the first case (of the convergent series, which happens whenever there are only finitely many players that chose $s^i = 1$) *Case 1* (since all players would like to choose $s^i = 1$ there), and the second case *Case 0* (all players would prefer to choose $s^i = 0$ here). This game has no Nash equilibria: Let

σ^i be a mixed strategy of player i, and denote by ξ^i the probability (under σ^i) that $s^i = 1$. By the Zero-One Law (mixed strategies are independent), either Case 1 happens almost surely, or Case 0 happens almost surely. In the former case, players strictly prefer their strategy $\xi^i = 1$ (i.e., $s^i = 1$ for sure); in the latter, $\xi^i = 0$ (i.e., $s^i = 0$).

A countably additive correlated equilibrium (with respect to Σ_0) is obtained in this game as follows: For each $j \in N$, let z_j be that element of S whose first j coordinates are 1 and all the rest are 0; i.e., $z_j = (z_j^i)_{i \in N}$ with $z_j^i = 1$ if $i \le j$ and $z_j^i = 0$ if $i > j$. Let p_1 be the probability measure on S with support $\{z_j : j \in N\}$ and $p_1(z_j) = 1/j - 1/(j+1)$ for all $j \in N$. Note that $p_1[s \in S : s^i = 1] = p_1[z_j : j \ge i] = 1/i$ for all $i \in N$ and that $p_1[\text{Case 1}] = 1$. Next, let p_0 be the product probability measure on S with marginals $p_0[s \in S : s^i = 1] = 1/i$ for all $i \in N$; note that p_0 [Case 0] $= 1$ (by the Zero-One Law, since $\Sigma 1/i = \infty$). Finally, put $p_2 = (p_1 + p_0)/2$. It is now easily checked that p_2 is indeed a correlated equilibrium: For every $i \in N$ and $s^i = 0, 1$, one has $p_2[\text{Case }1|s^i] = p_2[\text{Case }0|s^i] = 1/2$; therefore (3) is always satisfied (as an equality).

A noncountably additive correlated equilibrium (with respect to $\Sigma = 2^s$) is as follows (this is essentially an equilibrium of the type constructed in our proof of Theorem 2(i) below): Let $A \subset S$; if A contains only finitely many of the points $\{z_j\}_{j \in N}$ (defined in the previous paragraph), put $p_3(A) = 0$; if A contains all but finitely many of these points, put $p_3(A) = 1$. Now extend p_3 to a finitely additive probability measure on $(S, 2^s)$ (apply the Hahn–Banach Theorem[11]). It can now be easily checked that p_3 satisfies the following two properties: $p_3[s^i = 1] = p_3[z_j : j \ge i] = 1$ for each i; and $p_3[\text{Case 1}] = p_3[\Sigma_{i \in N} S^i < \infty] = p_3[z_j : j \in N] = 1$. It is now clear that p_3 is indeed a correlated equilibrium.

The equilibria in this example illustrate two points: First, that a (countably additive) correlated equilibrium may exist even when there is no Nash equilibrium. And second, that noncountably additive correlated equilibria may well be unintuitive (indeed, p_3-almost surely every player chooses 1, and also p_3-almost surely there are only finitely many 1's!). However, if the payoff functions are not continuous, countably additive correlated equilibria need not exist at all, as the next example will show.

Example 2. Identical to Example 1, except that in Case 1 (when there are finitely many 1's) we define $h^i(s) = s^i/i^2$ (instead of $h^i(s) = s^i$). Applying the same arguments used in Example 1 shows that there exists no

---

[11] The Axiom of Choice is used here.

Nash equilibrium, and also that p_3 is a (noncountably additive) correlated equilibrium.

Next, we prove that there exists no countably additive correlated equilibrium. Indeed, assume p is such an equilibrium. First, note that both Case 0 and Case 1 must have positive probability (otherwise, apply the same argument used to show that there is no Nash equilibrium). Now condition (3) for $r^i = 1$ and $t^i = 0$ is

$$p[\text{Case 1 \& } r^i = 1]/i^2 \geq p[\text{Case 0 \& } r^i = 1].$$

This implies that $p[\text{Case 0 \& } r^i = 1] \leq 1/i^2$; hence $p[\text{Case 0 \& } r^i = 1 \text{ infinitely often}] = 0$ (by the Borel–Cantelli Lemma, since the series $\Sigma 1/i^2$ converges; here is the only use of the countable additivity of p). Thus $p[\text{Case 0}] = 0$, a contradiction. $\square$

Proof of Theorem 2. (i) Consider finite sets $T = \prod_{i \in N} T^i \subset S$ such that $T^i \subset S^i$ for all i, and T^i is a singleton for all but finitely many players $i \in N$; we will call such a set T an "f-set." The game Γ_T obtained by replacing S^i with T^i for each $i \in N$ is equivalent to a finite game (only those i for which T^i is not a singleton are "real" players). By Theorem 1, every finite game has a correlated equilibrium; one therefore obtains a correlated equilibrium in Γ_T, which will be denoted by q_T. Clearly, we may regard q_T as a probability measure on S, with (finite) support T.

Consider the net $\{q_T : T \text{ an } f\text{-set}\}$, ordered by inclusion of the sets T. It belongs to the unit ball of $ba(S, \Sigma)$, the space of finitely additive measures on (S, Σ). By the Banach–Alaoglu Theorem (which is an easy consequence of Tychonoff's Theorem on the compactness of a product of compact spaces; see Dunford and Schwartz 1958, Theorem V.4.2), the unit ball is compact in the weak*-topology, which in this case is induced by $B(S, \Sigma)$, the space of bounded and measurable functions on (S, Σ) (cf. Dunford and Schwartz 1958, Theorem IV.5.1). Let p thus be a cluster point of $\{q_T\}$. We will show that p is a correlated equilibrium.

Indeed, fix $i \in N$ and $r^i, t^i \in S^i$. Define a real function f on S by: $f(s) = h^i(s) - h^i(s^{-i}, t^i)$ for $s \in S^{-i} \times \{r^i\}$ (i.e., when $s^i = r^i$), and $f(s) = 0$ otherwise. The assumptions on h^i imply that f too is bounded and Σ-measurable. Given $\varepsilon > 0$ and i, there exists therefore an f-set T with $T^i \supset \{r^i, t^i\}$, such that $|\int f \, dp - \int f \, dq_T| < \varepsilon$. It is easily checked that $\int f \, dq_T \geq 0$ (since q_T satisfies (1), hence (3), for all "real" players in

Γ_T, in particular i, and $r^i, t^i \in T^i$). Therefore the inequality above implies $\int f dp > -\varepsilon$, hence ≥ 0, which is exactly (3) for p.

(ii) Replace ba(S, Σ) in the proof of (i) with rca(S, Σ_0), the space of regular countably additive measures on (S, Σ_0), which is the dual of $C(S)$, the space of continuous real functions on S (by Riesz's Representation Theorem; see Dunford and Schwartz 1958, Theorem IV.6.3). □

4. Infinitely Many Players and Strategies

In this section we deal with the general case, where the set of players as well as the strategy sets may be infinite. A "game" was defined in Section 3; here it is no longer assumed that the (pure) strategy sets are finite. When defining a correlated equilibrium, it turns out that condition (3) is no longer appropriate, since $p(r^i)$ may well be zero for all $r^i \in S^i$ (for example, when S^i is a continuum). A first attempt is therefore to require (3) also for subsets $R^i \subset S^i$ and not only for singletons $\{r^i\}$. This is however not yet satisfactory, since t^i there need not be fixed—it may well depend on r^i. To obtain the correct conditions, we therefore consider again the *extended game* described in Section 2: An element $r \in S$ is chosen, each player i is informed only of the ith coordinate r^i of r, and then the original game is played. A strategy of player i in the extended game is thus a function ζ^i from S^i into itself, that associates to each "recommendation" $r^i \in S^i$ a choice of action $\zeta^i(r^i) \in S^i$ in the original game. A correlated equilibrium is obtained if, when each player i uses the strategy $\zeta^i =$ identify, a Nash equilibrium results (in the extended game).

Formally, let Σ^i be an algebra on S^i, let Σ_0 be the σ-algebra on S generated by the product of the Σ^i's, and let $\Sigma \supset \Sigma_0$ be any algebra on S. We assume that, for each $i \in N$, the payoff function h^i is bounded and Σ-measurable. A *correlated equilibrium (with respect to $\{\Sigma^i\}_{i \in N}$ and Σ)* is a probability measure p on (S, Σ), such that, for all $i \in N$ and all Σ^i-measurable[12] functions $\zeta^i \colon S^i \to S^i$, the following inequality holds:

$$\int_S \left[h^i(s^{-i}, s^i) - h^i(s^{-i}, \zeta^i(s^i)) \right] dp(s) \geq 0. \tag{4}$$

The left-hand side in (4) is the difference between the payoff of player i (in the extended game) when he always follows the "recommendation," and his

---

[12] I.e., $\{s^i \in S^i \colon \zeta^i(s^i) \in A\} \in \Sigma^i$ for every $A \in \Sigma^i$.

payoff when he plays the strategy ζ^i instead. It is easy to see that, when S^i is a finite set, conditions (4) and (3) are equivalent: (3) is obtained by taking in (4) $\zeta^i(s^i) = s^i$ for all $s^i \neq r^i$ and $\zeta^i(r^i) = t^i$; vice versa, given a function ζ^i, sum the inequalities (3) over all r^i with $t^i = \zeta^i(r^i)$, to yield (4).

As we saw in the previous section, noncountably additive correlated equilibria may be quite "unreasonable." We will therefore deal here only with the existence of countably additive correlated equilibria (however, a result parallel to Theorem 2(i) may be obtained here too).

Theorem 3. *Assume that, for each $i \in N$, the space S^i is compact Hausdorff and the function h^i is continuous (where S is endowed with the product topology). Let Σ^i be the Borel σ-algebra on S^i, and let Σ be the Borel σ-algebra on[13] S. Then there exists a countably additive correlated equilibrium with respect to $\{\Sigma^i\}_{i \in N}$ and Σ.*

Proof.   The construction is similar to that used in the proof of Theorem 2. A set $T = \prod_{i \in N} T^i$ will be called an "f-set" if T^i is a finite subset of S^i for all $i \in N$, and moreover T^i is a singleton for all but finitely many i's. To each f-set T there corresponds an (essentially) finite game Γ_T (obtained by replacing S^i with T^i for all i); let q_T be a correlated equilibrium of Γ_T (whose existence follows from Theorem 1), regarded as a probability measure on S (with finite support T). The net $\{q_T : T$ an f-set$\}$ (ordered by inclusion of the sets T) belongs to the unit ball of rca(S) (the space of regular countably additive measures on S; recall that each S^i is Hausdorff, therefore S is Hausdorff, hence measures with finite support are regular). Again, the Banach–Alaoglu Theorem implies the existence of a cluster point p with respect to the topology induced by $C(S)$ (the space of continuous functions on S). We will show that p is indeed a correlated equilibrium.

Let $i \in N$ and let $\zeta^i : S^i \to S^i$ be a measurable function. If ζ^i were a continuous function, inequality (4) would easily be proved by an argument similar to that used in the Proof of Theorem 2 (here, p is a cluster point relative to continuous functions); since however ζ^i need not be continuous, we need more elaborate arguments. We will deal first with a special case, from which the general case will then easily follow.

A special case. There exist $t^i \in S^i$ and $R^i \in \Sigma^i$ such that $\zeta^i(s^i) = t^i$ for all $t^i \in R^i$ and $\zeta^i(s^i) = s^i$ otherwise. Fix $\varepsilon > 0$. The measure p is

---

[13]Note that $\Sigma \supset \Sigma_0$ (= product of the Σ^i's). A. S. Nowak has pointed out that the inclusion may be strict in the nonmetrizable case.

regular; therefore there exist sets $F, G \subset S$ such that: F is closed, G is open, $F \subset R^i \times S^{-i} \subset G$ and[14] $p(G \backslash F) < \varepsilon$. Define[15] $F^i = \text{proj}_i F$ and $G^i = S^i \backslash \text{proj}_i(S \backslash G)$, then F^i is closed and G^i is open (since S is compact), $F^i \subset R^i \subset G^i$ and[16] $p((G^i \backslash F^i) \times S^{-i} \le p(G \backslash F) < \varepsilon$. Next, apply Urisohn's Lemma (see Dunford and Schwartz 1958, Theorems I.5.2 and I.5.9), to obtain a continuous function $\varphi \colon S^i \to [0,1]$ such that $\varphi(S^i) = 1$ for all $s^i \in F^i$ and $\varphi(s^i) = 0$ for all $s^i \notin G^i$.

Define[17] $f(s) = \varphi(s^i)[h^i(s) - h^i(s^{-i}, t^i)]$; since $f \colon S \to \mathbb{R}$ is a continuous function, there exists an f-set T with $T^i \supset \{t^i\}$ such that (all the integrals are over S)

$$\left| \int f \, dp - \int f \, dq_T \right| < \varepsilon.$$

Next, let M be a bound on h^i; then

$$\left| \int f \, dp - \int [h^i(s) - h^i(s^{-i}, \zeta^i(s^i))] dp \right| < 4M\varepsilon,$$

since the integrands may differ only on $(G^i \backslash F^i) \times S^{-i}$. Finally,

$$\int f \, dq_T = \sum_{s^i \in T^i} \varphi(s^i) \sum_{s^{-i} \in T^{-i}} [h^i(s) - h^i(s^{-i}, t^i)] q_T(s) \ge 0,$$

since (1) holds for q_T and $\varphi \ge 0$. The last three displayed inequalities together imply that the left-hand side of (4) is $> -(4M + 1)\varepsilon$, hence, ε being arbitrary, that (4) holds.

The general case. Let $\varepsilon > 0$ be given. Since both S^i and S^{-i} are compact and the function h^i is continuous, it is straightforward to obtain a finite partition of S^i into disjoint measurable sets $\{A_k\}_{k=1,\dots,K}$ such that $|h^i(s^{-i}, s^i) - h^i(s^{-i}, t^i)| < \varepsilon$ for any s^i, t^i belonging to the same A_k and every $s^{-i} \in S^{-i}$. For each k, fix some $t_k^i \in A_k$; let $R_k^i = \{s^i \in S^i : \zeta^i(s^i) \in A_k\}$ and define a Σ^i-measurable function ζ_k^i as follows: $\zeta_k^i(s^i) = t_k^i$ for $s^i \in R_k^i$ and ζ_k^i is the identity outside R_k^i. Inequality (4) holds for ζ_k^i, since the special case applies to it. Summing up these inequalities for all k, we obtain

---

[14] The symbol $\backslash$ denotes set-theoretic subtraction.

[15] The projection from S onto S^i is denoted proj_i.

[16] What we have shown here is that in a compact space, the marginal of a regular measure is also regular.

[17] One may regard φ as a mixed strategy (more precisely, a behavioral strategy) of player i in the extended game: when the recommendation (to i) is s^i, he follows it (i.e., he plays s^i) with probability $1 - \varphi(s^i)$ and he plays t^i with probability $\varphi(s^i)$.

(4) for the admissible function η^i defined by $\eta^i(s^i) = t^i_k$ if $s^i \in R^i_k$ for some k. The construction above implies that $|h^i(s^{-i}, \zeta^i(s^i)) - h^i(s^{-i}, \eta^i(s^i))| < \varepsilon$ for all s, therefore the left-hand side of (4) for ζ^i is $> -\varepsilon$, hence ≥ 0. $\quad\square$

Remark. Under the assumptions of Theorem 3, one may prove—using a fixed-point theorem—the existence of a Nash equilibrium (e.g., see Fan 1952 [apply his Theorem 2] or Glicksberg 1952 [his proof in Section 2 may be easily applied to an arbitrary set of players]).

References

Aumann, R. J. (1974), "Subjectivity and Correlation in Randomized Strategies," *Journal of Mathematical Economics*, 1, 67–95.

————(1987), "Correlated Equilibrium as an Expression of Bayesian Rationality," *Econometrica*, 55, 1–18.

Dunford, N. and J. T. Schwartz (1958), *Linear Operators*, Vol. I. New York: Interscience.

Fan, K. (1952), "Fixed-Point Theorems in Locally Convex Topological Spaces," *Proceedings of the National Academy of Sciences of the United States of America*, 38, 121–126.

Glicksberg, I. L. (1952), "A Further Generalization of the Kakutani Fixed Point Theorem, with Application to Nash Equilibrium Points," *Proceedings of the American Mathematical Society*, 3, 170–174.

Peleg, B. (1969), "Equilibrium Points for Games with Infinitely Many Players," *Journal of the London Mathematical Society*, 44, 292–294.

Part II

Regret Matching

Chapter 2

A SIMPLE ADAPTIVE PROCEDURE LEADING TO CORRELATED EQUILIBRIUM*

Sergiu Hart and Andreu Mas-Colell

We propose a new and simple adaptive procedure for playing a game: "regret-matching." In this procedure, players may depart from their current play with probabilities that are proportional to measures of regret for not having used other strategies in the past. It is shown that our adaptive procedure guarantees that, with probability one, the empirical distributions of play converge to the set of correlated equilibria of the game.

1. Introduction

The leading noncooperative equilibrium notions for N-person games in strategic (normal) form are Nash equilibrium (and its refinements) and correlated equilibrium. In this paper we focus on the concept of correlated equilibrium.

A *correlated equilibrium*—a notion introduced by Aumann (1974)—can be described as follows: Assume that, before the game is played, each player receives a private signal (which does not affect the payoffs). The player may

Originally published in *Econometrica*, 68 (2000), 1127–1150.

*Received October 1998; revised June 1999.

Keywords: adaptive procedure; correlated equilibrium; no regret; regret matching; simple strategies.

Previous versions: March 1996 (handout); December 1996; November 1997; February 1998.

Research partially supported by grants of the U.S.–Israel Binational Science Foundation, the Israel Academy of Sciences and Humanities, the Spanish Ministry of Education, and the Generalitat de Catalunya.

We want to acknowledge the useful comments and suggestions of Robert Aumann, Antonio Cabrales, Dean Foster, David Levine, Alvin Roth, Reinhard Selten, Sylvain Sorin, an editor, the anonymous referees, and the participants at various seminars where this work was presented.

then choose his action in the game depending on this signal. A correlated equilibrium of the original game is just a Nash equilibrium of the game with the signals. Considering all possible signal structures generates all correlated equilibria. If the signals are (stochastically) independent across the players, it is a Nash equilibrium (in mixed or pure strategies) of the original game. But the signals could well be correlated, in which case new equilibria may obtain.

Equivalently, a correlated equilibrium is a probability distribution on N-tuples of actions, which can be interpreted as the distribution of play instructions given to the players by some "device" or "referee." Each player is given—privately—instructions for his own play only; the joint distribution is known to all of them. Also, for every possible instruction that a player receives, the player realizes that the instruction provides a best response to the random estimated play of the other players—assuming they all follow their instructions.

There is much to be said for correlated equilibrium. See Aumann (1974, 1987) for an analysis and foundational arguments in terms of rationality. Also, from a practical point of view, it could be argued that correlated equilibrium may be the most relevant noncooperative solution concept. Indeed, with the possible exception of well-controlled environments, it is hard to exclude a priori the possibility that correlating signals are amply available to the players, and thus find their way into the equilibrium.

This paper is concerned with dynamic considerations. We pose the following question: *Are there simple adaptive procedures always leading to correlated equilibrium?*

Foster and Vohra (1997) have obtained a procedure converging to the set of correlated equilibria. The work of Fudenberg and Levine (1999) led to a second one. We introduce here a procedure that we view as particularly simple and intuitive (see Section 4 for a comparative discussion of all these procedures). It does not entail any sophisticated updating, prediction, or fully rational behavior. Our procedure takes place in discrete time and it specifies that players adjust strategies probabilistically. This adjustment is guided by "regret measures" based on observation of past periods. Players know the past history of play of all players, as well as their own payoff matrix (but not necessarily the payoff matrices of the other players). Our Main Theorem is: The adaptive procedure generates trajectories of play that almost surely converge to the set of correlated equilibria.

The procedure is as follows: At each period, a player may either continue playing the same strategy as in the previous period, or switch to other strategies, with probabilities that are proportional to how much higher his accumulated payoff would have been had he always made that change in the past. More precisely, let U be his total payoff up to now; for each strategy k different from his last period strategy j, let $V(k)$ be the total payoff he would have received if he had played k every time in the past that he chose j (and everything else remained unchanged). Then only those strategies k with $V(k)$ larger than U may be switched to, with probabilities that are proportional to the differences $V(k) - U$, which we call the "regret" for having played j rather than k. These probabilities are normalized by a fixed factor, so that they add up to strictly less than 1; with the remaining probability, the same strategy j is chosen as in the last period.

It is worthwhile to point out three properties of our procedure. First, its simplicity; indeed, it is very easy to explain and to implement. It is not more involved than fictitious play (Brown 1951 and Robinson 1951); note that in the two-person zero-sum case, our procedure also yields the minimax value. Second, the procedure is *not* of the "best-reply" variety (such as fictitious play, smooth fictitious play [Fudenberg and Levine 1995, 1999] or calibrated learning [Foster and Vohra 1997]; see Section 4 for further details). Players do not choose only their "best" actions, nor do they give probability close to 1 to these choices. Instead, all "better" actions may be chosen, with probabilities that are proportional to the apparent gains, as measured by the regrets; the procedure could thus be called "*regret matching*." And third, there is "inertia." The strategy played in the last period matters: There is always a positive probability of continuing to play this strategy and, moreover, changes from it occur only if there is reason to do so.

At this point a question may arise: Can one actually guarantee that the smaller set of Nash equilibria is always reached? The answer is definitely "no." On the one hand, in our procedure, as in most others, there is a natural coordination device: the common history, observed by all players. It is thus reasonable to expect that, at the end, independence among the players will not obtain. On the other hand, the set of Nash equilibria is a mathematically complex set (a set of fixed points; by comparison, the set of correlated equilibria is a convex polytope), and simple adaptive procedures cannot be expected to guarantee the global convergence to such a set.

After this introductory section, in Section 2 we present the model, describe the adaptive procedure, and state our result (the Main Theorem). Section 3 is devoted to a "stylized variation" of the procedure of Section 2. It is a variation that lends itself to a very direct proof, based on Blackwell's (1956a) Approachability Theorem. This is a new instrument in this field, which may well turn out to be widely applicable.

Section 4 contains a discussion of the literature, together with a number of relevant issues. The proof of the Main Theorem is relegated to the Appendix.

2. The Model and Main Result

Let $\Gamma = (N, (S^i)_{i \in N}, (u^i)_{i \in N})$ be a finite N-person game in strategic (normal) form: N is the set of players, S^i is the set of strategies of player i, and $u^i: \prod_{i \in N} S^i \to \mathbb{R}$ is player i's payoff function. All sets N and S^i are assumed to be finite. Denote by $S := \prod_{i \in N} S^i$ the set of N-tuples of strategies; the generic element of S is $s = (s^i)_{i \in N}$, and s^{-i} denotes the strategy combination of all players except i, i.e., $s^{-i} = (s^{i'})_{i' \neq i}$. We focus attention on the following solution concept:

Definition. A probability distribution ψ on S is a *correlated equilibrium* of Γ if, for every $i \in N$, every $j \in S^i$, and every $k \in S^i$ we have[1]

$$\sum_{s \in S: s^i = j} \psi(s)[u^i(k, s^{-i}) - u^i(s)] \leq 0.$$

If in the above inequality we replace the right-hand side by an $\varepsilon > 0$, then we obtain the concept of a *correlated ε-equilibrium*.

Note that every Nash equilibrium is a correlated equilibrium. Indeed, Nash equilibria correspond to the special case where ψ is a product measure; that is, the play of the different players is independent. Also, the set of correlated equilibria is nonempty, closed, and convex, and even in simple games (e.g., "chicken") it may include distributions that are not in the convex hull of the Nash equilibrium distributions.

Suppose now that the game Γ is played repeatedly through time: $t = 1, 2, \ldots$. At time $t + 1$, given a history of play $h_t = (s_\tau)_{\tau=1}^t \in \prod_{\tau=1}^t S$, we postulate that each player $i \in N$ chooses $s_{t+1}^i \in S^i$ according to a probability distribution[2] $p_{t+1}^i \in \Delta(S^i)$ which is defined in the following way: For every two different strategies $j, k \in S^i$ of player i, suppose i were to replace strategy j, every time that it was played in the past, by strategy k; his payoff at time τ, for $\tau \leq t$, would become

$$W_\tau^i(j, k) := \begin{cases} u^i(k, s_\tau^{-i}), & \text{if } s_\tau^i = j, \\ u^i(s_\tau), & \text{otherwise.} \end{cases} \tag{2.1a}$$

---

[1] We write $\sum_{s \in S: s^i = j}$ for the sum over all N-tuples s in S whose ith coordinate s^i equals j.
[2] We write $\Delta(Q)$ for the set of probability distributions over a finite set Q.

The resulting difference in i's average payoff up to time t is then

$$D_t^i(j,k) := \frac{1}{t} \sum_{\tau=1}^t W_\tau^i(j,k) - \frac{1}{t} \sum_{\tau=1}^t u^i(s_\tau)$$

$$= \frac{1}{t} \sum_{\tau \leq t : s_\tau^i = j} [u^i(k, s_\tau^{-i}) - u^i(s_\tau)]. \qquad (2.1b)$$

Finally, denote

$$R_t^i(j,k) := [D_t^i(j,k)]^+ = \max\{D_t^i(j,k), 0\}. \qquad (2.1c)$$

The expression $R_t^i(j,k)$ has a clear interpretation as a measure of the (average) "regret" at period t for not having played, every time that j was played in the past, the different strategy k.

Fix $\mu > 0$ to be a large enough number.[3] Let $j \in S^i$ be the strategy last chosen by player i, i.e., $j = s_t^i$. Then the probability distribution $p_{t+1}^i \in \Delta(S^i)$ used by i at time $t+1$ is defined as

$$\begin{cases} p_{t+1}^i(k) := \frac{1}{\mu} R_t^i(j,k), & \text{for all } k \neq j, \\ p_{t+1}^i(j) := 1 - \sum_{k \in S^i : k \neq j} p_{t+1}^i(k). \end{cases} \qquad (2.2)$$

Note that the choice of μ guarantees that $p_{t+1}^i(j) > 0$; that is, there is always a positive probability of playing the same strategy as in the previous period. The play $p_1^i \in \Delta(S^i)$ at the initial period is chosen arbitrarily.[4]

Informally, (2.2) may be described as follows. Player i starts from a "reference point": his current actual play. His choice next period is governed by propensities to depart from it. It is natural therefore to postulate that, if a change occurs, it should be to actions that are perceived as being better, relative to the current choice. In addition, and in the spirit of adaptive behavior, we assume that all such better choices get positive probabilities; also, the better an alternative action seems, the higher the probability of

---

[3] The parameter μ is fixed throughout the procedure (independent of time and history). It suffices to take μ so that $\mu > 2M^i(m^i - 1)$ for all $i \in N$, where M^i is an upper bound for $|u^i(\cdot)|$ and m^i is the number of strategies of player i. Even better, we could let μ satisfy $\mu > (m^i - 1)|u^i(k, s^{-i}) - u^i(j, s^{-i})|$ for all $j, k \in S^i$, all $s^{-i} \in S^{-i}$, and all $i \in N$ (and moreover we could use a different μ^i for each player i).
[4] Actually, the procedure could start with any finite number of periods where the play is arbitrary.

choosing it next time. Further, there is also inertia: the probability of staying put (and playing the same action as in the last period) is always positive.

More precisely, the probabilities of switching to different strategies are proportional to their regrets relative to the current strategy. The factor of proportionality is constant. In particular, if the regrets are small, then the probability of switching from current play is also small.

For every t, let $z_t \in \Delta(S)$ be the empirical distribution of the N-tuples of strategies played up to time t. That is, for every[5] $s \in S$,

$$z_t(s) := \frac{1}{t}|\{\tau \leq t : s_\tau = s\}| \qquad (2.3)$$

is the relative frequency that the N-tuple s has been played in the first t periods. We can now state our main result.

Main Theorem. *If every player plays according to the adaptive procedure (2.2), then the empirical distributions of play z_t converge almost surely as $t \to \infty$ to the set of correlated equilibrium distributions of the game Γ.*

Note that convergence to the *set* of correlated equilibria does not imply that the sequence z_t converges to a *point*. The Main Theorem asserts that the following statement holds with probability one: For any $\varepsilon > 0$ there is $T_0 = T_0(\varepsilon)$ such that for all $t > T_0$ we can find a correlated equilibrium distribution ψ_t at a distance less than ε from z_t. (Note that this T_0 depends on the history; it is an "a.s. finite stopping time.") That is, the Main Theorem says that, with probability one, for any $\varepsilon > 0$, the (random) trajectory $(z_1, z_2, \ldots, z_t, \ldots)$ enters and then stays forever in the ε-neighborhood in $\Delta(S)$ of the set of correlated equilibria. Put differently: Given any $\varepsilon > 0$, there exists a constant (i.e., independent of history) $t_0 = t_0(\varepsilon)$ such that, with probability at least $1 - \varepsilon$, the empirical distributions z_t for *all* $t > t_0$ are in the ε-neighborhood of the set of correlated equilibria. Finally, let us note that because the set of correlated equilibria is nonempty and compact, the statement "the trajectory (z_t) converges to the set of correlated equilibria" is equivalent to the statement "the trajectory (z_t) is such that for any $\varepsilon > 0$ there is $T_1 = T_1(\varepsilon)$ with the property that z_t is a correlated ε-equilibrium for all $t > T_1$."

We conclude this section with a few comments (see also Section 4):

(1) Our adaptive procedure (2.2) requires player i to know his own payoff matrix (but not those of the other players) and, at time $t+1$, the history h_t;

---

[5]We write $|Q|$ for the number of elements of a finite set Q.

actually, the empirical distribution z_t of $(s_1, s_2, \ldots, s_t)$ suffices. In terms of computation, player i needs to keep record of the time t together with the $m^i(m^i - 1)$ numbers $D_t^i(j, k)$ for all $j \neq k$ in S^i (and update these numbers every period).

(2) At every period the adaptive procedure that we propose randomizes only over the strategies that exhibit positive regret relative to the most recently played strategy. Some strategies may, therefore, receive zero probability. Suppose that we were to allow for trembles. Specifically, suppose that at every period we put a $\delta > 0$ probability on the uniform tremble (each strategy thus being played with probability at least δ/m^i). It can be shown that in this case the empirical distributions z_t converge to the set of correlated ε-equilibria (of course, ε depends on δ, and it goes to zero as δ goes to zero). In conclusion, unlike most adaptive procedures, ours does not rely on trembles (which are usually needed, technically, to get the "ergodicity" properties); moreover, our result is robust with respect to trembles.

(3) Our adaptive procedure depends only on one parameter,[6] μ. This may be viewed as an "inertia" parameter (see Sections 4(g) and 4(h)): A higher μ yields lower probabilities of switching. The convergence to the set of correlated equilibria is always guaranteed (for any large enough μ; see footnote 5), but the speed of convergence changes with μ.

(4) We know little about additional convergence properties for z_t. It is easy to see that the empirical distributions z_t either converge to a Nash equilibrium in pure strategies, or must be infinitely often outside the set of correlated equilibria (because, if z_t is a correlated equilibrium from some time on, then[7] all regrets are 0, and the play does not change). This implies, in particular, that interior (relative to $\Delta(S)$) points of the set of correlated equilibria that are not pure Nash equilibria are unreachable as the limit of some z_t (but it is possible that they are reachable as limits of a *subsequence* of z_t).

(5) There are other procedures enjoying convergence properties similar to ours: the procedures of Foster and Vohra (1997), of Fudenberg and Levine (1999), and of Theorem A in Section 3 below; see the discussion in Section 4. The delimitation of general classes of procedures converging to correlated equilibria seems, therefore, an interesting research problem.[8]

---

[6]Using a parameter μ (rather than a fixed normalization of the payoffs) was suggested to us by Reinhard Selten.

[7]See the Proposition in Section 3.

[8]See Hart and Mas-Colell (2001a) and Cahn (2004) for such results.

3. No Regret and Blackwell Approachability

In this section (which can be viewed as a motivational preliminary) we shall replace the adaptive procedure of Section 2 by another procedure that, while related to it, is more stylized. Then we shall analyze it by means of Blackwell's (1956a) Approachability Theorem, and prove that it yields convergence to the set of correlated equilibria. In fact, the Main Theorem stated in Section 2, and its proof in Appendix 1, were inspired by consideration and careful study of the result of this section. Furthermore, the procedure here is interesting in its own right (see, for instance, the Remark following the statement of Theorem A, and (d) in Section 4).

Fix a player i and recall the procedure of Section 2: At time $t + 1$ the transition probabilities, from the strategy played by player i in period t to the strategies to be played at $t+1$, are determined by the stochastic matrix defined by the system (2.2). Consider now an invariant probability vector $q_t^i = (q_t^i(j))_{j \in S^i} \in \Delta(S^i)$ for this matrix (such a vector always exists). That is, q_t^i satisfies

$$q_t^i(j) = \sum_{k \neq j} q_t^i(k) \frac{1}{\mu} R_t^i(k, j) + q_t^i(j) \left[1 - \sum_{k \neq j} \frac{1}{\mu} R_t^i(j, k) \right],$$

for every $j \in S^i$. By collecting terms, multiplying by μ, and formally letting $R_t^i(j, j) := 0$, the above expression can be rewritten as

$$\sum_{k \in S^i} q_t^i(k) R_t^i(k, j) = q_t^i(j) \sum_{k \in S^i} R_t^i(j, k), \tag{3.1}$$

for every $j \in S^i$.

In this section we shall assume that play at time $t + 1$ by player i is determined by a solution q_t^i to the system of equations (3.1); i.e., $p_{t+1}^i(j) := q_t^i(j)$. In a sense, we assume that player i at time $t + 1$ goes instantly to the invariant distribution of the stochastic transition matrix determined by (2.2). We now state the key result.

Theorem A. *Suppose that at every period $t + 1$ player i chooses strategies according to a probability vector q_t^i that satisfies (3.1). Then player i's regrets $R_t^i(j, k)$ converge to zero almost surely for every j, k in S^i with $j \neq k$.*

Remark. Note that—in contrast to the Main Theorem, where every player uses (2.2)—no assumption is made in Theorem A on how players different from i choose their strategies (except for the fact that for every t, given the history up to t, play is independent among players). In the terminology of Fudenberg and Levine (1999, 1998), the adaptive procedure of this section is "(universally) calibrated." For an extended discussion of this issue, see Section 4(d).

What is the connection between regrets and correlated equilibria? It turns out that a necessary and sufficient condition for the empirical distributions to converge to the set of correlated equilibria is precisely that all regrets converge to zero. More generally, we have the following proposition.

Proposition. *Let $(s_t)_{t=1,2,...}$ be a sequence of plays (i.e., $s_t \in S$ for all t) and let[9] $\varepsilon \geq 0$. Then:* $\limsup_{t \to \infty} R_t^i(j,k) \leq \varepsilon$ *for every $i \in N$ and every $j,k \in S^i$ with $j \neq k$, if and only if the sequence of empirical distributions z_t (defined by (2.3)) converges to the set of correlated ε-equilibria.*

Proof.   For each player i and every $j \neq k$ in S^i we have

$$D_t^i(j,k) = \frac{1}{t} \sum_{\tau \leq t : s_\tau^i = j} [u^i(k, s_\tau^{-i}) - u^i(j, s_\tau^{-i})]$$

$$= \sum_{s \in S : s^i = j} z_t(s)[u^i(k, s^{-i}) - u^i(j, s^{-i})].$$

On any subsequence where z_t converges, say $z_{t'} \to \psi \in \Delta(S)$, we get

$$D_{t'}^i(j,k) \to \sum_{s \in S : s^i = j} \psi(s)[u^i(k, s^{-i}) - u^i(j, s^{-i})].$$

The result is immediate from the definition of a correlated ε-equilibrium and (2.1c).   □

Theorem A and the Proposition immediately imply the following corollary.

Corollary. *Suppose that at each period $t+1$ every player i chooses strategies according to a probability vector q_t^i that satisfies (3.1). Then the empirical distributions of play z_t converge almost surely as $t \to \infty$ to the set of correlated equilibria of the game Γ.*

Before addressing the formal proof of Theorem A, we shall present and discuss Blackwell's Approachability Theorem.

The basic setup contemplates a decision-maker i with a (finite) action set S^i. For a finite indexing set L, the decision-maker receives an $|L|$-dimensional

---

[9]Note that both $\varepsilon > 0$ and $\varepsilon = 0$ are included.

vector payoff $v(s^i, s^{-i}) \in \mathbb{R}^L$ that depends on his action $s^i \in S^i$ and on some external action s^{-i} belonging to a (finite) set S^{-i} (we will refer to $-i$ as the "opponent"). The decision problem is repeated through time. Let $s_t = (s_t^i, s_t^{-i}) \in S^i \times S^{-i}$ denote the choices at time t; of course, both i and $-i$ may use randomizations. The question is whether the decision-maker i can guarantee that the time average of the vector payoffs, $D_t := (1/t) \sum_{\tau \le t} v(s_\tau) \equiv (1/t) \sum_{\tau \le t} v(s_\tau^i, s_\tau^{-i})$, approaches a predetermined set (in $\mathbb{R}^L$).

Let $\mathscr{C}$ be a convex and closed subset of $\mathbb{R}^L$. The set $\mathscr{C}$ is *approachable* by the decision-maker i if there is a procedure[10] for i that guarantees that the average vector payoff D_t approaches the set $\mathscr{C}$ (i.e.,[11] $\mathrm{dist}(D_t, \mathscr{C}) \to 0$ almost surely as $t \to \infty$), regardless of the choices of the opponent $-i$. To state Blackwell's result, let $w_{\mathscr{C}}$ denote the *support function* of the convex set $\mathscr{C}$, i.e., $w_{\mathscr{C}}(\lambda) := \sup\{\lambda \cdot c : c \in \mathscr{C}\}$ for all λ in $\mathbb{R}^L$. Given a point $x \in \mathbb{R}^L$ which is not in $\mathscr{C}$, let $F(x)$ be the (unique) point in $\mathscr{C}$ that is closest to x in the Euclidean distance, and put $\lambda(x) := x - F(x)$; note that $\lambda(x)$ is an outward normal to the set $\mathscr{C}$ at the point $F(x)$.

Blackwell's Approachability Theorem. *Let $\mathscr{C} \subset \mathbb{R}^L$ be a convex and closed set, with support function $w_{\mathscr{C}}$. Then $\mathscr{C}$ is approachable by i if and only if for every $\lambda \in \mathbb{R}^L$ there exists a mixed strategy $q_\lambda \in \Delta(S^i)$ such that[12]*

$$\lambda \cdot v(q_\lambda, s^{-i}) \le w_{\mathscr{C}}(\lambda), \quad \text{for all } s^{-i} \in S^{-i}. \tag{3.2}$$

Moreover, the following procedure of i guarantees that $\mathrm{dist}(D_t, \mathscr{C})$ converges almost surely to 0 as $t \to \infty$: At time $t+1$, play $q_{\lambda(D_t)}$ if $D_t \notin \mathscr{C}$, and play arbitrarily if $D_t \in \mathscr{C}$.

We will refer to the condition for approachability given in the Theorem as the *Blackwell condition*, and to the procedure there as the *Blackwell procedure*. To get some intuition for the result, assume that D_t is not in $\mathscr{C}$, and let $\mathscr{H}(D_t)$ be the half-space of $\mathbb{R}^L$ that contains $\mathscr{C}$ (and not D_t) and is bounded by the supporting hyperplane to $\mathscr{C}$ at $F(D_t)$ with normal $\lambda(D_t)$; see Figure 1. When i uses the Blackwell procedure, it guarantees that $v(q_{\lambda(D_t)}, s^{-i})$ lies in $\mathscr{H}(D_t)$ for all s^{-i} in S^{-i} (by (3.2)). Therefore,

---

[10]In the repeated setup, we refer to a (behavior) strategy as a "procedure."

[11]$\mathrm{dist}(x, A) := \min\{\|x - a\| : a \in A\}$, where $\|\cdot\|$ is the Euclidean norm. Strictly speaking, Blackwell's definition of approachability requires also that the convergence of the distance to 0 be uniform over the procedures of the opponent; i.e., there is a procedure of i such that for every $\varepsilon > 0$ there is $t_0 \equiv t_0(\varepsilon)$ such that for any procedure of $-i$ we have $\Pr[\mathrm{dist}(D_t, \mathscr{C}) < \varepsilon \text{ for all } t > t_0] > 1 - \varepsilon$. The Blackwell procedure (defined in the next Theorem) guarantees this as well.

[12]$v(q, s^{-i})$ denotes the expected payoff, i.e., $\sum_{s^i \in S^i} q(s^i) v(s^i, s^{-i})$. Of course, only $\lambda \ne 0$ with $w_{\mathscr{C}}(\lambda) < \infty$ need to be considered in (3.2).

given D_t, the expectation of the next period payoff $E[v(s_{t+1})|D_t]$ will lie in the half-space $\mathscr{H}(D_t)$ for any pure choice s_{t+1}^{-i} of $-i$ at time $t+1$, and thus also for any randomized choice of $-i$. The expected average vector payoff at period $t+1$ (conditional on D_t) is

$$E[D_{t+1}|D_t] = \frac{t}{t+1}D_t + \frac{1}{t+1}E[v(s_{t+1})|D_t].$$

When t is large, $E[D_{t+1}|D_t]$ will thus be *inside* the circle of center $F(D_t)$ and radius $\|\lambda(D_t)\|$. Hence

$$\text{dist}(E[D_{t+1}|D_t], \mathscr{C}) \leq \|E[D_{t+1}|D_t] - F(D_t)\| < \|\lambda(D_t)\|$$
$$= \text{dist}(D_t, \mathscr{C})$$

(the first inequality follows from the fact that $F(D_t)$ is in $\mathscr{C}$). A precise computation shows that the distance not only decreases, but actually goes to zero.[13] For proofs of Blackwell's Approachability Theory, see[14] Blackwell (1956a), or Mertens, Sorin, and Zamir (1995, Theorem 4.3).

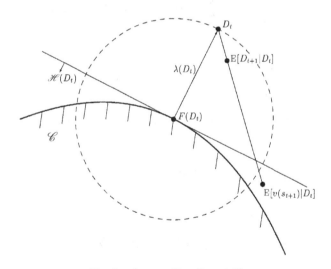

Fig. 1.   Approaching the set $\mathscr{C}$.

---

[13]Note that one looks here at *expected* average payoffs; the Strong Law of Large Numbers for Dependent Random Variables—see the Proof of Step M10 in the Appendix—implies that the *actual* average payoffs also converge to the set $\mathscr{C}$.

[14]The Blackwell condition is usually stated as follows: For every $x \notin \mathscr{C}$ there exists $q(x) \in \Delta(S^i)$ such that $[x - F(x)] \cdot [v(q(x), s^{-i}) - F(x)] \leq 0$, for all $s^{-i} \in S^{-i}$. It is easy to verify that this is equivalent to our formulation. We further note a simple way of stating the Blackwell result: A convex set $\mathscr{C}$ is approachable if and only if any half-space containing $\mathscr{C}$ is approachable.

We now prove Theorem A.

Proof of Theorem A.   As mentioned, the proof of this Theorem consists of an application of Blackwell's Approachability Theorem. Let

$$L := \{(j, k) \in S^i \times S^i : j \neq k\},$$

and define the vector payoff $v(s^i, s^{-i}) \in \mathbb{R}^L$ by letting its $(j, k) \in L$ coordinate be

$$[v(s^i, s^{-i})](j, k) := \begin{cases} u^i(k, s^{-i}) - u(j, s^{-i}), & \text{if } s^i = j, \\ 0, & \text{otherwise.} \end{cases}$$

Let $\mathscr{C}$ be the nonpositive orthant $\mathbb{R}^L_- := \{x \in \mathbb{R}^L : x \leq 0\}$. We claim that $\mathscr{C}$ is approachable by i. Indeed, the support function of $\mathscr{C}$ is given by $w_{\mathscr{C}}(\lambda) = 0$ for all $\lambda \in \mathbb{R}^L_+$ and $w_{\mathscr{C}}(\lambda) = \infty$ otherwise; so only $\lambda \in \mathbb{R}^L_+$ need to be considered. Condition (3.2) is

$$\sum_{(j,k)\in L} \lambda(j, k) \sum_{s^i \in S^i} q_\lambda(s^i)[v(s^i, s^{-i})](j, k) \leq 0,$$

or

$$\sum_{(j,k)\in L} \lambda(j, k) q_\lambda(j)[u^i(k, s^{-i}) - u^i(j, s^{-i})] \leq 0 \qquad (3.3)$$

for all $s^{-i} \in S^{-i}$. After collecting terms, the left-hand side of (3.3) can be written as

$$\sum_{j\in S^i} \alpha(j) u^i(j, s^{-i}), \qquad (3.4a)$$

where

$$\alpha(j) := \sum_{k\in S^i} q_\lambda(k)\lambda(k, j) - q_\lambda(j) \sum_{k\in S^i} \lambda(j, k). \qquad (3.4b)$$

Let $q_\lambda \in \Delta(S^i)$ be an invariant vector for the nonnegative $S^i \times S^i$ matrix with entries $\lambda(j, k)$ for $j \neq k$ and 0 for $j = k$ (such a q_λ always exists). That is, q_λ satisfies

$$\sum_{k\in S^i} q_\lambda(k)\lambda(k, j) = q_\lambda(j) \sum_{k\in S^i} \lambda(j, k), \qquad (3.5)$$

for every $j \in S^i$. Therefore $\alpha(j) = 0$ for all $j \in S^i$, and so inequality (3.3) holds true (as an equality[15]) for all $s^{-i} \in S^{-i}$. The Blackwell condition is thus satisfied by the set $\mathscr{C} = \mathbb{R}^L_-$.

Consider D_t, the average payoff vector at time t. Its (j,k)-coordinate is $(1/t) \sum_{\tau \le t} [v(s_\tau)](j,k) = D^i_t(j,k)$. If $D_t \notin \mathbb{R}^L_-$, then the closest point to D_t in $\mathbb{R}^L_-$ is $F(D_t) = [D_t]^-$ (see Figure 2), hence $\lambda(D_t) = D_t - [D_t]^- = [D_t]^+ = (R^i_t(j,k))_{(j,k) \in L}$, which is the vector of regrets at time t. Now the given strategy of i at time $t+1$ satisfies (3.1), which is exactly condition (3.5) for $\lambda = \lambda(D_t)$. Hence player i uses the Blackwell procedure for $\mathbb{R}^L_-$, which guarantees that the average vector payoff D_t approaches $\mathbb{R}^L_-$, or $R^i_t(j,k) \to 0$ a.s. for every $j \ne k$.   □

Remark. The proof of Blackwell's Approachability Theorem also provides bounds on the speed of convergence. In our case, one gets the following: The expectation $E[R^i_t(j,k)]$ of the regrets is of the order of $1/\sqrt{t}$, and the probability that z_t is a correlated ε-equilibrium for all $t > T$ is at least $1 - ce^{-cT}$ (for an appropriate constant $c > 0$ depending on ε; see Foster and Vohra 1999, Section 4.1). Clearly, a better speed of convergence[16] for

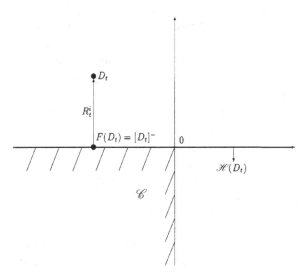

Fig. 2.   Approaching $\mathscr{C} = \mathbb{R}^L_-$.

---

[15]Note that this is precisely Formula (2) in the Proof of Theorem 1 in Hart and Schmeidler (1989); see Section 4(i).
[16]Up to a constant factor.

the expected regrets cannot be guaranteed, since, for instance, if the other players play stationary mixed strategies, then the errors are of the order $1/\sqrt{t}$ by the Central Limit Theorem.

4. Discussion

This section discusses a number of important issues, including links and comparisons to the relevant literature.

(a) Foster and Vohra. The seminal paper in this field of research is Foster and Vohra (1997). They consider, first, "forecasting rules"—on the play of others—that enjoy good properties, namely, "calibration." Second, they assume that each player best-replies to such calibrated forecasts. The resulting procedure leads to correlated equilibria. The motivation and the formulation are quite different from ours; nonetheless, their results are close to our results (specifically, to our Theorem A), since their calibrated forecasts are also based on regret measures.[17]

(b) Fudenberg and Levine. The next important paper is Fudenberg and Levine (1999) (see also their 1998 book). In that paper they offer a class of adaptive procedures, called "calibrated smooth fictitious play," with the property that for every $\varepsilon > 0$ there are procedures in the class that guarantee almost sure convergence to the set of correlated ε-equilibria (but the conclusion does not hold for $\varepsilon = 0$). The formal structure of these procedures is also similar to that of our Theorem A, in the sense that the mixed choice of a given player at time t is determined as an invariant probability vector of a transition matrix. However, the transition matrix (and therefore the stochastic dynamics) is different from the regret-based transition matrix of our Theorem A. To understand further the similarities and differences between the Fudenberg and Levine procedures and our own, the next two subsections, (c) and (d), contain a detour on the concepts of "universal consistency" and "universal calibration."

(c) Universal Consistency. The term "universal consistency" is due to Fudenberg and Levine (1995). The concept goes back to Hannan (1957), who

---

[17]These regrets are defined on an ε-grid on $\Delta(S^{-i})$, with ε going to zero as t goes to infinity. Therefore, at each step in their procedure one needs to compute the invariant vector for a matrix of an increasingly large size; by comparison, in our Theorem A the size of the matrix is fixed, $m^i \times m^i$.

proved the following result: There is a procedure (in the setup of Section 2) for player i that guarantees, no matter what the other players do, that

$$\limsup_{t \to \infty} \left[\max_{k \in S^i} \frac{1}{t} \sum_{\tau=1}^{t} u^i(k, s_\tau^{-i}) - \frac{1}{t} \sum_{\tau=1}^{t} u^i(s_\tau) \right] \leq 0 \quad \text{a.s.} \quad (4.1)$$

In other words, i's average payoff is, in the limit, no worse than if he were to play any *constant* strategy $k \in S^i$ for all $\tau \leq t$. This property of the Hannan procedure for player i is called *universal consistency* by Fudenberg and Levine (1995) (it is "universal" since it holds no matter how the other players play). Another universally consistent procedure was shown by Blackwell (1956b) to result from his Approachability Theorem (see also Luce and Raiffa 1957, pp. 482–483).

The adaptive procedure of our Theorem A is also universally consistent. Indeed, for each j in S', (4.1) is guaranteed even when restricted to those periods when player i chose that particular j; this being true for all j in S^i, the result follows. However, the application of Blackwell's Approachability Theorem in Section 3 suggests the following particularly simple procedure.

At time t, for each strategy k in S^i, let

$$D_t^i(k) := \frac{1}{t} \sum_{\tau=1}^{t} [u^i(k, s_\tau^{-i}) - u^i(s_\tau)], \quad (4.2a)$$

$$p_{t+1}^i(k) := \frac{[D_t^i(k)]^+}{\sum_{k' \in S^i} [D_t^i(k')]^+}, \quad (4.2b)$$

if the denominator is positive, and let $p_{t+1}^i \in \Delta(S^i)$ be arbitrary otherwise. The strategy of player i is then, at time $t + 1$, to choose k in S^i with probability $p_{t+1}^i(k)$. These probabilities are thus proportional to the "unconditional regrets" $[D_t^i(k)]^+$ (by comparison to the "conditional on j" regrets of Section 2). We then have the following theorem.

Theorem B. *The adaptive procedure* (4.2) *is universally consistent for player i.*

The proof of Theorem B is similar to the proof of Theorem A in Section 3 and is omitted.

Fudenberg and Levine (1995) propose a class of procedures that turn out to be universally ε-consistent[18]: "smooth fictitious play." Player i

---

[18]That is, the right-hand side of (4.1) is $\varepsilon > 0$ instead of 0.

follows a smooth fictitious play behavior rule if at time t he plays a mixed strategy $\sigma^i \in \Delta(S^i)$ that maximizes the sum of his expected payoff (with the actions of the remaining players distributed as in the empirical distribution up to t) and $\lambda v^i(\sigma^i)$, where $\lambda > 0$ and v^i is a strictly concave smooth function defined on i's strategy simplex, $\Delta(S^i)$, with infinite length gradient at the boundary of $\Delta(S^i)$. The result of Fudenberg and Levine is then that, given any $\varepsilon > 0$, there is a sufficiently small λ such that universal ε-consistency obtains for player i. Observe that, for small λ, smooth fictitious play is very close to fictitious play (it amounts to playing the best response with high probability and the remaining strategies with low but positive probability). The procedure is, therefore, clearly distinct from (4.2): In (4.2) all the better, even if not best, replies are played with significant probability; also, in (4.2) the inferior replies get zero probability. Finally, it is worth emphasizing that the tremble from best response is required for the Fudenberg and Levine result, since fictitious play is not guaranteed to be consistent. In contrast, the procedure of (4.2) has no trembles.

The reader is referred to Hart and Mas-Colell (2001a), where a wide class of universally consistent procedures is exhibited and characterized (including as special cases (4.2) as well as smooth fictitious play).

(d) Universal Calibration. The idea of "universal calibration," also introduced[19] by Fudenberg and Levine (1998, 1999), is that, again, regret measures go to zero irrespective of the other players' play. The difference is that, now, the set of regret measures is richer: It consists of regrets that are conditional on the strategy currently played by i himself. Recall the Proposition of Section 3: If such universally calibrated strategies are played by all players, then all regrets become nonpositive in the limit, and thus the convergence to the correlated equilibrium set is guaranteed.

The procedure of Theorem A is universally calibrated; so (up to ε) is the "calibrated smooth fictitious play" of Fudenberg and Levine (1999). The two procedures stand to each other as, in the unconditional version, Theorem B stands to "smooth fictitious play."

The procedure (2.2) of our Main Theorem is not universally calibrated. If only player i follows the procedure, we cannot conclude that all his regrets go to zero; adversaries who know the procedure used by player i could keep

---

[19]They actually call it "calibration"; we prefer the term "universal calibration," since it refers to *any* behavior of the opponents (as in their "[conditional] universal consistency").

his regrets positive.[20] Such sophisticated strategies of the other players, however, are outside the framework of our study—which deals with simple adaptive behavior. In fact, it turns out that the procedure of our Main Theorem is guaranteed to be calibrated not just against opponents using the same procedure, but also against a wide class of behaviors.[21]

We regard the simplicity of (2.2) as a salient point. Of course, if one needs to guarantee calibration even against sophisticated adversaries, one may have to give up on simplicity and resort to the procedure of Theorem A instead.

(e) Better-reply vs. Best-reply. Note that all the procedures in the literature reviewed above are best-reply-based: A player uses (almost) exclusively actions that are (almost) best-replies to a certain belief about his opponents. In contrast, our procedure gives significant probabilities to any actions that are just better (rather than best). This has the additional effect of making the behavior continuous, without need for approximations.

(f) Eigenvector Procedures. The procedure of our Main Theorem differs from all the other procedures leading to correlated equilibria (including that of our Theorem A) in an important aspect: It does not require the player to compute, at every step, an invariant (eigen-) vector for an appropriate positive matrix. Again, the simplicity[22] of (2.2) is an essential property when discussing unsophisticated behavior; this is the reason we have sought this result as our Main Theorem.

(g) Inertia. A specific and most distinctive feature by which the procedure of our Main Theorem differs from those of Theorem A and the other works mentioned above is that in the former the individual decisions privilege the most recent action taken: The probabilities used at period $t + 1$ are best thought of as propensities to depart from the play at t.

Viewed in this light, our procedure has significant inertial characteristics. In particular, there is a positive probability of moving from the strategy played at t only if there is another that appears better (in which case the

---

[20]At each time $t + 1$, let them play an $(N - 1)$-tuple of strategies that minimizes the expected (relative to p_{t+1}^i) payoff of player i; for an example, see Fudenberg and Levine (1998, Section 8.10).

[21]Namely, such that the dependence of any one choice of $-i$ on any one past choice of i is small, relative to the number of periods; see Cahn (2004).

[22]For a good test of the simplicity of a procedure, try to explain it verbally; in particular, consider the procedure of our Main Theorem vs. those requiring the computation of eigenvectors.

probabilities of playing the better strategies are proportional to the regrets relative to the period t strategy).[23]

(h) Friction. The procedure (2.2) exhibits "friction": There is always a positive probability of continuing with the period t strategy.[24] To understand the role played by friction,[25] suppose that we were to modify the procedure (2.2) by requiring that the switching probabilities be rescaled in such a way that a switch occurs if and only if there is at least one better strategy (i.e., one with positive regret). Then the result of the Main Theorem may not hold. For example, in the familiar two-person 2×2 coordination game, if we start with an uncoordinated strategy pair, then the play alternates between the two uncoordinated pairs. However, no distribution concentrated on these two pairs is a correlated equilibrium.

It is worth emphasizing that in our result the breaking away from a bad cycle, like the one just described, is obtained not by ergodic arguments but by the probability of staying put (i.e., by friction). What matters is that the diagonal of the transition matrix be positive, rather than that all the entries be positive (which, indeed, will not hold in our case).

(i) The Set of Correlated Equilibria. The set of correlated equilibria of a game is, in contrast to the set of Nash equilibria, geometrically simple: It is a convex set (actually, a convex polytope) of distributions. Since it includes the Nash equilibria we know it is nonempty. Hart and Schmeidler (1989) (see also Nau and McCardle 1990) provide an elementary (nonfixed point) proof of the nonemptiness of the set of correlated equilibria. This is done by using the Minimax Theorem. Specifically, Hart and Schmeidler proceed by associating to the given N-person game an auxiliary two-person zero-sum game. As it turns out, the correlated equilibria of the original game correspond to the maximin strategies of player I in the auxiliary game. More precisely, in the Hart–Schmeidler auxiliary game, player I chooses a distribution over N-tuples of actions, and player II chooses a pair of strategies for one of the N original players (interpreted as a play and a suggested deviation from it). The payoff to auxiliary player II is the expected gain

---

[23] It is worth pointing out that if a player's last choice was j, then the relative probabilities of switching to k or to k' do not depend only on the average utilities that would have been obtained if j had been changed to k or to k' in the past, but also on the average utility that was obtained in those periods by playing j itself (it is the magnitude of the increases in moving from j to k or to k' that matters).

[24] See Sanchirico (1996) and Section 4.6 in Fudenberg and Levine (1998) for a related point in a best-reply context.

[25] See Step M7 in the Proof of the Main Theorem in the Appendix.

of the designated original player if he were to follow the change suggested by auxiliary player II. In other words, it is the "regret" of that original player for not deviating. The starting point for our research was the observation that fictitious play applied to the Hart–Schmeidler auxiliary game must converge, by the result of Robinson (1951), and thus yield optimal strategies in the auxiliary game, in particular for player I—hence, correlated equilibria in the original game. A direct application of this idea does not, however, produce anything that is simple and separable across the N players (i.e., such that the choice of each player at time t is made independently of the other players' choices at t—an indispensable requirement).[26] Yet, our adaptive procedure is based on "no-regret" ideas motivated by this analysis and it is the direct descendant—several modifications later—of this line of research.[27]

(j) The Case of the Unknown Game. The adaptive procedure of Section 2 can be modified[28] to yield convergence to correlated equilibria also in the case where players neither know the game, nor observe the choices of the other players.[29] Specifically, in choosing play probabilities at time $t + 1$, a player uses information *only* on his own actual past play and payoffs (and *not* on the payoffs that would have been obtained if his past play had been different). The construction is based on replacing $D_t^i(j, k)$ (see (2.1b)) by

$$C_t^i(j,k) := \frac{1}{t} \left[\sum_{\tau \leq t : s_\tau^i = k} \frac{p_\tau^i(j)}{p_\tau^i(k)} u^i(s_\tau) - \sum_{\tau \leq t : s_\tau^i = j} u^i(s_\tau) \right].$$

Thus, the payoff that player i would have received had he played k rather than j is estimated by the actual payoffs he obtained when he did play k in the past.

For precise formulations, results and proofs, as well as further discussions, the reader is referred to Hart and Mas-Colell (2001b).

---

[26]This needed "decoupling" across the N original players explains why applying linear programming-type methods to reach the convex polytope of correlated equilibria is not a fruitful approach. The resulting procedures operate in the space of N-tuples of strategies S (more precisely, in $\Delta(S)$), whereas adaptive procedures should be defined for each player i separately (i.e., on $\Delta(S^i)$).

[27]For another interesting use of the auxiliary two-person zero-sum game, see Myerson (1997).

[28]Following a suggestion of Dean Foster.

[29]For similar constructions, see Baños (1968), Megiddo (1980), Foster and Vohra (1993), Auer *et al.* (1995), Roth and Erev (1995), Erev and Roth (1998), Camerer and Ho (1998), Marimon (1996, Section 3.4), and Fudenberg and Levine (1998, Section 4.8). One may view this type of result in terms of "stimulus-response" decision behavior models.

Appendix: Proof of the Main Theorem

This Appendix is devoted to the proof of the Main Theorem, stated in Section 2. The proof is inspired by the result of Section 3 (Theorem A). It is however more complex on account of our transition probabilities not being the invariant measures that, as we saw in Section 3, fitted so well with Blackwell's Approachability Theorem.

As in the standard proof of Blackwell's Approachability Theorem, the proof of our Main Theorem is based on a recursive formula for the distance of the vector of regrets to the negative orthant. However, our procedure (2.2) does not satisfy the Blackwell condition; it is rather a sort of iterative approximation to it. Thus, a simple one-period recursion (from t to $t + 1$) does not suffice, and we have to consider instead a multi-period recursion where a large "block" of periods, from t to $t + v$, is combined together. Both t and v are carefully chosen; in particular, t and v go to infinity, but v is relatively small compared to t.

We start by introducing some notation. Fix player i in N. For simplicity, we drop reference to the index i whenever this cannot cause confusion (thus we write D_t and R_t instead of D_t^i and R_t^i, and so on). Let $m := |S^i|$ be the number of strategies of player i, and let M be an upper bound on i's possible payoffs: $M \geq |u^i(s)|$ for all s in S. Denote $L := \{(j, k) \in S^i \times S^i : j \neq k\}$; then $\mathbb{R}^L$ is the $m(m-1)$-dimensional Euclidean space with coordinates indexed by L. For each $t = 1, 2, \dots$ and each (j, k) in L, put[30]

$$A_t(j, k) = 1_{\{s_t^i = j\}}[u^i(k, s_t^{-i}) - u^i(s_t)],$$

$$D_t(j, k) = \frac{1}{t} \sum_{\tau=1}^{t} A_\tau(j, k),$$

$$R_t(j, k) = D_t^+(j, k) \equiv [D_t(j, k)]^+.$$

We shall write A_t for the vector $(A_t(j, k))_{j \neq k} \in \mathbb{R}^L$; the same goes for D_t, D_t^+, R_t, and so on. Let $\Pi_t(\cdot, \cdot)$ denote the transition probabilities from t to $t + 1$ (these are computed after period t, based on h_t):

$$\Pi_t(j, k) := \begin{cases} \dfrac{1}{\mu} R_t(j, k), & \text{if } k \neq j, \\ 1 - \displaystyle\sum_{k' \neq j} \dfrac{1}{\mu} R_t(j, k'), & \text{if } k = j. \end{cases}$$

---

[30]We write 1_G for the indicator of the event G.

Thus, at time $t+1$ the strategy used by player i is to choose each $k \in S^i$ with probability $p_{t+1}^i(k) = \Pi_t(s_t^i, k)$. Note that the choice of μ guarantees that $\Pi_t(j,j) > 0$ for all $j \in S^i$ and all t. Finally, let

$$\rho_t := [\text{dist}(D_t, \mathbb{R}_-^L)]^2$$

be the squared distance (in $\mathbb{R}^L$) of the vector D_t to the nonpositive orthant $\mathbb{R}_-^L$. Since the closest point to D_t in $\mathbb{R}_-^L$ is[31] D_t^-, we have $\rho_t = \|D_t - D_t^-\|^2 = \|D_t^+\|^2 = \sum_{j \neq k}[D_t^+(j,k)]^2$.

It will be convenient to use the standard "O" notation: For two real-valued functions $f(\cdot)$ and $g(\cdot)$ defined on a domain X, "$f(x) = \mathrm{O}(g(x))$" means that there exists a constant $K < \infty$ such that $|f(x)| \leq Kg(x)$ for all x in[32] X. We write Pr for Probability, and E for Expectation. From now on, t, v, and w will denote positive integers; $h_t = (s_\tau)_{\tau \leq t}$ will be histories of length t; j, k, and s^i will be elements of S^i; s and s^{-i} will be elements of S and S^{-i}, respectively. Unless stated otherwise, all statements should be understood to hold "for all t, v, h_t, j, k, etc."; when histories h_t are concerned, only those that occur with positive probability are considered.

We divide the proof of the Main Theorem into 11 steps, M1–M11, which we now state formally; an intuitive guide follows.

- **Step M1:**

(i) $\mathrm{E}[(t+v)^2 \rho_{t+v}|h_t] \leq t^2 \rho_t + 2t \sum_{w=1}^{v} R_t \cdot \mathrm{E}[A_{t+w}|h_t] + \mathrm{O}(v^2)$; and

(ii) $(t+v)^2 \rho_{t+v} - t^2 \rho_t = \mathrm{O}(tv + v^2)$.

Define

$$\alpha_{t,w}(j, s^{-i}) := \sum_{k \in S'} \Pi_t(k, j) \Pr[s_{t+w} = (k, s^{-i})|h_t] - \Pr[s_{t+w} = (j, s^{-i})|h_t].$$

- **Step M2:**

$$R_t \cdot \mathrm{E}[A_{t+w}|h_t] = \mu \sum_{s^{-i} \in S^{-i}} \sum_{j \in S^i} \alpha_{t,w}(j, s^{-i}) u^i(j, s^{-i}).$$

- **Step M3:**

$$R_{t+v}(j,k) - R_t(j,k) = \mathrm{O}\left(\frac{v}{t}\right).$$

---

[31] We write $[x]^-$ for $\min\{x, 0\}$, and D_t^- for the vector $([D_t(j,k)]^-)_{(j,k) \in L}$.

[32] The domain X will usually be the set of positive integers, or the set of vectors whose coordinates are positive integers. Thus when we write, say, $f(t,v) = \mathrm{O}(v)$, it means $|f(t,v)| \leq Kv$ for all v and t. The constants K will always depend only on the game (through N, m, M, and so on) and on the parameter μ.

For each $t > 0$ and each history h_t, define an auxiliary stochastic process $(\hat{s}_{t+w})_{w=0,1,2,\ldots}$ with values in S as follows: The initial value is $\hat{s}_t = s_t$, and the transition probabilities are[33]

$$\Pr[\hat{s}_{t+w} = s|\hat{s}_t,\ldots,\hat{s}_{t+w-1}] := \prod_{i' \in N} \Pi_t^{i'}(\hat{s}_{t+w-1}^{i'}, s^{i'}).$$

(The $\hat{s}$-process is thus stationary: It uses the transition probabilities of period t at each period $t + w$, for all $w > 0$.)

- **Step M4:**

$$\Pr[s_{t+w} = s|h_t] - \Pr[\hat{s}_{t+w} = s|h_t] = O\left(\frac{w^2}{t}\right).$$

Define

$$\hat{\alpha}_{t,w}(j, s^{-i}) := \sum_{k \in S^i} \Pi_t(k, j)\Pr[\hat{s}_{t+w} = (k, s^{-i})|h_t] - \Pr[\hat{s}_{t+w} = (j, s^{-i})|h_t].$$

- **Step M5:**

$$\alpha_{t,w}(j, s^{-i}) - \hat{\alpha}_{t,w}(j, s^{-i}) = O\left(\frac{w^2}{t}\right).$$

- **Step M6:**

$$\hat{\alpha}_{t,w}(j, s^{-i}) = \Pr[\hat{s}_{t+w}^{-i} = s^{-i}|h_t][\Pi_t^{w+1} - \Pi_t^w](s_t^i, j),$$

where $\Pi_t^w \equiv (\Pi_t)^w$ is the wth power of the matrix Π_t, and $[\Pi_t^{w+1} - \Pi_t^w](s_t^i, j)$ denotes the (s_t^i, j) element of the matrix $\Pi_t^{w+1} - \Pi_t^w$.

- **Step M7:**

$$\hat{\alpha}_{t,w}(j, s^{-i}) = O(w^{-1/2}).$$

- **Step M8:**

$$E[(t + v)^2 \rho_{t+v}|h_t] \le t^2 \rho_t + O(v^3 + tv^{1/2}).$$

For each $n = 1, 2, \ldots$, let $t_n := \lfloor n^{5/3} \rfloor$ be the largest integer not exceeding $n^{5//3}$.

- **Step M9:**

$$E[t_{n+1}^2 \rho_{t_{n+1}}|h_{t_n}] \le t_n^2 \rho_{t_n} + O(n^2).$$

---

[33]We write $\Pi_t^{i'}$ for the transition probability matrix of player i' (thus Π_t is Π_t^i).

- **Step M10:**

$$\lim_{n \to \infty} \rho_{t_n} = 0 \quad \text{a.s.}$$

- **Step M11:**

$$\lim_{t \to \infty} R_t(j, k) = 0 \quad \text{a.s.}$$

We now provide an intuitive guide to the proof. The first step, M1(i), is our basic recursion equation. In Blackwell's Theorem, the middle term on the right-hand side vanishes (it is ≤ 0 by (3.2)). This is not so in our case; Steps M2–M8 are thus devoted to estimating this term. Step M2 yields an expression similar to (3.4), but here the coefficients α depend also on the moves of the other players. Indeed, given h_t, the choices s_{t+w}^i and s_{t+w}^{-i} are *not* independent when $w > 1$ (since the transition probabilities change with time). Therefore we replace the process $(s_{t+w})_{0 \leq w \leq v}$ by another process $(\hat{s}_{t+w})_{0 \leq w \leq v}$, with a *stationary* transition matrix (that of period t). For w small relative to t, the change in probabilities is small (see Steps M3 and M4), and we estimate the total difference (Step M5). Next (Step M6), we factor out the moves of the other players (which, in the $\hat{s}$-process, are independent of the moves of player i) from the coefficients $\hat{a}$. At this point we get the difference between the transition probabilities after w periods and after $w + 1$ periods (for comparison, in formula (3.4) we would replace both by the invariant distribution, so the difference vanishes). This difference is shown (Step M7) to be small, since w is large and the transition matrix has all its diagonal elements strictly positive.[34] Substituting in M1(i) yields the final recursive formula (Step M8). The proof is now completed (Steps M9–M11) by considering a carefully chosen subsequence of periods $(t_n)_{n=1,2,\dots}$.

The rest of this Appendix contains the proofs of the Steps M1–M11.

- **Proof of Step M1.** Because $D_t^- \in \mathbb{R}_-^L$ we have

$$\rho_{t+v} \leq \|D_{t+v} - D_t^-\|^2 = \left\| \frac{t}{t+v} D_t + \frac{1}{t+v} \sum_{w=1}^{v} A_{t+w} - D_t^- \right\|^2$$

$$= \frac{t^2}{(t+v)^2} \|D_t - D_t^-\|^2 + \frac{2t}{(t+v)^2} \sum_{w=1}^{v} (A_{t+w} - D_t^-) \cdot (D_t - D_t^-)$$

---

[34] For further discussion on this point, see the Proof of Step M7.

$$+ \frac{v^2}{(t+v)^2} \left\| \frac{1}{v} \sum_{w=1}^{v} A_{t+w} - D_t^- \right\|^2$$

$$\leq \frac{t^2}{(t+v)^2} \rho_t + \frac{2t}{(t+v)^2} \sum_{w=1}^{v} A_{t+w} \cdot R_t + \frac{v^2}{(t+v)^2} m(m-1)16M^2.$$

Indeed: $|u^i(s)| \leq M$, so $|A_{t+w}(j,k)| \leq 2M$ and $|D_t(j,k)| \leq 2M$, yielding the upper bound on the third term. As for the second term, note that $R_t = D_t^+ = D_t - D_t^-$ and $D_t^- \cdot D_t^+ = 0$. This gives the bound of (ii). To get (i), take conditional expectation given the history h_t (so ρ_t and R_t are known). □

• **Proof of Step M2.** We have

$$E[A_{t+w}(j,k)|h_t] = \sum_{s^{-i}} \phi(j,s^{-i})[u^i(k,s^{-i}) - u^i(j,s^{-i})],$$

where $\phi(j,s^{-i}) := \Pr[s_{t+w} = (j,s^{-i})|h_t]$. So

$$R_t \cdot E[A_{t+w}|h_t]$$

$$= \sum_{j} \sum_{k \neq j} R_t(j,k) \sum_{s^{-i}} \phi(j,s^{-i})[u^i(k,s^{-i}) - u^i(j,s^{-i})]$$

$$= \sum_{s^{-i}} \sum_{j} u^i(j,s^{-i}) \left[\sum_{k \neq j} R_t(k,j)\phi(k,s^{-i}) - \sum_{k \neq j} R_t(j,k)\phi(j,s^{-i}) \right]$$

(we have collected together all terms containing $u^i(j,s^{-i})$). Now, $R_t(k,j) = \mu \Pi_t(k,j)$ for $k \neq j$, and $\sum_{k \neq j} R_t(j,k) = \mu(1 - \Pi_t(j,j))$ by definition, so

$$R_t \cdot E[A_{t+w}|h_t] = \mu \sum_{s^{-i}} \sum_{j} u^i(j,s^{-i}) \left[\sum_{k} \Pi_t(k,j)\phi(k,s^{-i}) - \phi(j,s^{-i}) \right]$$

(note that the last sum is now over *all* k in S^i). □

• **Proof of Step M3.** This follows immediately from

$$(t+v)[D_{t+v}(j,k) - D_t(j,k)] = \sum_{w=1}^{v} A_{t+w}(j,k) - vD_t(j,k),$$

together with $|A_{t+w}(j,k)| \leq 2M$ and $|D_t(j,k)| \leq 2M$. □

• **Proof of Step M4.** We need the following Lemma, which gives bounds for the changes in the w-step transition probabilities as a function of changes in the 1-step transitions.

Lemma. *Let $(X_n)_{n \geq 0}$ and $(Y_n)_{n \geq 0}$ be two stochastic processes with values in a finite set B. Assume $X_0 = Y_0$ and*

$$|\Pr[X_n = b_n | X_0 = b_0, \ldots, X_{n-1} = b_{n-1}]$$
$$-\Pr[Y_n = b_n | Y_0 = b_0, \ldots, Y_{n-1} = b_{n-1}]| \leq \beta_n$$

for all $n \geq 1$ and all $b_0, \ldots, b_{n-1}, b_n \in B$. Then

$$|\Pr[X_{n+w} = b_{n+w} | X_0 = b_0, \ldots, X_{n-1} = b_{n-1}]$$
$$- \Pr[Y_{n+w} = b_{n+w} | Y_0 = b_0, \ldots, Y_{n-1} = b_{n-1}]| \leq |B| \sum_{r=0}^{w} \beta_{n+r}$$

for all $n \geq 1, w \geq 0$, and all $b_0, \ldots, b_{n-1}, b_{n+w} \in B$.

Proof. We write P_X and P_Y for the probabilities of the two processes $(X_n)_n$ and $(Y_n)_n$, respectively (thus $P_X[b_{n+w}|b_0, \ldots, b_{n-1}]$ stands for $\Pr[X_{n+w} = b_{n+w} | X_0 = b_0, \ldots, X_{n-1} = b_{n-1}]$, and so on). The proof is by induction on w.

$$P_X[b_{n+w}|b_0, \ldots, b_{n-1}]$$

$$= \sum_{b_n} P_X[b_{n+w}|b_0, \ldots, b_n] P_X[b_n|b_0, \ldots, b_{n-1}]$$

$$\leq \sum_{b_n} P_Y[b_{n+w}|b_0, \ldots, b_n] P_X[b_n|b_0, \ldots, b_{n-1}] + |B| \sum_{r=1}^{w} \beta_{n+r}$$

$$\leq \sum_{b_n} P_Y[b_{n+w}|b_0, \ldots, b_n](P_Y[b_n|b_0, \ldots, b_{n-1}] + \beta_n) + |B| \sum_{r=1}^{w} \beta_{n+r}$$

$$\leq P_Y[b_{n+w}|b_0, \ldots, b_{n-1}] + |B|\beta_n + |B| \sum_{r=1}^{w} \beta_{n+r}$$

(the first inequality is by the induction hypothesis). Exchanging the roles of X and Y completes the proof. $\quad\square$

We proceed now with the proof of Step M4. From t to $t + w$ there are $|N|w$ transitions (at each period, think of the players moving one after the other, in some arbitrary order). Step M3 implies that each transition probability for the $\hat{s}$-process differs from the corresponding one for the s-process by at most $O(w/t)$, which yields, by the Lemma, a total difference of $|N|w|S|O(w/t) = O(w^2/t)$. $\quad\square$

● **Proof of Step M5.** Immediate by Step M4.   □

● **Proof of Step M6.** Given h_t, the random variables $(\hat{s}^{i'}_{t+w})_w$ are independent over the different players i' in N; indeed, the transition probabilities are all determined at time t, and the players randomize independently. Hence:

$$\Pr[\hat{s}_{t+w} = (j, s^{-i})|h_t] = \Pr[\hat{s}^{-i}_{t+w} = s^{-i}|h_t]\Pr[\hat{s}^i_{t+w} = j|h_t],$$

implying that

$$\hat{\alpha}_{t,w}(j, s^{-i})$$

$$= \Pr[\hat{s}^{-i}_{t+w} = s^{-i}|h_t]\left[\sum_{k \in S^i} \Pi_t(k, j)\Pr[\hat{s}^i_{t+w} = k|h_t] - \Pr[\hat{s}^i_{t+w} = j|h_t]\right].$$

Now $\Pr[\hat{s}^i_{t+w} = j|h_t]$ is the probability of reaching j in w steps starting from s^i_t, using the transition probability matrix Π_t. Therefore $\Pr[\hat{s}^i_{t+w} = j|h_t]$ is the (s^i_t, j)-element of the wth power $\Pi^w_t \equiv (\Pi_t)^w$ of Π_t, i.e., $[\Pi^w_t](s^i_t, j)$. Hence

$$\hat{\alpha}_{t,w}(j, s^{-i}) = \Pr[\hat{s}^{-i}_{t+w} = s^{-i}|h_t]\left[\sum_{k \in S^i} \Pi_t(k, j)[\Pi^w_t](s^i_t, k) - [\Pi^w_t](s^i_t, j)\right]$$

$$= \Pr[\hat{s}^{-i}_{t+w} = s^{-i}|h_t][[\Pi^{w+1}_t](s^i_t, j) - [\Pi^w_t](s^i_t, j)],$$

completing the proof.   □

● **Proof of Step M7.** It follows from M6 using the following Lemma (recall that $\Pi_t(j, j) > 0$ for all $j \in S^i$).

Lemma. *Let* Π *be an* $m \times m$ *stochastic matrix with all of its diagonal entries positive. Then* $[\Pi^{w+1} - \Pi^w](j, k) = O(w^{-1/2})$ *for all* $j, k = 1, \ldots, m$.

Proof.[35] Let $\beta > 0$ be a lower bound on all the diagonal entries of Π, i.e., $\beta := \min_j \Pi(j, j)$. We can then write $\Pi = \beta I + (1 - \beta)\Lambda$, where Λ is also a

---

[35]If Π were a strictly positive matrix, then $\Pi^{w+1} - \Pi^w \to 0$ would be a standard result, because then Π^w would converge to the invariant matrix. However, we know only that the diagonal elements are positive. This implies that, if w is large, then with high probability there will be a positive fraction of periods when the process does not move. But this number is random, so the probabilities of going from j to k in w steps or in $w + 1$ steps should be almost the same (since it is like having r "stay put" transitions versus $r + 1$).

stochastic matrix. Now

$$\Pi^w = \sum_{r=0}^{w} \binom{w}{r} \beta^{w-r}(1-\beta)^r \Lambda^r,$$

and similarly for Π^{w+1}. Subtracting yields

$$\Pi^{w+1} - \Pi^w = \sum_{r=0}^{w+1} \gamma_r \binom{w}{r} \beta^{w-r}(1-\beta)^r \Lambda^r,$$

where $\gamma_r := \beta(w+1)/(w+1-r) - 1$. Now $\gamma_r > 0$ if $r > q := (w+1)(1-\beta)$, and $\gamma_r \leq 0$ if $r \leq q$; together with $0 \leq \Lambda^r(j,k) \leq 1$, we get

$$\sum_{r\leq q} \gamma_r \binom{w}{r} \beta^{w-r}(1-\beta)^r \leq [\Pi^{w+1} - \Pi^w](j,k) \leq \sum_{r>q} \gamma_r \binom{w}{r} \beta^{w-r}(1-\beta)^r.$$

Consider the left-most sum. It equals

$$\sum_{r\leq q} \binom{w+1}{r} \beta^{w+1-r}(1-\beta)^r - \sum_{r\leq q} \binom{w}{r} \beta^{w-r}(1-\beta)^r = G_{w+1}(q) - G_w(q),$$

where $G_n(\cdot)$ denotes the cumulative distribution function of a sum of n independent Bernoulli random variables, each one having the value 0 with probability β and the value 1 with probability $1 - \beta$. Using the normal approximation yields (Φ denotes the standard normal cumulative distribution function):

$$G_{w+1}(q) - G_w(q) = \Phi(x) - \Phi(y) + O\left(\frac{1}{\sqrt{(w+1)}}\right) + O\left(\frac{1}{\sqrt{w}}\right),$$

where

$$x := \frac{q - (w+1)(1-\beta)}{\sqrt{(w+1)\beta(1-\beta)}} \quad \text{and} \quad y := \frac{q - w(1-\beta)}{\sqrt{w\beta(1-\beta)}};$$

the two error terms $O((w + 1)^{-1/2})$ and $O(w^{-1/2})$ are given by the Berry–Esséen Theorem (see Feller 1965, Theorem XVI.5.1). By definition of q we have $x = 0$ and $y = O(w^{-1/2})$. The derivative of Φ is bounded, so $\Phi(x) - \Phi(y) = O(x - y) = O(w^{-1/2})$. Altogether, the left-most sum is $O(w^{-1/2})$. A similar computation applies to the right-most sum. $\qquad\square$

• **Proof of Step M8.** Steps M5 and M7 imply $\alpha_{t,w}(j, s^{-i}) = O(w^2/t + w^{-1/2})$. The formula of Step M2 then yields

$$R_t \cdot \mathrm{E}[A_{t+w}|h_t] = O\left(\frac{w^2}{t} + w^{-1/2}\right).$$

Adding over $w = 1, 2, \ldots, v$ (note that $\sum_{w=1}^{v} w^\lambda = O(v^{\lambda+1})$ for $\lambda \neq -1$) and substituting into Step M1(i) gives the result.   □

• **Proof of Step M9.** We use the inequality of Step M8 for $t = t_n$ and $v = t_{n+1} - t_n$. Because $v = \lfloor (n+1)^{5/3} \rfloor - \lfloor n^{5/3} \rfloor = O(n^{2/3})$, we have $v^3 = O(n^2)$ and $tv^{1/2} = O(n^{5/3+1/3}) = O(n^2)$, and the result follows.   □

• **Proof of Step M10.** We use the following result (see Loève 1978, Theorem 32.1.E):

Theorem (Strong Law of Large Numbers for Dependent Random Variables). *Let X_n be a sequence of random variables and b_n a sequence of real numbers increasing to ∞, such that the series $\sum_{n=1}^{\infty} \mathrm{var}(X_n)/b_n^2$ converges. Then*

$$\frac{1}{b_n}\sum_{v=1}^{n}[X_v - \mathrm{E}[X_v|X_1, \ldots, X_{v-1}]] \xrightarrow[n\to\infty]{} 0 \quad a.s.$$

We take $b_n := t_n^2$, and $X_n := b_n\rho_{t_n} - b_{n-1}\rho_{t_{n-1}} = t_n^2\rho_{t_n} - t_{n-1}^2\rho_{t_{n-1}}$. By Step M1(ii) we have $|X_n| \leq O(t_n v_n + v_n^2) = O(n^{7/3})$, thus $\sum_n \mathrm{var}(x_n)/b_n^2 = \sum_n O(n^{14/3})/n^{20/3} = \sum_n O(1/n^2) < \infty$. Next, Step M9 implies

$$(1/b_n) \sum_{v \leq n} \mathrm{E}[X_v|X_1, \ldots, X_{v-1}] \leq O\left(n^{-10/3} \sum_{v \leq n} v^2\right)$$
$$= O(n^{-10/3} n^3) = O(n^{-1/3}) \to 0.$$

Applying the Theorem above thus yields that ρ_{t_n}, which is nonnegative and equals $(1/b_n) \sum_{v \leq n} X_v$, must converge to 0 a.s.   □

• **Proof of Step M11.** Since $\rho_{t_n} = \sum_{j \neq k} [R_{t_n}(j, k)]^2$, the previous Step M10 implies that $R_{t_n}(j, k) \to 0$ a.s. $n \to \infty$, for all $j \neq k$. When $t_n \leq t \leq t_{n+1}$, we have $R_t(j, k) - R_{t_n}(j, k) = O(n^{-1})$ by the inequality of Step M3, so $R_t(j, k) \to 0$ a.s. $t \to \infty$.   □

References

Auer, P., N. Cesa-Bianchi, Y. Freund, and R. E. Schapire (1995), "Gambling in a Rigged Casino: The Adversarial Multi-Armed Bandit Problem," in *Proceedings of the 36th Annual Symposium on Foundations of Computer Science*, pp. 322–331.

Aumann, R. J. (1974), "Subjectivity and Correlation in Randomized Strategies," *Journal of Mathematical Economics*, 1, 67–96.

―――― (1987), "Correlated Equilibrium as an Expression of Bayesian Rationality," *Econometrica*, 55, 1–18.

Baños, A. (1968), "On Pseudo-Games," *The Annals of Mathematical Statistics*, 39, 1932–1945.

Blackwell, D. (1956a), "An Analog of the Minmax Theorem for Vector Payoffs," *Pacific Journal of Mathematics*, 6, 1–8.

―――― (1956b), "Controlled Random Walks," in *Proceedings of the International Congress of Mathematicians* 1954, Vol. III, ed. by E. P. Noordhoff. Amsterdam: North-Holland, pp. 335–338.

Brown, G. W. (1951), "Iterative Solutions of Games by Fictitious Play," in *Activity Analysis of Production and Allocation*, Cowles Commission Monograph 13, ed. by T. C. Koopmans. New York: Wiley, pp. 374–376.

Cahn, A. (2004), "General Procedures Leading to Correlated Equilibria," *International Journal of Game Theory*, 33, 21–40. [Chapter 6]

Camerer, C. and T.-H. Ho (1998), "Experience-Weighted Attraction Learning in Coordination Games: Probability Rules, Heterogeneity, and Time-Variation," *Journal of Mathematical Psychology*, 42, 305–326.

Erev, I. and A. E. Roth (1998), "Predicting How People Play Games: Reinforcement Learning in Experimental Games with Unique, Mixed Strategy Equilibria," *American Economic Review*, 88, 848–881.

Feller, W. (1965), *An Introduction to Probability Theory and Its Applications*, Vol. II, 2nd edition. New York: Wiley.

Foster, D. and R. V. Vohra (1993), "A Randomization Rule for Selecting Forecasts," *Operations Research*, 41, 704–709.

―――― (1997), "Calibrated Learning and Correlated Equilibrium," *Games and Economic Behavior*, 21, 40–55.

―――― (1998), "Asymptotic Calibration," *Biometrika*, 85, 379–390.

―――― (1999), "Regret in the Online Decision Problem," *Games and Economic Behavior*, 29, 7–35.

Fudenberg, D. and D. K. Levine (1995), "Universal Consistency and Cautious Fictitious Play," *Journal of Economic Dynamics and Control*, 19, 1065–1089.

―――― (1998), *Theory of Learning in Games*. Cambridge, MA: The MIT Press.

―――― (1999), "Conditional Universal Consistency," *Games and Economic Behavior*, 29, 104–130.

Hannan, J. (1957), "Approximation to Bayes Risk in Repeated Play," in *Contributions to the Theory of Games*, Vol. III, Annals of Mathematics Studies 39, ed. by M. Dresher, A. W. Tucker, and P. Wolfe. Princeton: Princeton University Press, pp. 97–139.

Hart, S. and A. Mas-Colell (2001a), "A General Class of Adaptive Strategies," *Journal of Economic Theory*, 98, 26–54. [Chapter 3]

—— (2001b), "A Reinforcement Procedure Leading to Correlated Equilibrium," in *Economic Essays*, ed. by G. Debreu, W. Neuefeind, and W. Trockel. Berlin: Springer, pp. 181–200. [Chapter 4]

Hart, S. and D. Schmeidler (1989), "Existence of Correlated Equilibria," *Mathematics of Operations Research*, 14, 18–25. [Chapter 1]

Loève, M. (1978), *Probability Theory*, Vol. II, 4th edition. Berlin: Springer.

Luce, R. D. and H. Raiffa (1957), *Games and Decisions*. New York: Wiley.

Marimon, R. (1996), "Learning From Learning in Economics," in *Advances in Economic Theory*, ed. by D. Kreps. Cambridge, UK: Cambridge University Press.

Megiddo, N. (1980), "On Repeated Games with Incomplete Information Played by Non-Bayesian Players," *International Journal of Game Theory*, 9, 157–167.

Mertens, J.-F., S. Sorin, and S. Zamir (1995), "Repeated Games, Part A," CORE DP-9420 (mimeo).

Myerson, R. B. (1997), "Dual Reduction and Elementary Games," *Games and Economic Behavior*, 21, 183–202.

Nau, R. F. and K. F. McCardle (1990), "Coherent Behavior in Noncooperative Games," *Journal of Economic Theory*, 50, 424–444.

Robinson, J. (1951), "An Iterative Method of Solving a Game," *Annals of Mathematics*, 54, 296–301.

Roth, A. E. and I. Erev (1995), "Learning in Extensive-Form Games: Experimental Data and Simple Dynamic Models in the Intermediate Term," *Games and Economic Behavior*, 8, 164–212.

Sanchirico, C. W. (1996), "A Probabilistic Model of Learning in Games," *Econometrica*, 64, 1375–1393.

Chapter 3

A GENERAL CLASS OF ADAPTIVE STRATEGIES*

Sergiu Hart and Andreu Mas-Colell

We exhibit and characterize an entire class of simple adaptive strategies, in the repeated play of a game, having the Hannan-consistency property: in the long run, the player is guaranteed an average payoff as large as the best-reply payoff to the empirical distribution of play of the other players; i.e., there is no "regret." Smooth fictitious play (Fudenberg and Levine 1995) and regret matching (Hart and Mas-Colell 2000) are particular cases. The motivation and application of the current paper come from the study of procedures whose empirical distribution of play is, in the long run, (almost) a correlated equilibrium. For the analysis we first develop a generalization of Blackwell's (1956) approachability strategy for games with vector payoffs.

1. Introduction

Consider a game repeated through time. We are interested in strategies of play which, while simple to implement, generate desirable outcomes. Such strategies, typically consisting of moves in "improving" directions, are usually referred to as adaptive.

Originally published in *Journal of Economic Theory*, 98 (2001), 26–54.

*Received April 19, 1999; final version received August 16, 2000.
JEL classification: C7, D7, C6.
Keywords: adaptive strategies; approachability; correlated equilibrium; fictitious play; regret; regret matching; smooth fictitious play.
The research was partially supported by grants of the Israel Academy of Sciences and Humanities, the Spanish Ministry of Education, the Generalitat de Catalunya, CREI, and the EU-TMR Research Network.
The authors thank the referees and editors for their useful comments.

In Hart and Mas-Colell (2000) we presented simple adaptive strategies with the property that, if used by all players, the empirical distribution of play is, in the long run, (almost) a correlated equilibrium of the game; for other procedures leading to correlated equilibria, see Foster and Vohra (1997) and Fudenberg and Levine (1999).[1] From this work we are led—for reasons we will comment upon shortly—to the study of a concept originally introduced by Hannan (1957). A strategy of a player is called *Hannan-consistent* if it guarantees that his long-run average payoff is as large as the highest payoff that can be obtained (i.e., the one-shot best-reply payoff) against the empirical distribution of play of the other players. In other words, a strategy is Hannan-consistent if, given the play of the others, there is no regret in the long run for not having played (constantly) any particular action. As a matter of terminology, the *regret* of player i for an action[2] k at period t is the difference in his average payoff up to t that results from replacing his actual past play by the constant play of action k. Hannan-consistency thus means that all regrets are nonpositive, as t goes to infinity.

In this paper we concentrate on the notion of Hannan-consistency, rather than on its stronger conditional version which characterizes convergence to the set of correlated equilibria (see Hart and Mas-Colell 2000). This is just to focus on essentials. The extension to the conditional setup is straightforward; see Section 5 below.[3]

Hannan-consistent strategies have been obtained by several authors: Hannan (1957), Blackwell (1956b) (see also Luce and Raiffa 1957, pp. 482–483), Foster and Vohra (1993, 1998), Fudenberg and Levine (1995), Freund and Schapire (1999), and Hart and Mas-Colell (2000), Section 4(c).[4] The strategy of Fudenberg and Levine (1995) (as well as those of Hannan 1957, Foster and Vohra 1993, 1998, and Freund and Schapire 1999) is a smoothed-out version of fictitious play. (We note that fictitious play—which may be stated as "at each period play an action with maximal regret"—is by itself

---

[1] For these and the other topics discussed in this paper, the reader is referred also to the book of Fudenberg and Levine (1998) and the survey of Foster and Vohra (1999) (as well as to the other papers in the special issue of *Games and Economic Behavior* 29 (1999).

[2] Think of this as the "regret for not having played k in the past."

[3] Note that conditional regrets have been used by Foster and Vohra (1997) to prove the existence of calibrated forecasts.

[4] See also Baños (1968) and Megiddo (1980) and, in the computer science literature, Littlestone and Warmuth (1994), Auer *et al.* (1995), and the book of Borodin and El-Yaniv (1998).

not Hannan-consistent.) In contrast, the strategy of Hart and Mas-Colell (2000a, Section 4(c)), called "regret matching," prescribes, at each period, play probabilities that are proportional to the (positive) regrets. That is, if we write $D(k)$ for the regret of i for action k at time t (as defined above) and $D_+(k)$ for the *positive regret* (i.e., $D(k)$ when $D(k) > 0$ and 0 when $D(k) \leq 0$), then the probability of playing action k at period $t+1$ is simply $D_+(k)/\sum_{k'} D_+(k')$.

Clearly, a general examination is called for. Smooth fictitious play and regret matching should be but particular instances of a whole class of adaptive strategies with the Hannan-consistency property. In this paper we exhibit and characterize such a class. It turns out to contain, in particular, a large variety of new, simple adaptive strategies.

In Hart and Mas-Colell (2000), we have introduced Blackwell's (1956a) approachability theory for games with vector payoffs as the appropriate basic tool for the analysis: the vector payoff is simply the vector of regrets. In this paper, therefore, we proceed in two steps. First, in Section 2, we generalize Blackwell's result: Given an approachable set (in vector payoff space), we find the class of ("directional") strategies that guarantee that the set is approached. We defer the specifics to that section. Suffice it to say that Blackwell's strategy emerges as the particular quadratic case of a continuum of strategies where continuity and, interestingly, integrability feature decisively.

Second, in Section 3, we apply the general theory to the regret framework and derive an entire class of Hannan-consistent strategies. A feature common to them all is that, in the spirit of bounded rationality, they aim at "better" rather than "best" play. We elaborate on this aspect, and carry out an explicit discussion of fictitious play in Section 4. Section 5 discusses a number of extensions, including conditional regrets and correlated equilibria.

2. The Approachability Problem

2.1. *Model and Main Theorem*

In this section we will consider games where a player's payoffs are *vectors* (rather than, as in standard games, scalar real numbers), as introduced by Blackwell (1956a). This setting may appear unnatural at first. However, it has turned out to be quite useful: the coordinates may represent different commodities or contingent payoffs in different states of the world (when there is incomplete information), or, as we will see below (in Section 3), regrets in a standard game.

Formally, we are given a game in strategic form played by a player i against an opponent $-i$ (which may be Nature and/or the other players). The *action* sets are the finite[5] sets S^i for player i and S^{-i} for $-i$. The payoffs are *vectors* in some Euclidean space. We denote the payoff function by[6] $A : S \equiv S^i \times S^{-i} \to \mathbb{R}^m$; thus $A(s^i, s^{-i}) \in \mathbb{R}^m$ is the payoff vector when i chooses s^i and $-i$ chooses s^{-i}. As usual, A is extended bilinearly to mixed actions; thus[7] $A : \Delta(S^i) \times \Delta(S^{-i}) \to \mathbb{R}^m$.

Let time be discrete, $t = 1, 2, \ldots$, and denote by $s_t = (s_t^i, s_t^{-i}) \in S^i \times S^{-i}$ the actions chosen by i and $-i$, respectively, at time t. The payoff vector in period t is $a_t := A(s_t)$, and $\bar{a}_t := (1/t) \sum_{\tau \le t} a_\tau$ is the average payoff vector up to t. A *strategy*[8] *for player i* assigns to every history of play $h_{t-1} = (s_\tau)_{\tau \le t-1} \in (S)^{t-1}$ a (randomized) choice of action $\sigma_t^i \equiv \sigma_t^i(h_{t-1}) \in \Delta(S^i)$ at time t, where $[\sigma_t^i(h_{t-1})](s^i)$ is, for each s^i in S^i, the probability that i plays s^i at period t following the history h_{t-1}.

Let $\mathscr{C} \subset \mathbb{R}^m$ be a convex and closed[9] set. The set $\mathscr{C}$ is *approachable* by player i (cf. Blackwell 1956a; see Remark 3 below) if there is a strategy of i such that, no matter what $-i$ does,[10] $\text{dist}(\bar{a}_t, \mathscr{C}) \to 0$ almost surely as $t \to \infty$. Blackwell's result can then be stated as follows:

---

[5]See however Remark 2 below. Also, we always assume that S^i contains at least two elements.

[6]$\mathbb{R}$ is the real line, and $\mathbb{R}^m$ is the m-dimensional Euclidean space. For $x = (x_k)_{k=1}^m$ and $y = (x_k)_{k=1}^m$ in $\mathbb{R}^m$, we write $x \ge y$ when $x_k \ge y_k$ for all k, and $x \gg y$ when $x_k > y_k$ for all k. The nonnegative, nonpositive, positive, and negative orthants of $\mathbb{R}^m$ are, respectively, $\mathbb{R}_+^m := \{x \in \mathbb{R}^m : x \ge 0\}$, $\mathbb{R}_-^m := \{x \in \mathbb{R}^m : x \le 0\}$, $\mathbb{R}_{++}^m := \{x \in \mathbb{R}^m : x \gg 0\}$, and $\mathbb{R}_{--}^m := \{x \in \mathbb{R}^m : x \ll 0\}$.

[7]Given a finite set Z, we write $\Delta(Z)$ for the set of probability distributions on Z, i.e., the $(|Z| - 1)$-dimensional unit simplex $\Delta(Z) := \{p \in \mathbb{R}_+^Z : \sum_{z \in Z} p(z) = 1\}$ (the notation $|Z|$ stands for the cardinality of the set Z).

[8]We use the term "action" for a one-period choice and the term "strategy" for a multi-period choice.

[9]A set is approachable if and only if its closure is approachable; we thus assume without loss of generality that the set $\mathscr{C}$ is closed.

[10]We emphasize that the strategies of the opponents $(-i)$ are not in any way restricted; in particular, they may randomize and, furthermore, correlate their actions. All the results in this paper hold against all possible strategies of $-i$, and thus, *a fortiori*, for any specific class of strategies (like independent mixed strategies, and so on). Moreover, requiring independence over $j \ne i$ will not increase the set of strategies of i that guarantee approachability, since the worst $-i$ can do may always be taken to be pure (and thus independent).

Blackwell's Approachability Theorem. (1) *A convex and closed set $\mathscr{C}$ is approachable if and only if every half-space $\mathscr{H}$ containing $\mathscr{C}$ is approachable.*

(2) *A half-space $\mathscr{H}$ is approachable if and only if there exists a mixed action of player i such that the expected vector payoff is guaranteed to lie in $\mathscr{H}$; i.e., there is $\sigma^i \in \Delta(S^i)$ such that $A(\sigma^i, s^{-i}) \in \mathscr{H}$ for all $s^{-i} \in S^{-i}$.*

The condition for $\mathscr{C}$ to be approachable may be restated as follows (since, clearly, it suffices to consider in (1) only "minimal" half-spaces containing $\mathscr{C}$): For every $\lambda \in \mathbb{R}^m$ there exists $\sigma^i \in \Delta(S^i)$ such that

$$\lambda \cdot A(\sigma^i, s^{-i}) \le w(\lambda) := \sup\{\lambda \cdot y : y \in \mathscr{C}\} \quad \text{for all } s^{-i} \in S^{-i} \qquad (2.1)$$

(w is the "support function" of $\mathscr{C}$; note that only those $\lambda \ne 0$ with $w(\lambda) < \infty$ matter for (2.1)). Furthermore, the strategy constructed by Blackwell that yields approachability uses, at each step t where the current average payoff $\bar{a}_{t-1}$ is not in $\mathscr{C}$, a mixed choice σ^i_t satisfying (2.1) for that vector $\lambda \equiv \lambda(\bar{a}_{t-1})$ which goes to $\bar{a}_{t-1}$ from that point y in $\mathscr{C}$ that is closest to $\bar{a}_{t-1}$ (see Fig. 1). To get some intuition, note that the next-period expected payoff vector $b := \mathrm{E}[a_t | h_{t-1}]$ lies in the half-space $\mathscr{H}$, and thus satisfies $\lambda \cdot b \le w(\lambda) < \lambda \cdot \bar{a}_{t-1}$, which implies that

$$\lambda \cdot (\mathrm{E}[\bar{a}_t | h_{t-1}] - \bar{a}_{t-1}) = \lambda \cdot \left(\frac{1}{t}b + \frac{t-1}{t}\bar{a}_{t-1} - \bar{a}_{t-1}\right) = \frac{1}{t}\lambda \cdot (b - \bar{a}_{t-1}) < 0.$$

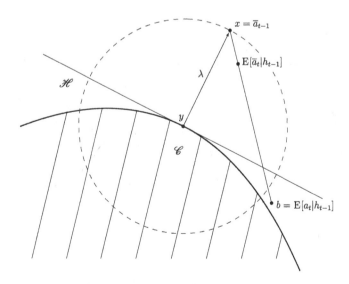

Fig. 1.   Approaching the set $\mathscr{C}$ by Blackwell's strategy.

Therefore the expected average payoff $E[\bar{a}_t|h_{t-1}]$ moves from $\bar{a}_{t-1}$ in the "general direction" of $\mathscr{C}$; in fact, it is closer than $\bar{a}_{t-1}$ to $\mathscr{C}$. Hence $E[\bar{a}_t|h_{t-1}]$ converges to $\mathscr{C}$, and so does the average payoff $\bar{a}_t$ (by the Law of Large Numbers).

Fix an approachable convex and closed set $\mathscr{C}$. We will now consider general strategies of player i which—like Blackwell's strategy above—are defined in terms of a *directional mapping*, that is, a function $\Lambda : \mathbb{R}^m \backslash \mathscr{C} \rightarrow \mathbb{R}^m$ that associates to every $x \notin \mathscr{C}$ a corresponding "direction" $\Lambda(x)$. Given such a mapping Λ, a strategy of player i is called a Λ-*strategy* if, whenever $\bar{a}_{t-1}$ does not lie in $\mathscr{C}$, it prescribes using at time t a mixed action σ_t^i that satisfies

$$\Lambda(\bar{a}_{t-1}) \cdot A(\sigma_t^i, s^{-i}) \leq w(\Lambda(\bar{a}_{t-1})) \quad \text{for all } s^{-i} \in S^{-i} \qquad (2.2)$$

(see Fig. 2: a Λ-strategy guarantees that, when $x = \bar{a}_{t-1} \notin \mathscr{C}$, the next-period expected payoff vector $b = E[a_t|h_{t-1}]$ lies in the smallest half-space $\mathscr{H}$ with normal $\Lambda(x)$ that contains $\mathscr{C}$); notice that there is no requirement when $\bar{a}_{t-1} \in \mathscr{C}$. We are interested in finding conditions on the mapping Λ such that, if player i uses a Λ-strategy, then the set $\mathscr{C}$ is guaranteed to be approached, no matter what $-i$ does.

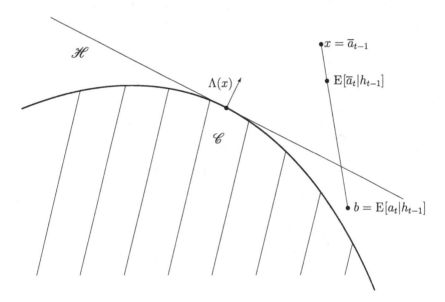

Fig. 2.   A Λ-strategy.

We introduce three conditions on a directional mapping Λ, relative to the given set $\mathscr{C}$.

(D1) Λ is continuous.

(D2) Λ is integrable, namely there exists a Lipschitz function[11] $P : \mathbb{R}^m \to \mathbb{R}$ such that $\nabla P(x) = \phi(x)\Lambda(x)$ for almost every $x \notin \mathscr{C}$, where $\phi : \mathbb{R}^m \backslash \mathscr{C} \to \mathbb{R}_{++}$ is a continuous positive function.

(D3) $\Lambda(x) \cdot x > w(\Lambda(x))$ for all $x \notin \mathscr{C}$.

See Fig. 3. The geometric meaning of (D3) is that the point x is strictly separated from the set $\mathscr{C}$ by $\Lambda(x)$. Note that (D3) implies that all λ with $w(\lambda) = \infty$, as well as $\lambda = 0$, are not allowable directions. Also, observe that the combination of (D1) and (D2) implies that P is continuously differentiable on $\mathbb{R}^m \backslash \mathscr{C}$ (see Clarke 1983, Corollary to Proposition 2.2.4 and Theorem 2.5.1). We will refer to the function P as the *potential* of Λ.

The main result of this section is:

Theorem 2.1. *Suppose that player i uses a Λ-strategy, where Λ is a directional mapping satisfying (D1), (D2), and (D3) for the approachable convex and closed set $\mathscr{C}$. Then the average payoff vector is guaranteed to approach*

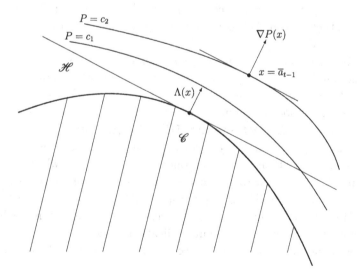

Fig. 3.   The directional mapping Λ and level sets of the potential P.

---

[11]Note that P is defined on the whole space $\mathbb{R}^m$.

the set $\mathscr{C}$; that is, $\mathrm{dist}(\bar{a}_t, \mathscr{C}) \to 0$ *almost surely as* $t \to \infty$, *for any strategy of* $-i$.

Before proving the theorem (in the next subsection), we state a number of comments.

Remarks. 1. The conditions (D1)–(D3) are independent of the game A (they depend on $\mathscr{C}$ only). That is, given a directional mapping Λ satisfying (D1)–(D3), a Λ-strategy is guaranteed to approach $\mathscr{C}$ for *any* game A for which $\mathscr{C}$ is approachable (of course, the specific choice of action depends on A, according to (2.2)). It is in this sense that we refer to the Λ-strategies as "universal."

2. The action sets S^i and S^{-i} need not be finite; as we will see in the proof, it suffices for the range of A to be bounded.

3. As in Blackwell's result, our proof below yields *uniform* approachability: For every ε there is $t_0 \equiv t_0(\varepsilon)$ such that $\mathrm{E}[\mathrm{dist}(\bar{a}_t, \mathscr{C})] < \varepsilon$ for all $t > t_0$ and all strategies of $-i$ (i.e., t_0 is independent of the strategy of $-i$).

4. The conditions on P are invariant under strictly increasing monotone transformations (with positive derivative); that is, only the level sets of P matter.

5. If the potential P is a convex function and $\mathscr{C} = \{y : P(y) \leq c\}$ for some constant c, then (D3) is automatically satisfied: $P(x) > P(y)$ implies $\nabla P(x) \cdot x > \nabla P(x) \cdot y$.

6. Given a norm $\| \cdot \|$ on $\mathbb{R}^m$, consider the resulting "distance from $\mathscr{C}$" function $P(x) := \min_{y \in \mathscr{C}} \|x - y\|$. If P is a smooth function (which is always the case when either the norm is smooth—i.e., the corresponding unit ball has a smooth boundary—or when the boundary of $\mathscr{C}$ is smooth), then the mapping $\Lambda = \nabla P$ satisfies (D1)–(D3) (the latter by the previous Remark 5). In particular, the Euclidean l_2-norm yields precisely the Blackwell strategy, since then $\nabla P(x)$ is proportional to $x - y(x)$, where $y(x) \in \mathscr{C}$ is the point in $\mathscr{C}$ closest to x. The l_p-norm is smooth for $1 < p < \infty$; therefore it yields strategies that guarantee approachability for *any* approachable set $\mathscr{C}$. However, if the boundary of $\mathscr{C}$ is not smooth—for instance, when $\mathscr{C}$ is an orthant, an important case in applications—then (D1) is *not* satisfied in the extreme cases $p = 1$ and $p = \infty$ (see Fig. 4; more on these two cases below).

7. When $\mathscr{C} = \mathbb{R}^m_-$ and P is given by (D2), condition (D3) becomes $\nabla P(x) \cdot x > 0$ for every $x \notin \mathscr{C}$, which means that P is increasing along any ray from the origin that goes outside the negative orthant.

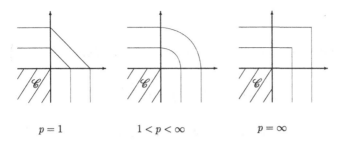

$$p = 1 \qquad\qquad 1 < p < \infty \qquad\qquad p = \infty$$

Fig. 4.   The l_p-potential for an orthant $\mathscr{C}$.

2.2. Proof of Theorem 2.1

We begin by proving two auxiliary results. The first applies to functions Q that satisfy conditions similar to but stronger than (D1)–(D3); the second allows us to reduce the general case to such a Q. The set $\mathscr{C}$, the mappings Λ and P, and the strategy of i (which is a Λ-strategy) are fixed throughout. Also, let K be a convex and compact set containing in its interior the range of A (recall that S is finite).

Lemma 2.2. *Let $Q : \mathbb{R}^m \to \mathbb{R}$ be a continuously differentiable function that satisfies*

(i) $Q(x) \geq 0$ *for all x;*
(ii) $Q(x) = 0$ *for all $x \in \mathscr{C}$;*
(iii) $\nabla Q(x) \cdot x - w(\nabla Q(x)) \geq Q(x)$ *for all $x \in K\backslash\mathscr{C}$; and*
(iv) $\nabla Q(x)$ *is nonnegatively proportional to $\Lambda(x)$ (i.e., $\nabla Q(x) = \phi(x)\Lambda(x)$ where $\phi(x) \geq 0$) for all $x \notin \mathscr{C}$.*

Then $\lim_{t\to\infty} Q(\bar{a}_t) = 0$ a.s. for any strategy of $-i$.

Proof.   We have $\bar{a}_t - \bar{a}_{t-1} = (1/t)(a_t - \bar{a}_{t-1})$; thus, writing x for $\bar{a}_{t-1}$,

$$Q(\bar{a}_t) = Q(x) + \nabla Q(x) \cdot \frac{1}{t}(a_t - x) + \mathrm{o}\left(\frac{1}{t}\right), \qquad (2.3)$$

since Q is (continuously) differentiable. Moreover, the remainder $\mathrm{o}(1/t)$ is uniform, since all relevant points lie in the compact set K. If $x \notin \mathscr{C}$ then player i plays at time t so that

$$\nabla Q(x) \cdot \mathrm{E}[a_t|h_{t-1}] \leq w(\nabla Q(x)) \qquad (2.4)$$

(by (2.2) and (iv)); if $x \in \mathscr{C}$ then $\nabla Q(x) = 0$ (by (i) and (ii)), and (2.4) holds too. Taking conditional expectation in (2.3) and then substituting

(2.4) yields

$$E[Q(\bar{a}_t)|h_{t-1}] \le Q(x) + \frac{1}{t}w(\nabla Q(x)) - (\nabla Q(x) \cdot x) + o\left(\frac{1}{t}\right)$$

$$\le Q(x) - \frac{1}{t}Q(x) + o\left(\frac{1}{t}\right),$$

where we have used (iii) when $x \notin \mathscr{C}$ and (i), (ii) when $x \in \mathscr{C}$. Thus

$$E[Q(\bar{a}_t)|h_{t-1}] \le \frac{t-1}{t}Q(\bar{a}_{t-1}) + o\left(\frac{1}{t}\right).$$

This may be rewritten as[12]

$$E[\zeta_t|h_{t-1}] \le o(1), \tag{2.5}$$

where $\zeta_t := tQ(\bar{a}_t) - (t-1)Q(\bar{a}_{t-1})$. Hence $\limsup_{t\to\infty}(1/t)\sum_{\tau \le t}$ $E[\zeta_\tau|h_{\tau-1}] \le 0$. The Strong Law of Large Numbers for Dependent Random Variables (see Loève 1978, Theorem 32.1.E) implies that $(1/t)\sum_{\tau \le t}(\zeta_\tau - E[\zeta_\tau|h_{\tau-1}]) \to 0$ a.s. as $t \to \infty$ (note that the ζ_t's are uniformly bounded, as can be immediately seen from equation (2.3), $\zeta_t = Q(\bar{a}_{t-1}) + \nabla Q(\bar{a}_{t-1}) \cdot (a_t - \bar{a}_{t-1}) + o(1)$, and from the fact that everything happens in the compact set K). Therefore $\limsup_{t\to\infty}(1/t)\sum_{\tau \le t}\zeta_\tau \le 0$. But $0 \le Q(\bar{a}_t) = (1/t)\sum_{\tau \le t}\zeta_\tau$, so $\lim_{t\to\infty}Q(\bar{a}_t) = 0$. $\qquad\square$

Lemma 2.3. *The function P satisfies:*

(c1) *If the boundary of $\mathscr{C}$ is connected, then there exists a constant c such that*

$$P(x) = c, \quad \text{if } x \in \text{bd } \mathscr{C};$$
$$P(x) > c, \quad \text{if } x \notin \mathscr{C}.$$

(c2) *If the boundary of $\mathscr{C}$ is not connected, then there exists a $\lambda \in \mathbb{R}^m\backslash\{0\}$ such that[13] $\mathscr{C} = \{x \in \mathbb{R}^m : -w(-\lambda) \le \lambda \cdot x \le w(\lambda)\}$ (where $w(\lambda) < \infty$*

---

[12]Recall that the remainder term $o(1/t)$ was uniform; that is, for every $\varepsilon > 0$ there is $t_0(\varepsilon)$ such that $o(1) < \varepsilon$ is guaranteed for all $t > t_0(\varepsilon)$.

[13]This is a general fact about convex sets: The only case where the boundary of a convex closed set $\mathscr{C} \subset \mathbb{R}^m$ is not path-connected is when $\mathscr{C}$ is the set of points lying between two parallel hyperplanes. We prove this in Steps 1–3 below, independently of the function P.

and $w(-\lambda) < \infty$), and there are constants c_1 and c_2 such that

$$
\begin{aligned}
P(x) &= c_1, &&\text{if } x \in \text{bd } \mathscr{C} &&\text{and} &&\lambda \cdot x = w(\lambda); \\
P(x) &= c_2, &&\text{if } x \in \text{bd } \mathscr{C} &&\text{and} &&(-\lambda) \cdot x = w(-\lambda); \\
P(x) &> c_1, &&\text{if } x \notin \mathscr{C} &&\text{and} &&\lambda \cdot x > w(\lambda); \\
P(x) &> c_2, &&\text{if } x \notin \mathscr{C} &&\text{and} &&(-\lambda) \cdot x > w(-\lambda).
\end{aligned}
$$

Proof. Let $x_0, x_1 \in \text{bd } \mathscr{C}$ and denote by λ_j, for $j = 0, 1$, an outward unit normal to $\mathscr{C}$ at x_j; thus $\|\lambda_j\| = 1$ and $\lambda_j \cdot x_j = w(\lambda_j)$.

• **Step 1.** If $\lambda_1 \neq -\lambda_0$, we claim that there is a path on bd $\mathscr{C}$ connecting x_0 and x_1, and moreover that $P(x_0) = P(x_1)$. Indeed, there exists a vector[14] $z \in \mathbb{R}^m$ such that $\lambda_0 \cdot z > 0$ and $\lambda_1 \cdot z > 0$. The straight line segment connecting x_0 and x_1 lies in $\mathscr{C}$; we move it in the direction z until it reaches the boundary of $\mathscr{C}$. That is, for each $\eta \in [0, 1]$, let $y(\eta) := \eta x_1 + (1 - \eta)x_0 + \alpha(\eta)z$, where $\alpha(\eta) := \max\{\beta : \eta x_1 + (1 - \eta)x_0 + \beta z \in \mathscr{C}\}$; this maximum exists by the choice of z. Note that $y(\cdot)$ is a path on bd $\mathscr{C}$ connecting x_0 and x_1.

It is easy to verify that $\alpha(0) = \alpha(1) = 0$ and that $\alpha : [0, 1] \to \mathbb{R}_+$ is a concave function, and thus differentiable a.e. For each $k = 1, 2, \ldots$, define $y_k(\eta) := y(\eta) + (1/k)z$; then $y_k(\cdot)$ is a path in $\mathbb{R}^m \backslash \mathscr{C}$, the region where P is continuously differentiable. Let $\bar{\eta} \in (0, 1)$ be a point of differentiability of $\alpha(\cdot)$, thus also of $y(\cdot)$, $y_k(\cdot)$, and $P(y_k(\cdot))$; we have $dP(y_k(\bar{\eta}))/d\eta = \nabla P(y_k(\bar{\eta})) \cdot y_k'(\bar{\eta}) = \nabla P(y_k(\bar{\eta})) \cdot y'(\bar{\eta})$. By (D3), $\nabla P(y_k(\bar{\eta})) \cdot y_k(\bar{\eta}) > w(\nabla P(y_k(\bar{\eta}))) \geq \nabla P(y_k(\bar{\eta})) \cdot y(\eta)$ for any $\eta \in [0, 1]$ (the second inequality since $y(\eta) \in \mathscr{C}$). Thus, for any accumulation point q of the bounded[15] sequence $(\nabla P(y_k(\bar{\eta})))_{k=1}^\infty$, we get $q \cdot y(\bar{\eta}) \geq q \cdot y(\eta)$ for all $\eta \in [0, 1]$. Therefore $q \cdot y(\eta)$ is maximized at $\eta = \bar{\eta}$, which implies that $q \cdot y'(\bar{\eta}) = 0$. This holds for *any* accumulation point q; hence $\lim_{k \to \infty} dP(y_k(\bar{\eta}))/d\eta = 0$ for almost every $\bar{\eta}$. Therefore

$$
P(x_1) - P(x_0) = P(y(1)) - P(y(0)) = \lim_k [P(y_k(1)) - P(y_k(0))]
$$

$$
= \lim_k \int_0^1 \frac{dP(y_k(\eta))}{d\eta} d\eta = \int_0^1 \lim_k \frac{dP(y_k(\eta))}{d\eta} d\eta = 0
$$

(again, P is Lipschitz, so $dP(y_k(\eta))/d\eta$ are uniformly bounded).

---

[14] Take for instance $z = \lambda_0 + \lambda_1$.

[15] Recall that P is Lipschitz.

• **Step 2.** If $\lambda_1 = -\lambda_0$ and there is another boundary point x_2 with outward unit normal λ_2 different from both $-\lambda_0$ and $-\lambda_1$, then we get paths on bd $\mathscr{C}$ connecting x_0 to x_2 and x_1 to x_2, and also $P(x_0) = P(x_2)$ and $P(x_1) = P(x_2)$—thus we get the same conclusion as in Step 1.

• **Step 3.** If $\lambda_1 = -\lambda_0$ and no x_2 and λ_2 as in Step 2 exist, it follows that the unit normal to every point on the boundary of $\mathscr{C}$ is either λ_0 or $-\lambda_0$; thus $\mathscr{C}$ is the set bounded between the two parallel hyperplanes $\lambda_0 \cdot x = w(\lambda_0)$ and $-\lambda_0 \cdot x = w(-\lambda_0)$. In particular, the boundary of $\mathscr{C}$ is not connected, and we are in case (c2). Note that in this case when x_0 and x_1 lie on the same hyperplane then $P(x_0) = P(x_1)$ by Step 1 (since $\lambda_1 = \lambda_0 \neq -\lambda_0$).

• **Step 4.** If it is case (c1)—thus not (c2)—then the situation of Step 3 is not possible; thus for any two boundary points x_0 and x_1 we get $P(x_0) = P(x_1)$ by either Step 1 or Step 2.

• **Step 5.** Given $x \notin \mathscr{C}$, let $x_0 \in$ bd $\mathscr{C}$ be the point in $\mathscr{C}$ that is closest to x. Then the line segment from x to x_0 lies outside $\mathscr{C}$, i.e., $y(\eta) := \eta x + (1 - \eta)x_0 \notin \mathscr{C}$ for all $\eta \in (0,1]$. By (D3) and $x_0 \in \mathscr{C}$, it follows that $\nabla P(y(\eta)) \cdot y(\eta) > w(\nabla P(y(\eta))) \geq \nabla P(y(\eta)) \cdot x_0$, or, after dividing by $\eta > 0$, that $\nabla P(y(\eta)) \cdot (x - x_0) > 0$, for all $\eta \in (0,1]$. Hence $P(x) - P(x_0) = \int_0^1 \nabla P(y(\eta)) \cdot y'(\eta) \, d\eta = \int_0^1 \nabla P(y(\eta)) \cdot (x - x_0) \, d\eta > 0$, showing that $P(x) > c$ in Case (c1) and $P(x) > c_1$ or $P(x) > c_2$ in Case (c2). $\qquad\square$

We can now prove the main result of this section.

Proof of Theorem 2.1. First, use Lemma 2.3 to replace P by P_1 as follows: When the boundary of $\mathscr{C}$ is connected (Case (c1)), define $P_1(x) := (P(x) - c)^2$ for $x \notin \mathcal{C}$ and $P_1(x) := 0$ for $x \in \mathscr{C}$; when the boundary of $\mathscr{C}$ is not connected (Case (c2)), define $P_1(x) := (P(x) - c_1)^2$ for $x \notin \mathscr{C}$ with $\lambda \cdot x > w(\lambda)$, $P_1(x) := (P(x) - c_2)^2$ for $x \notin \mathscr{C}$ with $(-\lambda) \cdot x > w(-\lambda)$, and $P_1(x) := 0$ for $x \in \mathscr{C}$. It is easy to verify that: P_1 is continuously differentiable; $\nabla P_1(x)$ is positively proportional to $\nabla P(x)$ and thus to $\Lambda(x)$ for $x \notin \mathscr{C}$; $P_1(x) \geq 0$ for all x; and $P_1(x) = 0$ if and only if $x \in \mathscr{C}$.

Given $\varepsilon > 0$, let $k \geq 2$ be a large enough integer such that

$$\frac{\nabla P_1(x) \cdot x - w(\nabla P_1(x))}{P_1(x)} \geq \frac{1}{k} \qquad (2.6)$$

for all x in the compact set $K \cap \{x : P_1(x) \geq \varepsilon\}$ (the minimum of the above ratio is attained and it is positive by (D3)). Put[16] $Q(x) := ([P_1(x) - \varepsilon]_+)^k$.

---

[16]We write $[z]_+$ for the positive part of z, i.e., $[z]_+ := \max\{z, 0\}$.

Then Q is continuously differentiable (since $k \geq 2$) and it satisfies all the conditions of Lemma 2.2. To check (iii): When $Q(x) = 0$ we have $\nabla Q(x) = 0$, and when $Q(x) > 0$ we have

$$\nabla Q(x) \cdot x - w(\nabla Q(x)) = k(P_1(x) - \varepsilon)^{k-1}[\nabla P_1(x) \cdot x - w(\nabla P_1(x))]$$
$$\geq (P_1(x) - \varepsilon)^{k-1} P_1(x)$$
$$\geq Q(x).$$

(the first inequality follows from (2.6)).

By Lemma 2.2, it follows that the Λ-strategy guarantees a.s. $\lim_{t\to\infty} Q(\bar{a}_t) = 0$, or $\limsup_{t\to\infty} P_1(\bar{a}_t) \leq \varepsilon$. Since $\varepsilon > 0$ is arbitrary, this yields a.s. $\lim_{t\to\infty} P_1(\bar{a}_t) = 0$, or $\bar{a}_t \to \mathscr{C}$. □

Remark. P may be viewed (up to a constant, as in the definition of P_1 above) as a generalized distance to the set $\mathscr{C}$ (compare with Remark 6 in Section 2.1).

2.3. Counterexamples

In this subsection we provide counterexamples showing the indispensability of the conditions (D1)–(D3) for the validity of Theorem 2.1. The first two examples refer to (D1), the third to (D2), and the last one to (D3).

Example 2.4. *The role of* (D1): Consider the following two-dimensional vector payoff matrix A

	C1	C2
R1	$(0,-1)$	$(0,1)$
R2	$(1,0)$	$(-1,0)$

Let i be the Row player and $-i$ the Column player. The set $\mathscr{C} := \mathbb{R}^2_-$ is approachable by the Row player since $w(\lambda) < \infty$ whenever $\lambda \geq 0$, and then the mixed action $\sigma^{\text{Row}}(\lambda) := (\lambda_1/(\lambda_1 + \lambda_2), \lambda_2/(\lambda_1 + \lambda_2))$ of the Row player yields $\lambda \cdot A(\sigma^{\text{Row}}(\lambda), \gamma) = 0 = w(\lambda)$ for any action γ of the Column player.

We define a directional mapping Λ_∞ on $\mathbb{R}^2 \backslash \mathbb{R}^2_-$,

$$\Lambda_\infty(x) := \begin{cases} (1,0), & \text{if } x_1 > x_2; \\ (0,1), & \text{if } x_1 \leq x_2. \end{cases}$$

Clearly Λ_∞ is *not continuous*, i.e., it does not satisfy (D1); it does however satisfy (D3) and (D2) (with $P(x) = \max\{x_1, x_2\}$, the l_∞-potential; see Remark 6 in Section 2.1). Consider a Λ_∞-strategy for the Row player that, when $x := \bar{a}_{t-1} \notin \mathscr{C}$, plays $\sigma^{\text{Row}}(\Lambda_\infty(x))$ at time t; that is, he plays R1 when $x_1 > x_2$, and R2 when $x_1 \leq x_2$. Assume that the Column player plays[17] C2 when $x_1 > x_2$, and C1 when $x_1 \leq x_2$. Then, starting with, say, $a_1 = (0,1) \notin \mathscr{C}$, the vector payoff a_t will always be either $(0,1)$ or $(1,0)$, thus on the line $x_1 + x_2 = 1$, so the average $\bar{a}_t$ does not converge to $\mathscr{C} = \mathbb{R}^2_-$.

Example 2.5. *The role of* $(D1)$, *again*: The same as in Example 2.4, but now the directional mapping is Λ_1, defined on $\mathbb{R}^2 \backslash \mathbb{R}^2_-$ by

$$\Lambda_1(x) := \begin{cases} (1,1), & \text{if } x_1 > 0 \quad \text{and} \quad x_2 > 0; \\ (1,0), & \text{if } x_1 > 0 \quad \text{and} \quad x_2 \leq 0; \\ (0,1), & \text{if } x_1 \leq 0 \quad \text{and} \quad x_2 > 0. \end{cases}$$

Again, the mapping Λ_1 is *not continuous*—it does not satisfy (D1)—but it satisfies (D3) and (D2) (with $P(x) := [x_1]_+ + [x_2]_+$, the l_1-potential). Consider a Λ_1-strategy for the Row player where at time t he plays $\sigma^{\text{Row}}(\Lambda_1(x))$ when $x := \bar{a}_{t-1} \notin \mathscr{C}$, and assume that the Column player plays C1 when $x_1 \leq 0$ and $x_2 > 0$, and plays C2 otherwise. Thus, if $x \notin \mathscr{C}$ then a_t is:

- $(0,1)$ or $(-1,0)$ with equal probabilities, when $x_1 > 0$ and $x_2 > 0$;
- $(0,1)$, when $x_1 > 0$ and $x_2 \leq 0$;
- $(1,0)$, when $x_1 \leq 0$ and $x_2 > 0$.

In all cases the second coordinate of a_t is nonnegative; therefore, if we start with, say, $a_1 = (0,1) \notin \mathscr{C}$, then, inductively, the second coordinate of $\bar{a}_{t-1}$ will be strictly positive, so that $\bar{a}_{t-1} \notin \mathscr{C}$ for all t. But then $\mathrm{E}[a_t | h_{t-1}] \in \mathcal{D} := \text{conv}\{(-1/2, 1/2), (0,1), (1,0)\}$, and $\mathcal{D}$ is disjoint from $\mathscr{C}$ and at a positive distance from it. Therefore $(1/t) \sum_{\tau \leq t} \mathrm{E}[a_\tau | h_{\tau-1}] \in \mathcal{D}$ and so, by the Strong Law of Large Numbers, $\lim \bar{a}_t = \lim(1/t) \sum_{\tau \leq t} a_\tau \in \mathcal{D}$ too (a.s.), so $\bar{a}_t$ does not approach[18] $\mathscr{C}$.

To get some intuition, consider the deterministic system where a_t is replaced by $\mathrm{E}[a_t | h_{t-1}]$. Then the point $(0, 1/3)$ is a stationary point for this dynamic. Specifically (see Fig. 5), if $\bar{a}_{t-1}$ is on the line segment joining $(-1/2, 1/2)$ with $(1,0)$, then $\mathrm{E}[\bar{a}_t | h_{t-1}]$ will be there too, moving towards

---

[17]In order to show that the strategy of the Row player does not guarantee approachability to $\mathscr{C}$, we exhibit one strategy of the Column player for which $\bar{a}_t$ does not converge to $\mathscr{C}$.
[18]One way to see this formally is by a separation argument: Let $f(x) := x_1 + 3x_2$; then $\mathrm{E}[f(a_t)|h_{t-1}] \geq 1$, so $\liminf f(\bar{a}_t) = \liminf(1/t) \sum_{\tau \leq t} f(a_t) \geq 1$, whereas $f(x) \leq 0$ for all $x \in \mathscr{C}$.

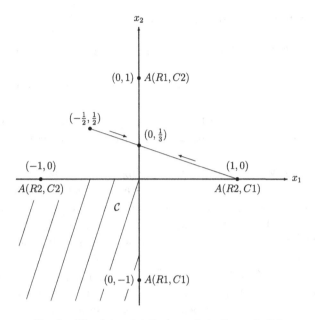

Fig. 5. The deterministic dynamic in Example 2.4.

$(-1/2, 1/2)$ when $\bar{a}_{t-1}$ is in the positive orthant and towards $(1, 0)$ when it is in the second orthant.

Example 2.6. *The role of* $(D2)$: Consider the following two-dimensional vector payoff matrix A:

	C1	C2	C3	C4
R1	$(0, 1)$	$(0, 0)$	$(0, -1)$	$(0, 0)$
R2	$(-1, 0)$	$(0, 0)$	$(1, 0)$	$(0, 0)$
R3	$(0, 0)$	$(0, -1)$	$(0, 0)$	$(0, 1)$
R4	$(0, 0)$	$(-1, 0)$	$(0, 0)$	$(1, 0)$

Again, the Row player is i and the Column player is $-i$. Let $\mathscr{C} := \{(0, 0)\}$. For every $\lambda \in \mathbb{R}^2 \setminus \{(0, 0)\}$, put $\mu_1 := |\lambda_1|/(|\lambda_1| + |\lambda_2|)$ and $\mu_2 := |\lambda_2|/(|\lambda_1| + |\lambda_2|)$, and define a mixed action $\sigma^{\mathrm{Row}}(\lambda)$ for the Row player and a pure action $c(\lambda)$ for the Column player, as follows:

- If $\lambda_1 \geq 0$ and $\lambda_2 \geq 0$ then $\sigma^{\mathrm{Row}}(\lambda) := (\mu_1, \mu_2, 0, 0)$ and $c(\lambda) := C1$;
- If $\lambda_1 < 0$ and $\lambda_2 \geq 0$ then $\sigma^{\mathrm{Row}}(\lambda) := (0, 0, \mu_1, \mu_2)$ and $c(\lambda) := C2$;
- If $\lambda_1 < 0$ and $\lambda_2 < 0$ then $\sigma^{\mathrm{Row}}(\lambda) := (\mu_1, \mu_2, 0, 0)$ and $c(\lambda) := C3$;
- If $\lambda_1 \geq 0$ and $\lambda_2 < 0$ then $\sigma^{\mathrm{Row}}(\lambda) := (0, 0, \mu_1, \mu_2)$ and $c(\lambda) := C4$.

It is easy to verify that in all four cases:

(a1) $\lambda \cdot A(\sigma^{\text{Row}}(\lambda), \gamma) \le 0 = w(\lambda)$ for any action γ of the column player; and

(a2) $A(\sigma^{\text{Row}}(\lambda), c(\lambda)) = (|\lambda_1| + |\lambda_2|)^{-1}\widehat{\lambda}$, where $\widehat{\lambda} := (-\lambda_2, \lambda_1)$.

Condition (a1) implies, by (2.1), that $\mathscr{C}$ is approachable by the Row player; and Condition (a2) means that $A(\sigma^{\text{Row}}(\lambda), c(\lambda))$ is $90°$ counterclockwise from λ.

Consider now the directional mapping Λ given by $\Lambda(x) := (x_1 + \alpha x_2, x_2 - \alpha x_1)$, where $\alpha > 0$ is a fixed constant.[19] Then (D1) and (D3) hold (for the latter, we have $x \cdot \Lambda(x) = (x_1)^2 + (x_2)^2 > 0 = w(\Lambda(x))$ for all $x \notin \mathscr{C}$), but *integrability* (D2) *is not satisfied*. To see this, assume that $\nabla P(x) = \phi(x)\Lambda(x)$ for all $x \notin \mathscr{C}$, where $\phi(x) > 0$ is a continuous function. Consider the path $y(\eta) := (\sin \eta, \cos \eta)$ for $\eta \in [0, 2\pi]$. We have $0 = P(y(2\pi)) - P(y(0)) = \int_0^{2\pi}(dP(y(\eta))/d\eta)d\eta = \int_0^{2\pi} \nabla P(y(\eta)) \cdot y'(\eta)d\eta$. But the integrand equals $(\phi(y(\eta)))(\sin \eta + \alpha \cos \eta, \cos \eta - \alpha \sin \eta) \cdot (\cos \eta, -\sin \eta) = \alpha\phi(y(\eta))$ which is everywhere positive, a contradiction.

We claim that if the Row player uses the Λ-strategy where at time t he plays[20] $\sigma^{\text{Row}}(\Lambda(\bar{a}_{t-1}))$, and if the Column player chooses $c(\Lambda(\bar{a}_{t-1}))$, then the distance to the set $\mathscr{C} = \{(0, 0)\}$ does not approach 0. Indeed, let $b_t := \mathrm{E}[a_t | h_{t-1}]$; then the strategies played imply that the vector $b_t = A(\sigma^{\text{Row}}(\Lambda(\bar{a}_{t-1})), c(\Lambda(\bar{a}_{t-1})))$ is perpendicular to $\Lambda(\bar{a}_{t-1})$ and makes an acute angle with the vector $\bar{a}_{t-1}$. Specifically,[21] $b_t \cdot \bar{a}_{t-1} = \beta \|b_t\| \|\bar{a}_{t-1}\|$, where $\beta := \alpha/\sqrt{1+\alpha^2} \le 1$. Therefore

$$(\mathrm{E}[t\|\bar{a}_t\| \,|\, h_{t-1}])^2 \ge \|\mathrm{E}[t\bar{a}_t | h_{t-1}]\|^2$$

$$= (t-1)^2\|\bar{a}_{t-1}\|^2 + 2(t-1)b_t \cdot \bar{a}_{t-1} + \|b_t\|^2$$

---

[19]This will provide a counterexample for any value of α, showing that it is not an isolated phenomenon.

[20]For all λ except those on the two axes (i.e., $\lambda_1 = 0$ or $\lambda_2 = 0$), the mixed action $\sigma^{\text{Row}}(\lambda)$ is uniquely determined by (2.2). If the Row player were to play in these exceptional cases another mixed action satisfying (2.2), it is easy to verify that the Column player could respond appropriately so that $\mathscr{C}$ is not approached. Thus, no Λ-strategy guarantees approachability.

[21]Writing b for b_t; x for $\bar{a}_{t-1}$; and λ for $\Lambda(\bar{a}_{t-1}) = \Lambda(x)$, we have $x = (1+\alpha^2)^{-1}(\lambda + \alpha\widehat{\lambda})$ (recall the definition of Λ and invert); thus $b \cdot x = (|\lambda_1| + |\lambda_2|)^{-1}\widehat{\lambda} \cdot (1+\alpha^2)^{-1}(\lambda + \alpha\widehat{\lambda}) = \alpha(1+\alpha^2)^{-1}(|\lambda_1| + |\lambda_2|)^{-1}\|\lambda\|^2$ (we have used (a2), $\lambda \cdot \widehat{\lambda} = 0$, and $\|\lambda\| = \|\widehat{\lambda}\|$). Now $\|b\| = (|\lambda_1| + |\lambda_2|)^{-1}\|\lambda\|$ and $\|x\| = (1+\alpha^2)^{-1/2}\|\lambda\|$, thus completing the proof.

$$\geq (t-1)^2\|\bar{a}_{t-1}\|^2 + 2(t-1)\beta\|b_t\|\|\bar{a}_{t-1}\| + \beta^2\|b_t\|^2$$
$$= ((t-1)\|\bar{a}_{t-1}\| + \beta\|b_t\|)^2.$$

Now $\|b_t\| \geq 1/\sqrt{2}$ (for instance, see (a2)), and we have thus obtained

$$E[t\|\bar{a}_t\| - (t-1)\|\bar{a}_{t-1}\|\mid h_{t-1}] \geq \beta/\sqrt{2} > 0,$$

from which it follows that $\liminf \|\bar{a}_t\| = \liminf(1/t)\sum_{\tau\leq t}[\tau\|\bar{a}_\tau\| - (\tau-1)\|\bar{a}_{\tau-1}\|] \geq \beta/\sqrt{2} > 0$ a.s., again by the Strong Law of Large Numbers. Thus the distance of the average payoff vector $\bar{a}_t$ from the set $\mathscr{C} = \{(0,0)\}$ is, with probability one, bounded away from 0 from some time on.

To get some intuition for this result, note that the direction of movement from $\bar{a}_{t-1}$ to $E[\bar{a}_t\mid h_{t-1}]$ is at a fixed angle $\theta \in (0,\pi/2)$ from $\bar{a}_{t-1}$, which, if the dynamic were deterministic, would generate a counterclockwise spiral that goes away from $(0,0)$.

Example 2.7. *The role of* $(D3)$: Consider the two-dimensional vector payoff matrix A

R1	$(0,1)$
R2	$(-1,0)$

(where i is the Row player, and $-i$ has one action). The set $\mathscr{C} := \mathbb{R}^2_-$ is approachable by the Row player (by playing "R2 forever"). Consider the directional mapping Λ defined on $\mathbb{R}^2 \backslash \mathbb{R}^2_-$ by $\Lambda(x) := (1,0)$. Then (D1) and (D2) are satisfied (with $P(x) := x_1$), but (D3) is not: $\Lambda(0,1) \cdot (0,1) = 0 = w(\Lambda(0,1))$. Playing "R1 forever" is a Λ-strategy, but the payoff is $(0,1) \notin \mathscr{C}$.

3. Regrets

3.1. *Model and preliminary results*

In this section we consider standard N-person games in strategic form (with *scalar* payoffs for each player). The set of players is a finite set N, the action set of each player i is a finite set S^i, and the payoff function of i is $u^i : S \to \mathbb{R}$, where $S := \Pi_{j\in N}S^j$; we will denote this game $\langle N, (S^i)_i, (u^i)_i\rangle$ by Γ.

As in the previous section, the game is played repeatedly in discrete time $t = 1, 2, \ldots$; denote by $s^i_t \in S^i$ the choice of player i at time t, and put $s_t = (s^i_t)_{i\in N} \in S$. The payoff of i in period t is $U^i_t := u^i(s_t)$, and $\overline{U}^i_t := (1/t)\sum_{\tau\leq t}U^i_\tau$ is his average payoff up to t.

Fix a player $i \in N$. Following Hannan (1957), we consider the *regrets* of player i, namely, for each one of his actions $k \in S^i$, the change in his average payoff if he were always to choose k (while no one else makes any change in his realized actions):

$$D_t^i(k) := \frac{1}{t} \sum_{\tau=1}^{t} u^i(k, s_\tau^{-i}) - \overline{U}_t^i = u^i(k, z_t^{-i}) - \overline{U}_t^i,$$

where $z_t^{-i} \in \Delta(S^{-i})$ is the empirical distribution of the actions chosen by the other players in the past.[22] A strategy of player i is called *Hannan-consistent* if, as t increases, all regrets are guaranteed—no matter what the other players do—to become almost surely nonpositive in the limit; that is, with probability one, $\limsup_{t \to \infty} D_t^i(k) \leq 0$ for all $k \in S^i$.

Following Hart and Mas-Colell (2000), it is useful to view the regrets of i as an m-dimensional vector payoff, where $m := |S^i|$. We thus define $A \equiv A^i : S \to \mathbb{R}^m$, the *i-regret vector-payoff game associated to* Γ, by

$$A_k(s^i, s^{-i}) := u^i(k, s^{-i}) - u^i(s^i, s^{-i}) \quad \text{for all } k \in S^i,$$

and

$$A(s^i, s^{-i}) := (A_k(s^i, s^{-i}))_{k \in S^i},$$

for all $s = (s^i, s^{-i}) \in S^i \times S^{-i} = S$. Rewriting the regret as

$$D_t^i(k) = \frac{1}{t} \sum_{\tau \leq t} [u^i(k, s_\tau^{-i}) - u^i(s_\tau^i, s_\tau^{-i})]$$

shows that the vector of regrets at time t is just the average of the A vector payoffs in the first t periods: $D_t^i = (1/t) \sum_{\tau \leq t} A(s_\tau)$. The existence of a Hannan-consistent strategy in Γ is thus equivalent to the approachability by player i of the nonpositive orthant $\mathbb{R}_-^{S^i}$ in the vector-payoff game A, and a strategy is Hannan-consistent if and only if it guarantees that $\mathbb{R}_-^{S^i}$ is approached.

We now present two important results that apply in all generality to the regret setup.

Proposition 3.1. *For any (finite) N-person game Γ, the nonpositive orthant $\mathbb{R}_-^{S^i}$ is approachable by player i in the i-regret vector-payoff associated game.*

---

[22]That is, $z_t^{-i}(s^{-i}) := |\{\tau \leq t : s_\tau^{-i} = s^{-i}\}|/t$ for each $s^{-i} \in S^{-i}$.

This Proposition follows immediately from the next one. Observe that the approachability of $\mathbb{R}^{S^i}_-$ is equivalent, by the Blackwell condition (2.1), to the following: For every $\lambda \in \Delta(S^i)$ there exists $\sigma^i(\lambda) \in \Delta(S^i)$, a mixed action of player i, such that

$$\lambda \cdot A(\sigma^i(\lambda), s^{-i}) \leq 0 \quad \text{for all } s^{-i} \in S^{-i} \tag{3.1}$$

(indeed, $w(\lambda)$ equals 0 for $\lambda \geq 0$ and it is infinite otherwise). That is, the expected regret obtained by playing $\sigma^i(\lambda)$ lies in the half-space (through the origin) with normal λ. In this regret setup, the mixture $\sigma^i(\lambda)$ may actually be chosen in a simple manner:

Proposition 3.2. *For any (finite) N-person game Γ and every $\lambda \in \Delta(S^i)$, condition (3.1) is satisfied by $\sigma^i(\lambda) = \lambda$.*

Proof. Given $\lambda \in \Delta(S^i)$, a $\sigma^i \equiv (\sigma^i_k)_{k \in S^i} \in \Delta(S^i)$ satisfies (3.1) if and only if

$$\sum_{k \in S^i} \lambda_k \sum_{j \in S^i} \sigma^i_j [u^i(k, s^{-i}) - u^i(j, s^{-i})] \leq 0 \tag{3.2}$$

for all $s^{-i} \in S^{-i}$. This may be rewritten as

$$\sum_{k \in S^i} u^i(k, s^{-i}) \left(\lambda_k \sum_{j \in S^i} \sigma^i_j - \sigma^i_k \sum_{j \in S^i} \lambda_j \right) = \sum_{k \in S^i} u^i(k, s^{-i})(\lambda_k - \sigma^i_k) \leq 0.$$

Therefore, by choosing σ^i so that all coefficients in the square brackets vanish—that is, by choosing $\sigma^i_k = \lambda_k$—we guarantee (3.2) and thus (3.1) for all s^{-i}. □

3.2. Regret-based strategies

The general theory of Section 2 is now applied to the regret situation. A *stationary regret-based* strategy for player i is a strategy of i such that the choices depend only on i's regret vector; that is, for every history h_{t-1}, the mixed action of i at time t is a function[23] of D^i_{t-1}

---

[23]Note that the time t does not matter: the strategy is "stationary."

only: $\sigma_t^i = \sigma^i(D_{t-1}^i) \in \Delta(S^i)$. The main result of this section is:

Theorem 3.3. *Consider a stationary regret-based strategy of player i given by a mapping $\sigma^i : \mathbb{R}^{S^i} \to \Delta(S^i)$ that satisfies the following:*

(R1) *There exists a continuously differentiable function $P : \mathbb{R}^{S^i} \to \mathbb{R}$ such that $\sigma^i(x)$ is positively proportional to $\nabla P(x)$ for every $x \notin \mathbb{R}_-^{S^i}$; and*

(R2) $\sigma^i(x) \cdot x > 0$ *for every $x \notin \mathbb{R}_-^{S^i}$.*

Then this strategy is Hannan-consistent for any (finite) N-person game.

Proof.   Apply Theorem 2.1 for $\mathscr{C} = \mathbb{R}_-^{S^i}$ together with Propositions 3.1 and 3.2: (D1) and (D2) yield (R1), and (D3) yields (R2).   □

We have thus obtained a wide class of strategies that are Hannan-consistent. It is noteworthy that these are "universal" strategies: the mapping σ^i is independent of the game (see also the "variable game" case in Section 5).

Condition (R2) says that when $D_{t-1}^i \notin \mathbb{R}_-^{S^i}$—i.e., when some regret is positive—the mixed choice σ_t^i of i satisfies $\sigma_t^i \cdot D_{t-1}^i > 0$. This is equivalent to

$$u^i(\sigma_t^i, z_{t-1}^{-i}) > \overline{U}_{t-1}^i. \tag{3.3}$$

That is, the expected payoff of i from playing σ_t^i against the empirical distribution z_{t-1}^{-i} of the actions chosen by the other players in the past is higher than his realized average payoff. Thus σ_t^i is a *better reply*, where "better" is relative to the obtained payoff. By comparison, fictitious play always chooses an action that is a *best reply* to the empirical distribution z_{t-1}^{-i}. For more on this "better vs. best" issue, see Section 4.2 below and Hart and Mas-Colell (2000, Section 4(e)).

We now describe a number of interesting special cases, in order of increasing generality.

1. l_2-*potential:* $P(x) = (\sum_{k \in S^i}([x_k]_+)^2)^{1/2}$. This yields (after normalization) $\Lambda(x) = (1/\|[x]_+\|_1)[x]_+$ for $x \notin \mathbb{R}_-^{S^i}$, and the resulting strategy is $\sigma_t^i(k) = [D_{t-1}^i(k)]_+ / \sum_{k' \in S^i}[D_{t-1}^i(k')]_+$ when $D_{t-1}^i \notin \mathbb{R}_-^{S^i}$. This is the Hannan-consistent strategy introduced in Hart and Mas-Colell (2000, Theorem B), where the play probabilities are proportional to the positive regrets.

2. l_p-*potential:* $P(x) = (\sum_{k \in S^i}([x_k]_+)^p)^{1/p}$ for some $1 < p < \infty$. This yields $\sigma_t^i(k) = ([D_{t-1}^i(k)]_+)^{p-1} / \sum_{k' \in S^i}([D_{t-1}^i(k')]_+)^{p-1}$; i.e., play

probabilities that are proportional to a fixed positive power $(p - 1 > 0)$ of the positive regrets.

3. *Separable potential*: A *separable* strategy is one where σ_t^i is proportional to a vector whose kth coordinate depends *only* on the kth regret; i.e., σ_t^i is proportional to a vector of the form $(\psi_k(D_{t-1}^i(k)))_{k \in S^i}$. Conditions (R1) and (R2) result in the following requirements:[24] For each k in S^i, the function $\psi_k : \mathbb{R} \to \mathbb{R}$ is continuous; $\psi_k(x_k) > 0$ for $x_k > 0$; and $\psi_k(x_k) = 0$ for $x_k \le 0$. The corresponding potential is $P(x) = \sum_{k \in S^i} \Psi_k(x_k)$, where $\Psi_k(x) := \int_{-\infty}^{x} \psi_k(y) dy$. Note that, unlike the previous two cases, the functions ψ_k may differ for different k, and they need not be monotonic (thus a higher regret may not lead to a higher probability).

Finally, observe that in all of the above cases, actions with negative or zero regret are never chosen. This need no longer be true in the general (nonseparable) case; see Section 4.2 below.

3.3. *Counterexamples*

The counterexamples of Section 2.3 translate easily into the regret setup.

- *The role of "better"* (R2). Consider the one-person game

R1	0
R2	1

  The resulting regret game is given in Example 2.7. The strategy of playing "R1 forever" satisfies condition (R1) but not condition (R2) (or (3.3)), and it is indeed not Hannan-consistent.
- *The role of continuity in* (R1). Consider the simplest two-person coordination game (a well-known stumbling block for many strategies)

	C1	C2
R1	$(1,1)$	$(0,0)$
R2	$(0,0)$	$(1,1)$

  The resulting regret game for the Row player is precisely the vector-payoff game of Examples 2.4 and 2.5, where we looked at the approachability question for the nonpositive orthant. The two strategies we considered there—which we have shown not to be

---

[24] Consider points x with $x_j = \pm\varepsilon$ for all $j \ne k$.

Hannan-consistent—are not continuous. They correspond to the l_∞-
and the l_1-potentials, respectively, which are not differentiable. (Note
in particular that the l_∞-case yields "fictitious play," which is further
discussed in Section 4.1 below.)

- *The role of integrability in* (R1). The vector-payoff game of our
  Example 2.6 can easily be seen to be a regret game. However, the
  approachable set there was not the nonpositive orthant. In order to
  get a counterexample to the result of Theorem 3.3 when integrability
  is not satisfied, one would need to resort to additional dimensions, that
  is, more than two strategies; we do not do it here, although it is plain
  that such examples are easy—though painful—to construct.

4. Fictitious Play and Better Play

4.1. *Fictitious play and smooth fictitious play*

As we have already pointed out, fictitious play may be viewed as a stationary regret-based strategy, corresponding to the l_∞-mapping (the directional
mapping generated by the l_∞-potential). It does not guarantee Hannan-
consistency (see Example 2.4 and Section 3.3); the culprit for this is the
lack of continuity (i.e., (D1)).

Before continuing the discussion it is useful to note a property of fictitious play: *The play at time t does not depend on the realized average payoff*
$\overline{U}_{t-1}^i$. Indeed, $\max_k D_{t-1}^i(k) = \max_k u^i(k, z_{t-1}^{-i}) - \overline{U}_{t-1}^i$, so an action $k \in S^i$
maximizes regret if and only if it maximizes the payoff against the empirical
distribution z_{t-1}^{-i} of the actions of $-i$. In the general approachability setup of
Section 2 (with $\mathscr{C} = \mathbb{R}_-^{S^i}$), this observation translates into the requirement
that the directional mapping Λ be invariant to adding the same constant
to all coordinates. That is, writing $e := (1, 1, \ldots, 1) \in \mathbb{R}^{S^i}$,

$$\Lambda(x) = \Lambda(y) \quad \text{for any } x, y \notin \mathbb{R}_-^{S^i} \text{ with } x - y = \alpha e \text{ for some scalar } \alpha. \quad (4.1)$$

Note that, as it should be, the l_∞-mapping satisfies this property (4.1).

Proposition 4.1. *A directional mapping Λ satisfies* (D2), (D3), *and* (4.1)
*for $\mathscr{C} = \mathbb{R}^m$ if and only if it is equivalent to the l_∞-mapping, i.e., its poten-
tial P satisfies $P(x) = \phi(\max_k x_k)$ for some strictly increasing function ϕ.*

Proof. Since $\mathscr{C} = \mathbb{R}^m$, the allowable directions are $\lambda \geq 0, \lambda \neq 0$. Thus
$\nabla P(x) \geq 0$ for a.e. $x \notin \mathbb{R}_-^m$ by (D2), implying that the limit of $\nabla P(x) \cdot x$
is ≤ 0 as x approaches the boundary of $\mathbb{R}_-^m$. But $\nabla P(x) \cdot x > 0$ for a.e.
$x \notin \mathbb{R}_-^m$ by (D3), implying that the limit of $\nabla P(x) \cdot x$ is in fact 0 as x

approaches bd $\mathbb{R}^m_-$. Because P is Lipschitz, it follows that P is constant on bd $\mathbb{R}^m_-$, i.e., $P(x) = P(0)$ for every $x \in$ bd $\mathbb{R}^m_-$. By (D3) again we have $P(x) > P(0)$ for all $x \notin \mathbb{R}^m_-$. Adding to this the invariance condition (4.1) implies that the level sets of P are all translates by multiples of e of bd $\mathbb{R}^m_- = \{x \in \mathbb{R}^m : \max_k x_k = 0\}$. □

Since the l_∞-mapping does not guarantee that $\mathscr{C} = \mathbb{R}^m_-$ is approached (again, see Example 2.4 and Section 3.3), we have:

Corollary 4.2. *There is no stationary regret-based strategy that satisfies* (R1) *and* (R2) *and is independent of realized average payoff.*

The import of the Corollary (together with the indispensability of conditions (R1) and (R2), as shown by the counterexamples in Section 3.3) is that one cannot simultaneously have independence of realized payoffs and guarantee Hannan-consistency in every game.

We must weaken one of the two properties. One possibility is to weaken the consistency requirement to ε-*consistency*, $\limsup_t D^i_t(k) \leq \varepsilon$ for all k. Fudenberg and Levine (1995) propose a smoothing of fictitious play that— like fictitious play itself—is independent of realized payoffs. In essence, their function P is convex, smooth, and satisfies the property that its level sets are obtained from each other by translations along the $e = (1, \ldots, 1)$ direction[25] (see Fig. 6). The level set of P through 0 is therefore smooth; it is very close to the boundary of the negative orthant but unavoidably distinct from it. The resulting strategy approaches $\mathscr{C} = \{x : P(x) \leq P(0)\}$ (recall Remark 5 in Section 2.1: a set of the form $\{x : P(x) \leq c\}$, for constant c, is approachable when $c \geq P(0)$—since it contains $\mathbb{R}^{S^i}_-$—and is not approachable when $c < P(0)$—since it does not contain 0). The set $\mathscr{C}$ is strictly larger than $\mathbb{R}^{S^i}_-$; it is an ε-neighborhood of the negative orthant $\mathbb{R}^{S^i}_-$. Thus one obtains only ε-consistency.[26,27]

---

[25]Specifically, $P(x) = \max_{\sigma^i \in \Delta(S^i)} \{\sigma^i \cdot x + \varepsilon v(\sigma^i)\}$, where $\varepsilon > 0$ is small and v is a strictly differentiably concave function, with gradient vector approaching infinite length as one approaches the boundary of $\Delta(S^i)$.

[26]Other smoothings have been proposed, including Hannan (1957), Foster and Vohra (1993, 1998), and Freund and Schapire (1999) (in the latter—which corresponds to exponential smoothing—the strategy is nonstationary, i.e., it depends not only on the point in regret space but also on the time t; of course, nonstationary strategies where ε decreases with t may yield exact consistency).

[27]Smooth fictitious play may be equivalently viewed, in our framework, as first taking a set $\mathscr{C}$ that is close to the negative orthant and has a smooth boundary, and then using the l_∞-distance from $\mathscr{C}$ as a potential (recall Remark 6 in Section 2.1).

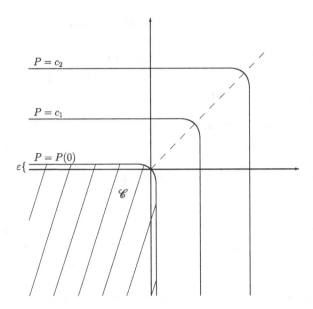

Fig. 6.   The level sets of the potential of smooth fictitious play.

The other possibility is to allow the strategy to depend also on the realized payoffs. Then there are strategies that are close to fictitious play and guarantee Hannan-consistency in any game. Take, for instance, the l_p-potential strategy for large enough p (see Section 3.2).[28]

4.2. *Better play*

All the examples presented until now satisfy an additional natural requirement, namely, that only actions with positive regret are played (provided, of course, that there are such actions). Formally, consider a stationary regret-based strategy of player i that is given by a mapping $\sigma^i : \mathbb{R}^{S^i} \to \Delta(S^i)$ (see Theorem 3.3); we add to (R1) and (R2) the following condition:[29]

(R3) For every $x \notin \mathbb{R}^{S^i}_{--}$,   if $x_k < 0$ then $[\sigma^i(x)]_k = 0$.

---

[28]This amounts to smoothing the norm and keeping $\mathscr{C}$ equal to the negative orthant, whereas the previous construction smoothed the boundary of $\mathscr{C}$ and kept the l_∞-norm. These are the two "dual" ways of generating a smooth potential (again, see Remark 6 in Section 2.1).

[29]Note that the condition needs to be satisfied not only for $x \notin \mathbb{R}^{S^i}_-$, but also for $x \in \text{bd } \mathbb{R}^{S^i}_-$.

Since x is the i-regret vector, (R3) means that σ^i gives probability 1 to the set of actions with nonnegative regret (unless all regrets are negative, in which case there is no requirement[30]). This may be rewritten as

$$[\sigma_t^i]_k > 0 \quad \text{only if } u^i(k, z_{t-1}^{-i}) \geq \overline{U}_{t-1}^i. \tag{4.2}$$

That is, only those actions k are played whose payoff against the empirical distribution z_{t-1}^{-i} of the opponents' actions is at least as large as the actual realized average payoff $\overline{U}_{t-1}^i$; in short, only the "better actions."[31] For an example where (R3) is *not* satisfied, see Fig. 7.

The l_p-potential strategies, for $1 < p < \infty$, and in fact all separable strategies (see Section 3.2) essentially[32] satisfy (R3). Fictitious play (with the l_∞-potential) also satisfies (R3): The action chosen is a "best" one (rather than just "better"). At the other extreme, the l_1-potential strategy gives equal probability to *all* better actions, so it also satisfies (R3). (However, these last two do not satisfy (R1)).

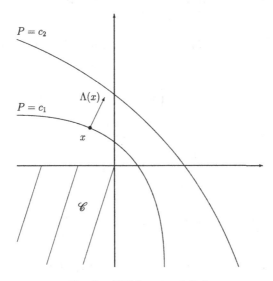

Fig. 7. (R3) is not satisfied.

---

[30] See Footnote 34 below.

[31] Observe that (R2) (or (3.3)) is a requirement on the average over all played actions k, whereas (R3) (or (4.2)) applies to each such k separately.

[32] The condition in (R3) is automatically satisfied in these cases for $x \notin \mathbb{R}^{S^i}$; one needs to impose it explicitly for $x \in \text{bd } \mathbb{R}_-^{S^i}$ (where, until now, we had no requirements).

Using condition (R3) yields the following result, which is a generalization of Theorem A of Monderer *et al.* (1997) for fictitious play:

Proposition 4.3. *Consider a stationary regret-based strategy of player i given by a mapping* $\sigma^i : \mathbb{R}^{S^i} \to \Delta(S^i)$ *that satisfies*[33] *(R3). Then, in any (finite) N-person game, the maximal regret of i is always nonnegative:*

$$\max_{k \in S^i} D_t^i(k) \geq 0 \quad \text{for all } t.$$

Proof. The proof is by induction, starting with $D_0^i(k) = 0$ for all k. Assume that $\max_k D_{t-1}^i(k) \geq 0$ (or, $D_{t-1}^i \notin \mathbb{R}_{--}^{S^i}$). By (R3), $D_{t-1}^i(k) \geq 0$ for any k chosen at time t; since $A_k(k, s_t^{-i}) = 0$ it follows that $D_t^i(k) = (1/t)((t-1)D_{t-1}^i(k) + t0) \geq 0$. □

Thus, the vector of regrets never enters the negative orthant.[34] Recall that the result of Theorem 3.3 is that the vector of regrets approaches the nonpositive orthant. To combine the two, we define *better play* as any stationary regret-based strategy of player i that is given by a mapping σ^i satisfying (R1)–(R3). We thus have

Corollary 4.4. *In any (finite) N-person game, if player i uses a better play strategy, then his maximal regret converges to 0 a.s.*

$$\lim_{t \to \infty} \max_{k \in S^i} D_t^i(k) = \lim_{t \to \infty} \left(\max_{k \in S^i} u^i(k, z_t^{-i}) - \overline{U}_t^i \right) = 0 \text{ a.s.}$$

That is, the average payoff $\overline{U}_t^i$ of player i up to time t is close, as $t \to \infty$, to i's best-reply payoff against the empirical distribution of the other players' actions. In particular, in a two-person zero-sum game we obtain the following.

Corollary 4.5. *In any (finite) two-person zero-sum game, if both players use better-play strategies, then:*

(i) *For each player, the empirical distribution of play converges to the set of optimal actions.*[35]

(ii) *The average payoff converges to the value of the game.*

---

[33] Note that (R1) and (R2) are not assumed.

[34] Therefore, at every period there are always actions with nonnegative regret—out of which the next action is chosen (and so the condition $x \notin \mathbb{R}_{--}^{S^i}$ in (R3) always holds).

[35] That is, the set of mixed actions that guarantee the value.

Proof. Let 1 be the maximizer and 2 the minimizer, and denote by v the minimax value of the game. Then $\max_{k \in S^1} u^1(k, z_t^2) \geq v$, so by Corollary 4.4 we have $\liminf_t \overline{U}_t^1 \geq v$. The same argument for player 2 yields the opposite inequality, thus $\lim_t \overline{U}_t^1 = v$. Therefore $\lim_t \max_{k \in S^1} u^1(k, z_t^2) = v$ (apply the Corollary again); hence any limit point of the sequence z_t^2 must be an optimal action of player 2; similarly for player 1. □

Thus, better play enjoys the same properties as fictitious play in two-person zero-sum games (for fictitious play, see Robinson 1951 for the convergence to the set of optimal strategies, and see Monderer *et al.* 1997, Theorem B, and Rivière 1997 for the convergence of the average payoff).

5. Discussion and Extensions

In this section we discuss a number of extensions of our results.

5.1. *Conditional regrets*

As stated in the Introduction, we have been led to the "no regret" Hannan-consistency property from considerations of "no conditional regret" that correspond to correlated equilibria (see Hart and Mas-Colell 2000). Given two actions k and j of player i, the *conditional regret* from j to k is the change that would have occurred in the average payoff of i if he had played action k in all those periods where he did play j (and everything else is left unchanged). That is,

$$DC_t^i(j, k) := \frac{1}{t} \sum_{\tau \leq t : s_\tau^i = j} [u^i(k, s_\tau^{-i}) - u^i(s_\tau)]$$

for every[36] $j, k \in S^i$. The vector of regrets DC_t^i is now in $\mathbb{R}^L$, where $L := S^i \times S^i$, and the empirical distribution of actions up to time t constitutes a correlated equilibrium if and only if $DC_t^i \leq 0$. Thus, the set to be approached is the nonpositive orthant $\mathbb{R}_-^L$. The corresponding game with vector payoffs A is defined as follows: the (j, k) coordinate of the vector payoff $A(s^i, s^{-i}) \in \mathbb{R}^L$ is $u^i(k, s^{-i}) - u^i(j, s^{-i})$ when $s^i = j$, and it is 0 otherwise; hence $DC_t^i = (1/t) \sum_{\tau \leq t} A(s_\tau)$.

---

[36]Note that each Hannan regret $D_t^i(k)$ is the sum of the conditional regrets $DC_t^i(j, k)$ over $j \neq k$. Thus the set of distributions of N-tuples of actions that satisfy the Hannan no-regret conditions includes the set of correlated equilibrium distributions. The inclusion is, in general, strict (the two sets coincide when every player has only two actions).

As in Propositions 3.1 and 3.2 (see Hart and Mas-Colell 2000, Section 3), it can easily be verified that:

- $\mathscr{C} = \mathbb{R}^L_-$ is always approachable.
- For every $\lambda \in \mathbb{R}^L_+$, the Blackwell approachability condition for $\mathscr{C} = \mathbb{R}^L_-$ ((2.1) or (3.1)) holds for any mixed action $\sigma^i = (\sigma^i_k)_{k \in S^i} \in \Delta(S^i)$ that satisfies

$$\sum_{j \in S^i} \sigma^i_j \lambda(j,k) = \sigma^i_k \sum_{j \in S^i} \lambda(k,j) \quad \text{for all } k \in S^i. \tag{5.1}$$

Viewing λ as an $S^i \times S^i$ matrix, condition (5.1) says that σ^i is an invariant vector for the (nonnegative) matrix λ.
- For every $\lambda \in \mathbb{R}^L_+$, there exists a $\sigma^i \in \Delta(S^i)$ satisfying (5.1).

Applying Theorem 2.1 yields a large class of strategies. For example (as in Section 3.2), if P is the l_p-potential for some $1 < p < \infty$, then σ^i is an invariant vector of the matrix of the $p-1$ powers of the nonnegative regrets.[37] In the more general separable case, σ^i is an invariant vector of the matrix whose (j,k) coordinate is $\psi_{(j,k)}(DC^i_{t-1}(j,k))$, where $\psi_{(j,k)}$ is any real continuous function that vanishes for $x \leq 0$ and is positive for $x > 0$. As in Hart and Mas-Colell (2000, Theorem A), if every player uses a strategy in this class (of course, different players may use different types of strategies), then the empirical distribution of play converges to the set of correlated equilibria of Γ.

Since finding invariant vectors is by no means a simple matter, in Hart and Mas-Colell (2000) much effort is devoted to obtaining simple adaptive procedures, which use the matrix of regrets as a one-step transition matrix. To do the same here, one would use instead the matrix $\Lambda(DC^i_{t-1})$.

5.2. Variable game

We noted in Section 3.2 that our strategies are game-independent. This allows us to consider the case where, at each period, a different game is being played (for example, a stochastic game). The strategy set of player i is the same set S^i in all games, but he does not know which game is currently being

---

[37] For $p = 2$ we get the matrix of regrets—which yields precisely Theorem A of Hart and Mas-Colell (2000).

played. All our results—in particular, Theorem 3.3—continue to hold[38] provided player i is told, *after* each period t, which game was played at time t and what were the chosen actions s_t^{-i} of the other players. Indeed, as in Section 3, i can then compute the vector $a := A(s_t^i, s_t^{-i}) \in \mathbb{R}^{S^i}$, update his regret vector—$D_t^i = (1/t)((t-1)D_{t-1}^i + a)$—and then play $\sigma^i(D_t^i)$ in the next period, where σ^i is any mapping satisfying (R1) and (R2).

5.3. *Unknown game*

When the player does not know the (fixed) game Γ that is played and is told, at each stage, only his own realized payoff (but not the choices of the other players)—in what may be referred to as a "stimulus–response" model—Hannan-consistency may nonetheless be obtained (see Foster and Vohra 1993, 1998, Auer *et al.* 1995, Fudenberg and Levine 1998, Section 4.8, Hart and Mas-Colell 2000, Section 4(j), and also Baños 1968, and Megiddo 1980 for related work). For instance, one can replace the regrets—which cannot be computed here—with appropriate estimates.[39]

References

Auer P., N. Cesa-Bianchi, Y. Freund, and R. E. Schapire (1995), "Gambling in a Rigged Casino: The Adversarial Multiarmed Bandit Problem," in *Proceedings of the 36th Annual Symposium on Foundations of Computer Science*, pp. 322–331.

Baños, A. (1968), "On Pseudo-Games," *The Annals of Mathematical Statistics*, 39, 1932–1945.

Blackwell, D. (1956a), "An Analog of the Minmax Theorem for Vector Payoffs," *Pacific Journal of Mathematics*, 6, 1–8.

———— (1956b), "Controlled Random Walks," in *Proceedings of the International Congress of Mathematicians* 1954, Vol. III, ed. by E. P. Noordhoff. Amsterdam: North-Holland, pp. 335–338.

Borodin, A. and R. El-Yaniv (1998), *Online Computation and Competitive Analysis*. Cambridge, UK: Cambridge University Press.

Clarke, F. (1983), *Optimization and Nonsmooth Analysis*. New York: Wiley.

Foster, D. and R. V. Vohra (1993), "A Randomized Rule for Selecting Forecasts," *Operations Research*, 41, 704–709.

———— (1997), "Calibrated Learning and Correlated Equilibrium," *Games and Economic Behavior*, 21, 40–55.

———— (1998), "Asymptotic Calibration," *Biometrika*, 85, 379–390.

---

[38] Assuming the payoffs of i are uniformly bounded.

[39] Specifically, use $\hat{D}_t^i(k) := (1/t) \sum_{\tau \leq t : s_\tau^i = k} (1/[\sigma_\tau^i]_k) U_\tau^i - \overline{U}_t^i$ instead of $D_t^i(k)$ (see Hart and Mas-Colell 2001, Section 3(c)).

_____ (1999), "Regret in the Online Decision Problem," *Games and Economic Behavior*, 29, 7–35.

Freund, Y. and R. E. Schapire (1999), "Adaptive Game Playing Using Multiplicative Weights," *Games and Economic Behavior*, 29, 79–103.

Fudenberg, D. and D. K. Levine (1995), "Universal Consistency and Cautious Fictitious Play," *Journal of Economic Dynamics and Control*, 19, 1065–1090.

_____ (1998), *Theory of Learning in Games*. Cambridge, MA: MIT Press.

_____ (1999), "Conditional Universal Consistency," *Games and Economic Behavior*, 29, 104–130.

Hannan, J. (1957), "Approximation to Bayes Risk in Repeated Play," in *Contributions to the Theory of Games*, Vol. III (*Annals of Mathematics Studies* 39), ed. by M. Dresher, A. W. Tucker and P. Wolfe. Princeton: Princeton University Press, pp. 97–139.

Hart, S. and A. Mas-Colell (2000), "A Simple Adaptive Procedure Leading to Correlated Equilibrium," *Econometrica*, 68, 1127–1150. [Chapter 2]

_____ (2001), "A Reinforcement Procedure Leading to Correlated Equilibrium," in *Economic Essays: A Festschrift for Werner Hildenbrand*, ed. by G. Debreu, W. Neuefeind, and W. Trockel. Berlin: Springer, pp. 181–200. [Chapter 4].

Littlestone, N. and M. K. Warmuth (1994), "The Weighted Majority Algorithm," *Information and Computation*, 108, 212–261.

Loève, M. (1978), *Probability Theory*, Vol. II, 4th Edition. New York: Springer.

Luce, R. D. and H. Raiffa (1957), *Games and Decisions*. New York: Wiley.

Megiddo, N. (1980), "On Repeated Games with Incomplete Information Played by Non-Bayesian Players," *International Journal of Game Theory*, 9, 157–167.

Monderer, D., D. Samet, and A. Sela (1997), "Belief Affirming in Learning Processes," *Journal of Economic Theory*, 73, 438–452.

Rivière, P. (1997), "Quelques Modèles de Jeux d'Evolution," Ph.D. thesis, Université Paris 6.

Robinson, J. (1951), "An Iterative Method of Solving a Game," *Annals of Mathematics*, 54, 296–301.

Chapter 4

A REINFORCEMENT PROCEDURE LEADING TO CORRELATED EQUILIBRIUM*

Sergiu Hart and Andreu Mas-Colell

We consider repeated games where at any period each player knows only his set of actions and the stream of payoffs that he has received in the past. He knows neither his own payoff function, nor the characteristics of the other players (how many there are, their strategies and payoffs). In this context, we present an adaptive procedure for play—called "modified-regret matching"—which is interpretable as a stimulus-response or reinforcement procedure, and which has the property that any limit point of the empirical distribution of play is a correlated equilibrium of the stage game.

1. Introduction

Werner Hildenbrand has repeatedly emphasized the usefulness, conceptual and technical, of carrying out equilibrium analysis by means of distributional notions. For social situations modeled as N-person games, the concept of correlated equilibrium, an extension of Nash equilibrium introduced by Aumann, can be viewed in a most natural way as imposing equilibrium conditions on the distribution of action combinations of the different players. Correlated equilibrium will be the subject of the present paper,

Originally published in *Economic Essays: A Festschrift for Werner Hildenbrand*, ed. by G. Debreu, W. Neuefeind, and W. Trockel. Berlin: Springer (2001), pp. 181–200.

*Dedicated with great admiration to Werner Hildenbrand on his 65th birthday.
Previous versions of these results were included in the Center for Rationality Discussion Papers #126 (December 1996) and #166 (March 1998).
We thank Dean Foster for suggesting the use of "modified regrets." The research is partially supported by grants of the Israel Academy of Sciences and Humanities; the Spanish Ministry of Education; the Generalitat de Catalunya; CREI; and the EU-TMR Research Network.
See the note at the end of the chapter.

and the style of our analysis should not surprise anybody familiar with Hildenbrand's research.

This paper continues the study of Hart and Mas-Colell (2000), where the notion of correlated equilibrium and that of approachability by means of simple "rules of thumb" or "adaptive procedures" of play were linked. We showed there that if a game is repeatedly played and the players determine their stage probabilities of play according to a procedure called "regret matching," then the empirical distribution of play converges with probability one to the set of correlated equilibria of the given game. *Regret matching* means that if at period t player i has taken action j, then the probability of playing another action k at time $t+1$ will be proportional to the "regret for not having played k instead of j," which is defined simply as the increase, if any, in the average payoff that would result if all past plays of action j were replaced by action k (and everything else remained unaltered).

The implementation of regret matching by a player requires that player to observe the actions taken by *all* players in the past. But that is not all. The player should also be able to carry out the thought experiment of computing what his payoffs would have been, had his actions in the past been different from what they really were. For this he needs to know what game he is playing (or, at the very least, his own payoff function).

In this paper we will show that this level of sophistication and knowledge is not really necessary to obtain correlated equilibria. We shall modify the regret matching procedure in such a way that the play probabilities are determined from the *actual realizations only.* Specifically, each player only needs to know the payoffs he received in past periods. He need not know the game he is playing—neither his own payoff function nor the other players' payoffs; in fact, he may well be oblivious of the fact that there is a game at all. The procedure that we shall examine is a "stimulus-response" or "reinforcement" procedure, in the sense that a relatively high payoff at period t will tend to increase the probability of playing at period $t+1$ the same action that was played at t.

In Section 2 we present the basic model and state our results. Section 3 contains some further discussions, and the proofs are given in Section 4.

To summarize: We consider repeated games where each player knows only his set of available choices (but nothing about the other players), and observes only his own actually realized payoffs. We exhibit simple strategies of play whereby the long-run frequencies of the various action combinations being played are tantamount to a correlated equilibrium. Of course, this sophisticated "macro"-picture that emerges cannot be seen at the individual level. This is a conclusion, we trust, with which Werner Hildenbrand can truly sympathize.

2. Model and Results

We consider finite N-person games in strategic (or "normal") form. The set of players is a finite set N, the action[1] set of each player $i \in N$ is a finite set S^i, and the payoff function of i is $u^i : S \to \mathbb{R}$, where[2] $S := \Pi_{\ell \in N} S^\ell$. We will denote this game $\langle N, (S^i)_{i \in N}, (u^i)_{i \in N} \rangle$ by Γ.

A correlated equilibrium—a concept introduced by Aumann (1974)—is nothing other than a Nash equilibrium where each player may receive a private signal before the game is played (the signals do not affect the payoffs; and the players may base their choices on the signals received). This may be equivalently described as follows: Assume that the signal of each player i is in fact a "play recommendation" $s^i \in S^i$, where the N-tuple $s = (s^i)_{i \in N}$ is selected (by a "device" or "mediator") according to a commonly known probability distribution. A correlated equilibrium results if each player realizes that the best he can do is to follow the recommendation, provided that all the other players do likewise.

Equivalently, a probability distribution ψ on the set S of action N-tuples (i.e., $\psi(s) \geq 0$ for all $s \in S$ and $\sum_{s \in S} \psi(s) = 1$) is a *correlated equilibrium* of the game Γ if, for every player $i \in N$ and every two actions $j, k \in S^i$ of i, we have

$$\sum_{s \in S: s^i = j} \psi(s)(u^i(k, s^{-i}) - u^i(s)) \leq 0, \tag{1}$$

where $s^{-i} \in S^{-i} := \Pi_{\ell \neq i} S^\ell$ denotes the action combination of all players except i (thus $s = (s^i, s^{-i})$). The inequality (1) (after dividing the expression there by $\sum_{s \in S: s^i = j} \psi(s)$) means that when the recommendation to player i is to choose action j, then choosing k instead of j cannot yield a higher expected payoff to i. Finally, a *correlated ε-equilibrium* obtains when 0 is replaced by ε on the right-hand side of (1).

Now assume that the game Γ is played repeatedly in discrete time $t = 1, 2, \ldots$. Denote by $s^i_t \in S^i$ the (realized) choice of player i at time t, and put $s_t = (s^i_t)_{i \in N} \in S$. The payoff of player i in period t is denoted $U^i_t := u^i(s_t)$.

---

[1] To avoid confusion, we will use the term "action" for the one-shot game, and "strategy" for the multi-stage game.
[2] $\mathbb{R}$ denotes the real line.

The basic setup of this paper assumes that the information of each player i consists just of his set of available choices S^i. He does not know the game Γ; in fact, he does not know how many players there are besides himself,[3] what their choices are, and what the payoff functions (his own as well as the others') are. The only thing player i observes is his own realized payoffs and actions. That is, in determining his play probabilities at time $t + 1$, player i's information consists of his own realized payoffs $U_1^i, U_2^i, \ldots, U_t^i$ in the previous t periods, as well as his actual actions $s_1^i, s_2^i, \ldots, s_t^i$ there.

In Hart and Mas-Colell (2000), we introduced *regret matching procedures*. These are simple adaptive strategies where the play probabilities are proportional to the "regrets" for not having played other actions. Specifically, for any two distinct actions $j \neq k$ in S^i and every time t, the *regret* of i at t from j to k is

$$R_t^i(j, k) := [D_t^i(j, k)]^+ \equiv \max\{D_t^i(j, k), 0\}, \tag{2}$$

where

$$D_t^i(j, k) := \frac{1}{t} \sum_{\tau \leq t : s_\tau^i = j} u^i(k, s_\tau^{-i}) - u^i(s_\tau). \tag{3}$$

This is the change in the average payoff of i that would have resulted if he had played action k every time in the past that he actually chose j. Rewriting this as

$$D_t^i(j, k) = \frac{1}{t} \sum_{\tau \leq t : s_\tau^i = j} u^i(k, s_\tau^{-i}) - \frac{1}{t} \sum_{\tau \leq t : s_\tau^i = j} u^i(s_\tau)$$

shows that player i can compute the second term: It is just $(1/t) \sum_{\tau \leq t : s_\tau^i = j} U_\tau^i$. But i cannot compute the first term, since he knows neither his own payoff function u^i nor the choices s_τ^{-i} of the other players.

Therefore we replace the first term by an "estimate" that can be computed on the basis of the available information; namely, we define

$$C_t^i(j, k) := \frac{1}{t} \sum_{\tau \leq t : s_\tau^i = k} \frac{p_\tau^i(j)}{p_\tau^i(k)} U_\tau^i - \frac{1}{t} \sum_{\tau \leq t : s_\tau^i = j} U_\tau^i, \tag{4}$$

where p_τ^i denotes the play probabilities at time t (thus $p_\tau^i(j)$ is the probability that i chose j at period τ); these probabilities p_τ^i are defined below

---

[3] Perhaps none: it could be a game against Nature.

(they will indeed depend only on $U_1^i, \ldots, U_t^i$). As in (2), the *modified regrets* are then

$$Q_t^i(j, k) := [C_t^i(j, k)]^+ \equiv \max\{C_t^i(j, k), 0\}. \qquad (5)$$

In words, the modified regret for not having used k instead of j measures the difference (strictly speaking, its positive part) of the average payoff over the periods when k was used and the periods when j was used. In addition, the payoffs of these periods are normalized in a manner that, intuitively speaking, makes the length of the respective periods comparable.[4]

Next we define the play probabilities, based on these modified regrets. Recall the regret matching procedure of Hart and Mas-Colell (2000): If $j := s_t^i$ is the action chosen by i at time t, then the probability of switching at time $t+1$ to another action $k \neq j$ is proportional (with a fixed factor l/μ) to the regret from j to k; with the remaining probability, the same action j is chosen again. Here we shall make two changes relative to Hart and Mas-Colell (2000). First, we need to guarantee that the sum of the proposed probabilities does not exceed one; multiplying the modified regrets by a factor l/μ (which we still do) does not suffice because the modified regrets are not bounded.[5] Second, we need every action to be played with some minimal frequency.[6] Thus we shall require that the transitions "tremble" over every action.

Let therefore $\mu > 0$ be a sufficiently large number; similarly to Hart and Mas-Colell (2000), it suffices to take μ so that $\mu \geq 2M^i m^i$ for all i, where $m^i := |S^i|$ is[7] the number of actions of i, and M^i is an upper bound on $|u^i(s)|$ for all $s \in S$. Let $0 < \delta < 1$, and define the play probabilities of

---

[4]For instance, if the probability of k being chosen is always twice that of j, then in the long run k will be played twice as often as j. So, in order to compare the sum of the payoffs over those periods when j is played with the parallel sum when k is played, one needs to multiply the k-sum by 1/2. When the probabilities change over time, Formula (4) uses the correct normalization: The difference between the regrets and the modified regrets has conditional expectation equal to 0 (see Step 10(i) and the Proof of Step 11(iii) in Section 4).

[5]Notice that there is no *a priori* lower bound on the probabilities appearing in the denominators.

[6]Roughly speaking, one needs this exogenous statistical "noise" to be able to estimate the contingent payoffs using only the realized payoffs (recall the first term in the modified regret). In fact, this is quite delicate, since the actions of the other players are never observed.

[7]We write $|A|$ for the number of elements of a finite set A.

See the note at the end of the chapter.

player i at time $t + 1$ by

$$p_{t+1}^i(k) := (1 - \delta) \min \left\{ \frac{1}{\mu} Q_t^i(j, k), \frac{1}{m^i} \right\} + \delta \frac{1}{m^i}, \quad \text{for } k \neq j; \quad \text{and}$$

$$\tag{6}$$

$$p_{t+1}^i(j) := 1 - \sum_{k \in S^i : k \neq j} p_{t+1}^i(k),$$

where $j := s_t^i$ is the choice of i in (the previous) period t. As for the first period, the play probabilities at $t = 1$ are given by an arbitrary strictly positive vector $p_1^i \in \Delta(S^i)$; for simplicity, assume that $p_1^i(j) \geq \delta/m^i$ for all j in S^i.

Formula (6) says that p_{t+1}^i is a weighted average of two probability vectors (note that $p_{t+1}^i(j) = (1 - \delta)(1 - \sum_{k \neq j} \min\{Q_t^i(j, k)/\mu, 1/m^i\}) + \delta/m^i$). The first, with weight $1 - \delta$, is given by the modified regrets in a manner similar to Hart and Mas-Colell (2000, Formula (2.2))) (note that taking the minimum with $1/m^i$ guarantees that the resulting sum does not exceed $1 - 1/m^i$). The second term, with weight δ, is just the uniform distribution over S^i : each $k \in S^i$ has equal probability $1/m^i$. This "uniform tremble" guarantees that all probabilities at time $t+1$ are at least $\delta/m^i > 0$.

Putting (4), (5), and (6) together implies that p_{t+1}^i depends only on the previously realized payoffs $U_1^i, \ldots, U_t^i$ and play probabilities $p_1^i, \ldots, p_t^i$. Therefore, by recursion, p_{t+1}^i does indeed depend only on $U_1^i, \ldots, U_t^i$.

Let[8] $z_t \in \Delta(S)$ be the empirical distribution of play up to t; that is, for every $s \in S$,

$$z_t(s) := \frac{1}{t} |\{\tau \leq t : s_\tau = s\}|$$

is the relative frequency that the N-tuple of actions s has been played in the first t stages.

Our first result is:

Theorem 1. *For every $\varepsilon > 0$ there is a $\delta_0 \equiv \delta_0(\varepsilon) > 0$ such that for all $\delta < \delta_0$, if every player i plays according to adaptive procedure (6), then the empirical distributions of play z_t converge almost surely as $t \to \infty$ to the set of correlated ε-equilibria of the game Γ.*

---

[8]We write $\Delta(A) := \{p \in \mathbb{R}_+^A : \sum_{a \in A} p(a) = 1\}$ for the set of probability distributions over the finite set A.

Thus: for almost every trajectory, z_t is close to a correlated ε-equilibrium for all large enough t; equivalently, z_t is a correlated 2ε-equilibrium from some time on (this time may depend on the trajectory; the theorem states that, with probability one, it is finite). As in the Main Theorem of Hart and Mas-Colell (2000), note that the convergence is not to a *point*, but to a *set*; that is, the distance to the set goes to 0. Finally, we remark that both μ and δ can be taken to be different for different players.

To obtain in the limit correlated equilibria—rather than correlated *approximate* equilibria—we need δ to decrease to 0 as t increases. A simple way to do this is to replace δ in (6) by δ/t^γ, where γ is a number strictly between 0 and[9] $1/4$. Thus, we now define

$$p^i_{t+1}(k) := \left(1 - \frac{\delta}{t^\gamma}\right) \min\left\{\frac{1}{\mu}Q^i_t(j,k), \frac{1}{m^i}\right\} + \frac{\delta}{t^\gamma}\frac{1}{m^i}, \quad \text{for } k \neq j; \quad \text{and}$$

$$p^i_{t+1}(j) := 1 - \sum_{k \in S^i : k \neq j} p^i_{t+1}(k), \tag{7}$$

where, again, $j := s^i_t$ is the choice of i in period t. At $t = 1$, let p^i_1 be an arbitrary vector in $\Delta(S^i)$ with $p^i_1(j) \geq \delta/m^i$ for all j in S^i. The second result is:

Theorem 2. *If every player i plays according to adaptive procedure (7), then the empirical distributions of play z_t converge almost surely as $t \to \infty$ to the set of correlated equilibria of the game Γ.*

We have chosen the particular type of sequence $\delta_t := \delta/t^\gamma$ for simplicity and convenience only. What matters is, first, that δ_t converge to 0, and second, that it do so sufficiently slowly (otherwise the modified regrets C^i_t may become too large; recall that the probabilities p^i_t, which are bounded from below by $\delta_{t-1}m^i$, appear in the denominators). This explains the need for an upper bound on γ (it turns out that $\gamma < 1/4$ suffices; see the proof in Section 4). Moreover, we note that one may well take different sequences δ^i_t for different players i (cf. the Remark at the end of Section 4).

---

[9]The reason for this restriction will be explained below. See the note at the end of the chapter regarding formula (7).

3. Discussion

(a) Reinforcement and stimulus-response

The play procedures of this paper can be interpreted as "reinforcement" or "stimulus-response" procedures (see, for example, Bush and Mosteller 1955, Roth and Erev 1995, Borgers and Sarin 1995, Erev and Roth 1998, Fudenberg and Levine 1998, Section 4.8). Indeed, the behavior of our players is very far from the ideal of a sophisticated decision-maker who makes optimal decisions given his (more or less well-formed) beliefs about the environment. The behavior we postulate is, in fact, much closer to the model of a reflex-oriented individual who, from a very limited conception of the world in which he lives, simply reacts to stimuli by reinforcing those behaviors with "pleasurable" consequences.

In order to be specific and to illustrate the above point further, let us assume that the payoffs are positive and focus on the limit case of our procedure where probabilities are chosen in exact proportion to the modified regrets (5). Suppose that player i has played action j at period t. We then have for every $k \neq j$:

- if $Q_t^i(j, k) > 0$ then $Q_{t+1}^i(j, k) < Q_t^i(j, k)$; and
- if $Q_t^i(j, k) = 0$ then $Q_{t+1}^i(j, k) = 0$.

Thus all the modified regrets decrease (or stay null) from t to $t+1$. Hence the probability of choosing j at $t+1$ gets "reinforced" (i.e., increases relative to the same probability at t) by the occurrence of j at t, while the probabilities of the other actions $k \neq j$ decrease. Moreover, as can easily be seen from Definitions (4) and (5), the higher the payoff obtained at time t (when j was played), the greater this reinforcement. Finally, all these effects decrease with time—since we average over t in (4).

(b) Related work

There is by now a substantive body of work on not fully optimal behavior in repeated games (see for instance the book of Fudenberg and Levine 1998). In particular, strategies that lead to the set of correlated equilibria have been proposed by Foster and Vohra (1997) and by Fudenberg and Levine (1999). There is also an older tradition, beginning with Hannan (1957), that focuses on another form of regret (of an "unconditional" kind— see (c) below) and on strategies that asymptotically take them down to zero. Clearly, our work (here and in Hart and Mas-Colell 2000, 2001) belongs to these lines of research. Since the main difference between Hart and Mas-Colell (2000) and the present paper is that here the players do not know the game, we want to point out that this "unknown game case" has already been studied. Specifically, in the context of Hannan regrets, see Foster and Vohra (1993, 1998), Auer *et al.* (1995) and Fudenberg and Levine (1998, Section 4.8) (and also Baños 1968 and Megiddo 1980 for related work).

(c) Hannan-consistency

We say that a strategy of a player is *Hannan-consistent* if it guarantees that his long-run average payoff is as large as the highest payoff that can be obtained by playing a constant action; that is, it is no less than the one-shot best-reply payoff against the empirical distribution of play of the other players. Formally, the *Hannan regret* of player i at time t for action $k \in S^i$ is defined as

$$DH_t^i(k) := \frac{1}{t}\sum_{\tau=1}^{t} u^i(k, s_\tau^{-i}) - \frac{1}{t}\sum_{\tau=1}^{t} U_\tau^i = u^i(k, z_t^{-i}) - \frac{1}{t}\sum_{\tau=1}^{t} U_\tau^i,$$

where $z_t^{-i} \in \Delta(S^{-i})$ is the empirical distribution of the actions chosen by the other players in the past.[10,11] A strategy of player i is then called *Hannan-consistent* if, as t increases, all Hannan-regrets are guaranteed to become almost surely non-positive in the limit, no matter what the other players do; that is, with probability one, $\limsup_{t\to\infty} DH_t^i(k) \leq 0$ for all $k \in S^i$. The reader is referred to Hart and Mas-Colell (2001) for detailed discussions and results.

In the setup of the present paper, the "modified-regret matching" approach leads to a simple reinforcement strategy that is Hannan-consistent (recall Hart and Mas-Colell 2000, Theorem B): For every $k \in S^i$ define

$$CH_t^i(k) := \frac{1}{t} \sum_{\tau \leq t : s_\tau^i = k} \frac{1}{p_\tau^i(k)} U_\tau^i = \frac{1}{t}\sum_{\tau=1}^{t} U_\tau^i,$$

and

$$p_{t+1}^i(k) := (1 - \delta_t)\frac{[CH_t^i(k)]_+}{\sum_{j \in S^i}[CH_t^i(j)]_+} + \delta_t \frac{1}{m^i}. \tag{1}$$

---

[10]I.e., $z_t^{-i}(s^{-i}) := |\{\tau \leq t : s_\tau^{-i} = s^{-i}\}|/t$ for each $s^{-i} \in S^{-i}$.

[11]Note that $DH_t^i(k) = \sum_{j\neq k} D_t^i(j, k)$; we can thus refer to $DH_t^i(k)$ as the "unconditional regret for k," and to $D_t^i(j, k)$ as the "regret for k, conditional on j."

Here $\delta_t = \delta/t^\gamma$ for some $\delta > 0$ and $0 < \gamma < 1/2$; we take $p_{t+1} \in \Delta(S^i)$ to be arbitrary for $t = 0$ and whenever the denominator vanishes. We have

Theorem 3. *The strategy* (8) *is Hannan-consistent.*

The proof of this theorem is parallel to, and simpler than, the proof of Theorem 2 below, and therefore omitted.

4. Proof

In this section we will prove Theorems 1 and 2 of Section 2 together. Let

$$\delta_t := \frac{\delta}{t^\gamma}$$

where $\delta > 0$ and $0 \le \gamma < 1/4$. For Theorem 1 take $\gamma = 0$, and for Theorem 2, $\gamma > 0$.

We introduce some notations, in addition to those of the previous sections. Fix player i in N; for simplicity, we drop reference to the index i whenever this does not create confusion (thus we write C_t instead of C_t^i, and so on). Recall that $m := |S^i|$ is the number of strategies of i, and M is an upper bound on the payoffs: $M \ge |u^i(s)|$ for all $s \in S$. Denote $L := \{(j,k) \in S^i \times S^i : j \ne k\}$; then $\mathbb{R}^L$ is the $m(m-1)$ Euclidean space with coordinates indexed by L.

For each $t = 1, 2, \ldots$ and each (j, k) in L, denote[12]

$$Z_t(j,k) := \frac{p_t^i(j)}{p_t^i(k)} 1_{\{s_t^i = k\}} - 1_{\{s_t^i = j\}};$$

$$B_t(j,k) := Z_t(j,k) u^i(s_t);$$

$$A_t(j,k) := 1_{\{s_t^i = j\}} (u^i(k, s_t^{-i}) - u^i(s_t)).$$

Thus, we have

$$C_t(j,k) = \frac{1}{t} \sum_{\tau \le t} B_\tau(j,k) \quad \text{and}$$

$$D_t(j,k) = \frac{1}{t} \sum_{\tau \le t} A_\tau(j,k).$$

---

[12] We write 1_E for the *indicator* of the event E (i.e., $1_E = 1$ if E occurs and $= 0$ otherwise).

We shall write B_t for the vector $(B_t(j,k))_{(j,k)\in L} \in \mathbb{R}^L$; and similarly for the other vectors A_t, C_t, and so on. Next, define

$$\Pi_t(j,k) := \begin{cases} (1-\delta_t)\min\left\{\dfrac{1}{\mu}C_t^+(j,k), \dfrac{1}{m}\right\} + \delta_t\dfrac{1}{m}, & \text{if } k \neq j, \\[2em] (1-\delta_t)\left(1 - \displaystyle\sum_{k'\neq j}\min\left\{\dfrac{1}{\mu}C_t^+(j,k'), \dfrac{1}{m}\right\}\right) + \delta_t\dfrac{1}{m}, & \text{if } k = j. \end{cases}$$

Note that $\Pi_t(j,\cdot) \in \Delta(S^i)$ for all $j \in S^i$; thus Π_t is a transition probability matrix on S^i. Both procedures (6) and (7) satisfy $p_{t+1}^i(k) = \Pi_t(s_t^i, k)$ for all k (where, again, $\gamma = 0$ corresponds to (6) and $\gamma > 0$ to (7)). Let

$$\rho_t := (\text{dist }(C_t, \mathbb{R}_-^L))^2 \equiv \sum_{j\neq k}(C_t^+(j,k))^2$$

be the squared Euclidean distance (in $\mathbb{R}^L$) of the vector C_t from the non-positive orthant $\mathbb{R}_-^L$.

We will use the standard "O" notation: For two real-valued functions $f(\cdot)$ and $g(\cdot)$ defined on a domain X, we take "$f(x) = O(g(x))$" to mean that there exists a constant $K < \infty$ such that $|f(x)| \leq Kg(x)$ for all x in X. From now on, t, v, w will denote positive integers; $h_t = (s_\tau)_{\tau\leq t}$ will be histories of length t; j, k, and s^i will be strategies of i (i.e., elements of S^i); s and s^{-i} will be elements of S and S^{-i}, respectively. Unless stated otherwise, all statements should be understood to hold "for all t, v, h_t, j, k, etc."; where histories h_t are concerned, only those that occur with positive probability are considered. Finally, Pr stands for Probability, E for Expectation and Var for Variance.

The proof of Theorems 1 and 2 will be divided into 14 steps. Step 14 shows that the regrets are "small" in the limit; the Proposition of Section 3 of Hart and Mas-Colell (2000) then implies that the empirical distributions are correlated approximate equilibria. We note that Steps 1–11 hold for *any* non-increasing sequence $\delta_t > 0$, whereas Steps 12–14 make use of the special form $\delta_t = \delta/t^\gamma$. A guide to the proof follows the statement of the steps.

• **Step 1:**

(i) $\mathrm{E}[(t+v)^2\rho_{t+v}|h_t] \leq t^2\rho_t + 2t\sum_{w=1}^v C_t^+ \cdot \mathrm{E}[B_{t+w}|h_t] + \mathrm{O}(\frac{v^2}{\delta_{t+v}^2})$.

(ii) $(t+v)^2\rho_{t+v} - t^2\rho_t = \mathrm{O}(\frac{tv+v^2}{\delta_{t+v}^2})$.

Define[13]

$$\beta_{t+w}(j) := \sum_{k \neq j} \frac{1}{\mu} C_t^+(k,j) p_{t+w}^i(k) - \sum_{k \neq j} \frac{1}{\mu} C_t^+(j,k) p_{t+w}^i(j).$$

- **Step 2:**

$$C_t^+ \cdot \mathrm{E}[B_{t+w}|h_t] = \mu \mathrm{E}\left[\sum_{j \in S^i} u^i(j, s_{t+w}^{-i}) \beta_{t+w}(j) \Big| h_t \right].$$

- **Step 3:**

    (i) $C_{t+v}(j,k) - C_t(j,k) = \mathrm{O}(\frac{v}{t\delta_{t+v}})$.

    (ii) $\Pi_{t+v}(j,k) - \Pi_t(j,k) = \mathrm{O}(\frac{v}{t\delta_{t+v}} + (\delta_t - \delta_{t+v}))$.

Define

$$\bar{Z}_t(j,k) := \frac{1}{t} \sum_{\tau=1}^{t} Z_\tau(j,k); \quad \text{and}$$

$$Y_t := \sum_{(j,k) \in L} |\bar{Z}(j,k)|.$$

- **Step 4:**

$$\Pi_t(j,k) - \frac{1}{\mu} C_t^+(j,k) = \mathrm{O}(\delta_t + Y_t) \quad \text{for all } j \neq k.$$

- **Step 5:**[14]

$$C_t^+ \cdot \mathrm{E}[B_{t+w}|h_t] = \mu \mathrm{E}\left[\sum_{j \in S^i} u^i(j, s_{t+w}^{-i})((\Pi_t)^2 - \Pi_t)(s_{t+w-1}^i, j) \Big| h_t \right]$$
$$+ \mathrm{O}\left(\delta_t + Y_t + \frac{w}{t\delta_{t+w}} \right).$$

For each $t > 0$ and each history h_t, we define an auxiliary stochastic process $(\hat{s}_{t+w})_{w=0,1,2,...}$ with values in S as follows: The initial state is $\hat{s}_t = s_t$, and

---

[13] Note that $\beta_{t+w}(j)$ is measurable with respect to h_{t+w} (actually, it depends only on h_{t+w-1} and s_{t+w}^{-i}, but not on s_{t+w}^i).

[14] $(\Pi_t)^2$ is the second power of the matrix Π_t (i.e., $\Pi_t\Pi_t$), and $((\Pi_t)^2 - \Pi_t)(k,j)$ is the (k,j)-element of the matrix $(\Pi_t)^2 - \Pi_t$.

the transition probabilities are[15]

$$\Pr[\hat{s}_{t+w} = s | \hat{s}_t, \dots, \hat{s}_{t+w-1}] := \prod_{\ell \in N} \Pi_t^\ell(\hat{s}_{t+w-1}^\ell, s^\ell).$$

That is, the $\hat{s}$-process is stationary: It uses the transition probabilities of period t at each period $t+1, t+2, \dots, t+w, \dots$.

- **Step 6:**

$$\Pr[s_{t+w} = s | h_t] - \Pr[\hat{s}_{t+w} = s | h_t] = O\left(\frac{w^2}{t\delta_{t+w}} + w(\delta_t - \delta_{t+w})\right).$$

- **Step 7:**

$$C_t^+ \cdot E[B_{t+w} | h_t] = \mu E\left[\sum_{j \in S^i} u^i(j, \hat{s}_{t+w}^{-i})((\Pi_t)^2 - \Pi_t)(\hat{s}_{t+w-1}^i, j) | h_t\right]$$

$$+ O\left(\delta_t + Y_t + \frac{w^2}{t\delta_{t+w}} + w(\delta_t - \delta_{t+w})\right).$$

- **Step 8:**

$$E\left[\sum_{j \in S^i} u^i(j, \hat{s}_{t+w}^{-i})((\Pi_t)^2 - \Pi_t)(\hat{s}_{t+w-1}^i, j) | h_t\right]$$

$$= \sum_{s^{-i} \in S^{-i}} \Pr[\hat{s}_{t+w}^{-i} = s^{-i} | h_t]((\Pi_t)^{w+1} - (\Pi_t)^w)(s_t^i, j) = O\left(\frac{1}{\sqrt{w}}\right)$$

- **Step 9:**

$$E[(t+v)^2 \rho_{t+v} | h_t] \leq$$

$$t^2 \rho_t + O\left(tv\delta_t + tvY_t + \frac{v^3}{\delta_{t+v}} + tv^2(\delta_t - \delta_{t+v}) + t\sqrt{v} + \frac{v^2}{\delta_{t+v}^2}\right).$$

- **Step 10:**

(i) $E[Z_t(j, k) | h_{t-1}] = 0.$

(ii) $\text{Var}[Z_t(j, k)] = O(\frac{1}{\delta_t}).$

---

[15]We write Π_t^ℓ for the transition probability matrix of player ℓ (thus Π_t is Π_t^i).

- **Step 11:**

  (i) $\lim_{t\to\infty} \bar{Z}_t(j,k) = 0$ a.s.

  (ii) $\lim_{t\to\infty} Y_t = 0$ a.s.

  (iii) $\lim_{t\to\infty} (C_t(j,k) - D_t(j,k)) = 0$ a.s.

Let ξ satisfy[16]

$$1 < \xi < \min\left\{\frac{2}{1+\gamma}, \frac{1}{4\gamma}\right\}; \tag{9}$$

such a ξ exists since $0 < \gamma < 1/4$. For each $n = 1, 2, \ldots$, let $t_n := \lfloor n^\xi \rfloor$ be the largest integer not exceeding n^ξ.

- **Step 12:** There exists $\eta < 2\xi - 1$ such that

$$\mathrm{E}[t_{n+1}^2 \rho_{t_{n+1}} | h_{t_n}] \le t_n^2 \rho_{t_n} + \mathrm{O}(\delta n^{2\xi - \xi\gamma - 1} + Y_{t_n} n^{2\xi - 1} + n^\eta).$$

- **Step 13:**

  (i) If $\gamma = 0$ then $\limsup_{n\to\infty} \rho_{t_n} = \mathrm{O}(\delta)$ a.s.

  (ii) If $\gamma > 0$ then $\lim_{n\to\infty} \rho_{t_n} = 0$ a.s.

- **Step 14:**

  (i) If $\gamma = 0$ then $\limsup_{t\to\infty} R_t(j,k) = \mathrm{O}(\sqrt{\delta})$ a.s.

  (ii) If $\gamma > 0$ then $\lim_{t\to\infty} R_t(j,k) = 0$ a.s.

We now provide a short intuitive overview of the steps of the proof. The proof is based on the Proof of the Main Theorem of Hart and Mas-Colell (2000) (see Steps M1–M11 in the Appendix there)—which in turn is inspired by the Approachability Theorem of Blackwell (1956)—with a number of additional steps needed to take care of the modifications. Most of the proof is devoted to showing that the modified regrets $Q_t \equiv C_t^+$ are small. From this one readily gets in Step 14 that the actual regrets $R_t \equiv D_t^+$ are also small, since the difference $C_t - D_t$ is a martingale converging almost surely to 0 (see Step 11(iii)). The main steps in the proof are as follows: We start with the basic recursion equation in Step 1(i) (similar to Hart and Mas-Colell 2000, Step M1(i)). Next, we estimate the "middle term" on the right-hand side of 1(i) by approximating the s-process with the $\hat{s}$-process, which is independent across players (Steps 2–7; parallel to Hart and

---

[16]When $\gamma = 0$, (9) is $1 < \xi < 2$.

Mas-Colell (2000, Steps M2–M6)). This leads to a formula similar to Hart and Mas-Colell (2000, Formula (3.4)), except that the invariant distribution $q\lambda$ in Hart and Mas-Colell (2000, (3.4b)) is replaced here by the transitions after w and $w+1$ periods, which are close to one another (Step 8; compare with Hart and Mas-Colell 2000, Step M7). Finally, we obtain the recursion formula in Step 9. On comparing this formula with the parallel one in Hart and Mas-Colell (2000, Step M8), we see that in Step 9 there are additional terms; one of them, which comes about because the modified regrets are not bounded, contains a random variable Y_t. Step 10, Step 11, and the Proof of Step 13 show that this term also goes to zero. Steps 12 and 13 complete the proof for the modified regrets, in a manner that is similar to Hart and Mas-Colell (2000, Steps M9–M11) (though more complicated, partly because of the Y_t terms). As we have indicated above, Step 14 yields the needed result for the actual regrets.

We now proceed to the proofs of Steps 1–14.

- **Proof of Step 1:** Because $C_t^- \in \mathbb{R}_-^L$ we have

$$\rho_{t+v} \le \|C_{t+w} - C_t^-\|^2 = \left\| \frac{t}{t+v} C_t + \frac{1}{t+v} \sum_{w=1}^{v} B_{t+w} - C_t^- \right\|^2$$

$$= \frac{t^2}{(t+v)^2} \|C_t - C_t^-\|^2 + \frac{2t}{(t+v)^2} \sum_{w=1}^{v} (B_{t+w} - C_t^-) \cdot (C_t - C_t^-)$$

$$+ \frac{v^2}{(t+v)^2} \left\| \frac{1}{v} \sum_{w=1}^{v} B_{t+w} - C_t^- \right\|^2$$

$$\le \frac{t^2}{(t+v)^2} \rho_t + \frac{2t}{(t+v)^2} \sum_{w=1}^{v} B_{t+w} \cdot C_t^+$$

$$+ \frac{v^2}{(t+v)^2} m(m-1) \frac{4M^2 m^2}{\delta_{t+v}^2}.$$

Indeed: For the second term, note that $C_t^- \cdot (C_t - C_t^-) = C_t^- \cdot C_t^+ = 0$. As for the third term, we have $|u^i(s)| \le M$ and $|Z_{t+w}(j,k)| \le m/\delta_{t+w} \le m/\delta_{t+v}$ for $w \le v$ (since the sequence δ_t is non-increasing); therefore $B_{t+w}(j,k)$ and $C_t(j,k)$ are each bounded by $2M m/\delta_{t+v}$. This yields (ii). To get (i), take the conditional expectation given h_t (thus ρ_t and C_t are fixed). □

- **Proof of Step 2:** Note that $B_{t+w}(j,k)$ vanishes except when $s_{t+w}^i = j,k$. We condition on h_{t+w-1} and s_{t+w}^{-i} (i.e., on the whole history

h_{t+w} *except* player i's choice at time $t + w$):

$$\mathrm{E}[B_{t+w}(j,k)|h_{t+w-1}, s_{t+w}^{-i} = s^{-i}]$$

$$= p_{t+w}^i(k)\frac{p_{t+w}^i(j)}{p_{t+w}^i(k)}u^i(k, s^{-i}) - p_{t+w}^i(j)u^i(j, s^{-i})$$

$$= p_{t+w}^i(j)u^i(k, s^{-i}) - u^i(j, s^{-i}))$$

Hence

$$C_t^+ \cdot \mathrm{E}[B_{t+w}|h_{t+w-1}, s_{t+w}^{-i} = s^{-i}]$$

$$= \sum_j \sum_{k \neq j} C_t^+(j,k)p_{t+w}^i(j)(u^i(k, s^{-i}) - u^i(j, s^{-i}))$$

$$= \sum_j u^i(j, s^{-i})\left(\sum_{k \neq j} C_t^+(k,j)p_{t+w}^i(k) - \sum_{k \neq j} C_t^+(j,k)p_{t+w}^i(j)\right)$$

(we have collected together all terms containing $u^i(j, s^{-i})$). Conditioning now on h_t yields the result. $\qquad\qquad\square$

- **Proof of Step 3:** We have

$$(t + v)|C_{t+v}(j,k) - C_t(j,k)| \leq \sum_{w=1}^{v} |B_{t+w}(j,k) - C_t(j,k)|.$$

Since both $B_{t+w}(j,k)$ and $C_t(j,k)$ are $\mathrm{O}(1/\delta_{t+v})$, we get $C_{t+v}(j,k) - C_t(j,k) = \mathrm{O}(v/(t\delta_{t+v}))$. For $j \neq k$, the difference between $\Pi_{t+v}(j,k)$ and $\Pi_t(j,k)$ is therefore at most $(1 - \delta_t)\mathrm{O}(v/(t\delta_{t+v})) + (\delta_t - \delta_{t+v})/m$. For $j = k$, it is at most m times this amount. $\qquad\qquad\square$

- **Proof of Step 4:** We distinguish two cases. First, when $(1/\mu)C_t^+(j,k) \leq 1/m$, we have

$$\Pi_t(j,k) - \frac{1}{\mu}C_t^+(j,k) = \delta_t\left(\frac{1}{m} - \frac{1}{\mu}C_t^+(j,k)\right),$$

and this is $\mathrm{O}(\delta_t)$ (it lies between 0 and δ_t/m).

Second, when $(1/\mu)C_t^+(j,k) \geq 1/m$, we have

$$\Pi_t(j,k) = \frac{1 - \delta_t}{m} + \frac{\delta_t}{m} = \frac{1}{m} \leq \frac{1}{\mu}C_t^+(j,k).$$

For the opposite inequality, note that $|Z_\tau(j,k)| \leq 2 + Z_\tau(j,k)$ (since the only possible negative value of $Z_\tau(j,k)$ is -1); thus

$$\frac{1}{\mu}C_t^+(j,k) \leq \frac{1}{\mu}|C_t(j,k)| \leq \frac{1}{\mu t}\sum_{\tau=1}^{t}|Z_\tau(j,k)||u^i(s_\tau)|$$

$$\leq \frac{2M}{\mu} + \frac{M}{\mu}\bar{Z}_t(j,k) \leq \frac{1}{m} + \frac{M}{\mu}\bar{Z}_t(j,k)$$

$$= \Pi_t(j,k) + \frac{M}{\mu}\bar{Z}_t(j,k)$$

(recall that $\mu \geq 2Mm$. $\square$

• **Proof of Step 5:** Denote s_{t+w-1}^i by r; then

$$\beta_{t+w}(j) = \sum_{k \neq j}\frac{1}{\mu}C_t^+(k,j)\Pi_{t+w-1}(r,k) - \sum_{k \neq j}\frac{1}{\mu}C_t^+(j,k)\Pi_{t+w-1}(r,j).$$

Also,

$$((\Pi_t)^2 - \Pi_t)(r,j) = (\Pi_t)^2(r,j) - \Pi_t(r,j)$$

$$= \sum_{k \in S^i}\Pi_t(k,j)\Pi_t(r,k) - \Pi_t(r,j)\sum_{k \in S^i}\Pi_t(j,k)$$

$$= \sum_{k \neq j}\Pi_t(k,j)\Pi_t(r,k) - \sum_{k \neq j}\Pi_t(j,k)\Pi_t(r,j)$$

(we have subtracted the j-term from both sums). Comparing the last expression with $\beta_{t+w}(j)$, we see that $((\Pi_t)^2 - \Pi_t)(r,j)$ is obtained by replacing each C_t^+/μ and each Π_{t+w-1} in $\beta_{t+w}(j)$ by Π_t. Thus

$$\beta_{t+w}(j) - ((\Pi_t)^2 - \Pi_t)(r,j) = \sum_{k \neq j}\left(\frac{1}{\mu}C_t^+(k,j) - \Pi_t(k,j)\right)\Pi_{t+w-1}(r,k)$$

$$+ \sum_{k \neq j}\Pi_t(k,j)(\Pi_{t+w-1}(r,k) - \Pi_t(r,k))$$

$$- \sum_{k \neq j}\left(\frac{1}{\mu}C_t^+(j,k) - \Pi_t(j,k)\right)$$

$$\times \Pi_{t+w-1}(r,j)$$

$$- \sum_{k \neq j} \Pi_t(j,k)(\Pi_{t+w-1}(r,j) - \Pi_t(r,j))$$

$$= O\left(\delta_t + Y_t + \frac{w}{t\delta_{t+w}}\right),$$

by the estimates of Steps 3(ii) and 4. It only remains to substitute this into the formula of Step 2. $\square$

• **Proof of Step 6:** We use the Lemma of Hart and Mas-Colell (2000, Step M4), which implies that if the 1-step transition probabilities of two Markov processes on S differ by at most β, then the w-step transition probabilities differ by at most $|S|w\beta$. Applying this to the $\hat{s}$- and s-processes, with β given by Step 3(ii), yields the result. $\square$

• **Proof of Step 7:** Replacing $(s_{t+w})_w$ by $(\hat{s}_{t+w})_w$ in the formula of Step 5 gives an additional error that is estimated in Step 6 (note that the two processes $(s_{t+w})_w$ and $(\hat{s}_{t+w})_w$ start from the same history h_t). $\square$

• **Proof of Step 8:** Given h_t, the random variables $\hat{s}_{t+w}^{-i}$ and $\hat{s}_{t+w-1}^i$ are independent, since the transition probabilities of the $\hat{s}$-process are all determined at time t, and the players randomize independently. Hence:

$$E\left[\sum_{j \in S^i} u^i(j, \hat{s}_{t+w}^{-i})((\Pi_t)^2 - \Pi_t)(\hat{s}_{t+w-1}^i, j)|h_t\right]$$

$$= \sum_{s^{-i} \in S^{-i}} \Pr[\hat{s}_{t+w}^{-i} = s^{-i}|h_t] \sum_{r \in S^i} \Pr[\hat{s}_{t+w-1}^i = r|h_t]((\Pi_t)^2 - \Pi_t)(r,j)$$

$$= \sum_{s^{-i}} \Pr[\hat{s}_{t+w}^{-i} = s^{-i}|h_t] \sum_r (\Pi_t)^{w-1}(s_t^i, r)((\Pi_t)^2 - \Pi_t)(r,j)$$

$$= \sum_{s^{-i}} \Pr[\hat{s}_{t+w}^{-i} = s^{-i}|h_t]((\Pi_t)^{w+1} - (\Pi_t)^w)(s_t^i, j).$$

The estimate of $O(1/\sqrt{w})$ is obtained by the Lemma of Hart and Mas-Colell (2000, Step M7, using $\min_j \Pi_t(j,j) \geq 1/m$ for all[17] t).

$\square$

---

[17]Which is based on a Central Limit Theorem estimate. Note that here (unlike the Main Theorem of Hart and Mas-Colell 2000) Π_t is a strictly positive stochastic matrix: All its entries are $\geq \delta_t/m > 0$. It can then be shown that $|(\Pi_t)^{w+1}(k,j) - (\Pi_t)^w(k,j)| \leq (1 - \delta_t/m)^w$. This alternative estimate can be used instead of $O(w^{-1/2})$ (but we then need $\gamma < 1/5$ rather than $\gamma < 1/4$).

• **Proof of Step 9:** Putting together the estimates of Steps 7 and 8 and recalling that $\sum_{w=1}^{v} w^\lambda = O(v^{\lambda+1})$ for $\lambda \neq -1$ yields

$$2^t \sum_{w=1}^{v} C_t^+ \cdot \mathrm{E}[B_{t+w}|ht] = O\left(tv\delta_t + tvY_t + \frac{tv^3}{\delta_{t+v}} + tv^2(\delta_t - \delta_{t+v}) + t\sqrt{v}\right).$$

Recalling the formula of Step 1(i) completes the proof.   □

• **Proof of Step 10:** Part (i) follows immediately from the definition of $Z_t(j,k)$:

$$\mathrm{E}[Z_t(j,k)|h_{t-1}] = \frac{p_t^i(j)}{p_t^i(k)} p_t^i(k) - p_t^i(j) = 0.$$

Therefore $\mathrm{E}[Z_t(j,k)] = 0$, and

$$\mathrm{Var}[Z_t(j,k)] = \mathrm{E}[Z_t^2(j,k)] = \mathrm{E}[\mathrm{E}[Z_t^2(j,k)|h_{t-1}]]$$
$$= \mathrm{E}\left[\frac{(p_t^i(j))^2}{p_t^i(k))^2} p_t^i(k) + (-1)^2 p_t^i(j)\right] \leq \mathrm{E}\left[\frac{1}{p_t^i(k)}\right] \leq \frac{m}{\delta_t},$$

which gives (ii).   □

• **Proof of Step 11:** We will use the following Strong Law of Large Numbers for Dependent Random Variables; see Loève (1978, Theorem 32.1.E):

Theorem 4. *Let* $(X_n)_{n=1,2,\ldots}$ *be a sequence of random variables and* $(b_n)_{n=1,2,\ldots}$ *a sequence of real numbers increasing to* ∞, *such that the series* $\sum_{n=1}^{\infty} \mathrm{Var}(X_n)/b_n^2$ *converges. Then*

$$\lim_{n\to\infty} \frac{1}{b_n} \sum_{\nu=1}^{n} (X_\nu - \mathrm{E}[X_\nu|X_1,\ldots,X_{\nu-1}]) = 0 \quad a.s.$$

In our case, we have by Step 10(ii) .

$$\sum_{t=1}^{\infty} \frac{1}{t^2} \mathrm{Var}[Z_t(j,k)] \leq \sum_{t=1}^{\infty} \frac{m}{t^2\delta_t} = \sum_{t=1}^{\infty} \frac{m}{\delta t^{2-\gamma}}.$$

This series converges, since $\gamma < 1/4 < 1$. Therefore

$$\frac{1}{t} \sum_{\tau \leq t} (Z_\tau(j,k) - \mathrm{E}[Z_\tau(j,k)|Z_1(j,k),\ldots,Z_{\tau-1}(j,k)]) \to 0 \quad a.s.$$

and thus, by Step 10(i), $\bar{Z}_t(j,k) \to 0$ a.s. This yields (i) and (ii).

To get (iii), note that $1_{\{s_t^i=k\}}u^i(s_t) = 1_{\{s_t^i=k\}}u^i(k, s_t^{-i})$, so

$$B_t(j,k) - A_t(j,k)$$

$$= \left(\frac{p_t^i(j)}{p_t^i(k)}1_{\{s_t^i=k\}} - 1_{\{s_t^i=j\}}\right)u^i(s_t) - 1_{\{s_t^i=j\}}(u^i(k, s_t^{-i}) - u^i(s_t))$$

$$= \left(\frac{p_t^i(j)}{p_t^i(k)}1_{\{s_t^i=k\}} - 1_{\{s_t^i=j\}}\right)u^i(k, s_t^{-i}) = Z_t(j,k)u^i(k, s_t^{-i}).$$

But s_t^i and s_t^{-i} are independent given h_{t-1}; therefore

$$\mathrm{E}[B_t(j,k) - A_t(j,k)|h_{t-1}] = \mathrm{E}[Z_t(j,k)|h_{t-1}]\mathrm{E}[u^i(k, s_t^{-i})|h_{t-1}] = 0,$$

since the first term is 0 by Step 10(i). Moreover,

$$\mathrm{Var}[B_t(j,k) - A_t(j,k)] = \mathrm{E}[Z_t^2(j,k)(u^i(k, s_t^{-i}))^2]$$

$$\leq M^2\mathrm{E}[Z_t^2(j,k)] = \mathrm{O}\left(\frac{1}{\delta_t}\right).$$

It follows that the series $\sum_t \mathrm{Var}[B_t(j,k) - A_t(j,k)]/t^2$ converges, implying that[18] $C_t(j,k) - D_t(j,k) = (1/t)\sum_{\tau\leq t}(B_\tau(j,k) - A_\tau(j,k)) \to 0$ a.s. $t \to \infty$ (argument as in the proof of (i) above).   □

• **Proof of Step 12:** Apply the inequality of Step 9 with $t = t_n = \lfloor n^\xi \rfloor$ and $v = t_{n+1} - t_n$. Then: $v = \mathrm{O}(n^{\xi-1})$; $\delta_t \approx \delta n^{-\xi\gamma}$; $\delta_{t+v} \approx \delta(n+1)^{-\xi\gamma} = \mathrm{O}(n^{-\xi\gamma})$; and[19] $\delta_t - \delta_{t+v} = \mathrm{O}(n^{-\xi\gamma-1})$. Therefore

$$\mathrm{E}[t_{n+1}^2\rho_{t_{n+1}}|h_{t_n}] \leq t_{n\rho t_n}^2 + \mathrm{O}(\delta n^{2\xi-\xi\gamma-1} + Y_{t_n}n^{2\xi-1})$$

$$+ \mathrm{O}(n^{3\xi+\xi\gamma-3} + n^{3\xi-\xi\gamma-3} + n^{(3\xi-1)/2} + n^{2\xi+2\xi\gamma-2}).$$

To complete the proof, note that the definition (9) of ξ implies that: $3\xi - \xi\gamma - 3 \leq 3\xi + \xi\gamma - 3 < 2\xi - 1$ since $\xi < 2/(1+\gamma)$; $(3\xi - 1)/2 < 2\xi - 1$ since $\xi > 1$; and $2\xi + 2\xi\gamma - 2 < 2\xi - 1$ since $\xi < 1/(4\gamma) \leq 1/(2\gamma)$. Therefore we take $\eta := \max\{3\xi + \xi\gamma - 3, (3\xi - 1)/2, 2\xi + 2\xi\gamma - 2\} < 2\xi - 1$.   □

---

[18]It is interesting to note that, while the regrets are invariant to a change of origin for the utility function (i.e., adding a constant to all payoffs), this is not so for the modified regrets. Nonetheless, Step 13 shows that the resulting difference is just a martingale converging a.s. to 0.

[19]When $\gamma = 0$ (and thus $\delta_t - \delta_{t+v} = 0$), this yields an (over)estimate of $\mathrm{O}(n^{-1})$, which will turn out however not to matter.

- **Proof of Step 13:** Let $b_n := t_n^2 \approx n^{2\xi}$ and $X_n := b_n \rho_{t_n} -$ $b_{n-1}\rho_{t_{n-1}} = t_n^2 \rho_{t_n} - t_{n-1}^2 \rho_{t_{n-1}}$. By Step 1(ii) we have $X_n = O((t_n v_n + v_n^2)/$ $\delta_{t_{n+1}}^2) = O(n^{2\xi+2\xi\gamma-1})$; thus $\sum_n \mathrm{Var}(X_n)/b_n^2 = \sum_n O(n^{4\xi+4\xi\gamma-2}/n^{4\xi}) = \sum_n O(n^{4\xi\gamma-2})$ converges (since $4\xi\gamma < 1$ by the choice of ξ).

Next, consider $(1/b_n)\sum_{\nu \leq n} E[X_\nu | X_1, \ldots, X_{\nu-1}]$. The inequality of Step 12 yields three terms: The first is $O(n^{-2\xi\delta} \sum_{\nu \leq n} \nu^{2\xi-\xi\gamma-1}) = O(\delta n^{-\xi\gamma})$; the second one converges to 0 a.s. as $t \to \infty$, since $Y_{t_n} \to 0$ by Step 10(ii) and Lemma 1 below (with $y_n = Y_{t_n}$ and $a_n = n^{2\xi-1}$; and the third one is $O(n^{\eta-(2\xi-1)}) \to 0$ since $\eta < 2\xi - 1$. Altogether, we get $O(\delta)$ when $\gamma = 0$, and 0 when $\gamma > 0$.

' The proof is completed by applying again the Strong Law of Large Numbers for Dependent Random Variables (Theorem 4) and noting that $0 < \rho_{t_n} = (1/b_n)\sum_{\nu \leq n} X_\nu$.

Lemma 1. *Assume: (i)* $y_n \to 0$ *as* $n \to \infty$; *(ii)* $a_n > 0$ *for all* n; *and (iii)* $\sum_{n=1}^{\infty} a_n = \infty$. *Then* $c_n := \sum_{\nu=1}^{n} a_\nu y_\nu / \sum_{\nu=1}^{n} a_\nu \to 0$ *as* $n \to \infty$.

Proof. Given $\varepsilon > 0$, let n_0 be such that $|y_n| < \varepsilon$ for all $n > n_0$. Then $c_n = \sum_{\nu \leq n_0} a_\nu y_\nu / \sum_{\nu \leq n} a_\nu + \sum_{n_0 < \nu \leq n} a_\nu y_\nu / \sum_{\nu \leq n} a_\nu$. The first term converges to 0 (since the numerator is fixed), and the second is bounded by ε. $\square$

- **Proof of Step 14:** When $t_n \leq t \leq t_{n+1}$, we have by Step 3(i): $C_t(j,k) - C_{t_n}(j,k) = O(v_n/(t_n\delta_{n+1})) = O(n^{\xi\gamma-1}) \to 0$, for all $j \neq k$. Thus $\limsup_{t \to \infty} \rho_t = \limsup_{n \to \infty} \rho_{t_n}$. Recalling Step 11(iii) completes the proof. $\square$

Remark. For simplicity, we have assumed that all players use the same sequence $(\delta_t)_t$. In the case of different sequences $(\delta_t^\ell)_t$ for the different players $\ell \in N$, with $\delta_t^\ell = \delta^\ell/t^{\gamma^\ell}$ for some $\delta^\ell > 0$ and $0 \leq \gamma^\ell < 1/4$, it is straightforward to check that the estimate of Step 6 becomes now $\sum_{\ell \in N} O(w^2/(t\delta_{t+w}^\ell) + w(\delta_t^\ell - \delta_{t+w}^\ell))$. Choosing ξ to satisfy $1 < \xi < \min\{2/(1 + \bar\gamma), 1/(4\bar\gamma)\}$, where $\bar\gamma := \max_{\ell \in N} \gamma^\ell$, yields, on the right-hand side of Step 12, $t_n^2 \rho_{t_n} + O(\delta^i n^{2\xi-\xi\gamma^i-1} + Y_{t_n} n^{2\xi-1}) + O(n^{2\xi-1})$, and Steps 13 and 14 go through with δ^i and γ^i instead of δ and γ, respectively. The final result in Step 14 becomes:

(i) If $\delta_t^i = \delta^i$ for all t, then $\limsup_t R_t^i(j,k) = O(\sqrt{\delta^i})$ a.s.
(ii) If $\delta_t^i \to 0$ (i.e. if $\gamma(i) > 0$), then $\lim_t R_t^i(j,k) = 0$ a.s.

Note. The authors wish to point out the following corrections to the originally published paper: $1/m^i$ replaces $1/(m^i - 1)$ in formulas (6) and (7), and $\mu \geq 2Mm^i$ replaces $\mu > 2M(m^i - 1)$ in the paragraph before (6); corresponding changes are made in the proof (mainly in Steps 4 and 8).

The reason for these corrections is that for the estimate $O(1/\sqrt{w})$ in Step 8 the diagonal elements of the matrix Π_t need to be bounded away from 0, *uniformly* in t (and, indeed, now $\Pi_t(j,j) \geq 1/m^i$ for all $j\varepsilon S^i$ and all $t \geq 1$).

References

Auer P., N. Cesa-Bianchi, Y. Freund, and R. E. Schapire (1995), "Gambling in a Rigged Casino: The Adversarial Multi-Armed Bandit Problem," in *Proceedings of the 36th Annual Symposium on Foundations of Computer Science*, pp. 322–331.

Aumann, R. J. (1974), "Subjectivity and Correlation in Randomized Strategies," *Journal of Mathematical Economics*, 1, 67–96.

Baños, A. (1968), "On Pseudo-Games," *The Annals of Mathematical Statistics*, 39, 1932–1945.

Blackwell, D. (1956), "An Analog of the Minmax Theorem for Vector Payoffs," *Pacific Journal of Mathematics*, 6, 1–8.

Borgers, T. and R. Sarin (1995), Naive Reinforcement Learning with Endogenous Aspirations, University College London (mimeo).

Bush, R. and F. Mosteller (1955), *Stochastic Models for Learning*. New York: Wiley.

Erev, I. and A. E. Roth (1998), "Predicting How People Play Games: Reinforcement Learning in Experimental Games with Unique, Mixed Strategies," *American Economic Review*, 88, 848–881.

Foster, D. and R. V. Vohra (1993), "A Randomized Rule for Selecting Forecasts," *Operations Research*, 41, 704–709.

—— (1997), "Calibrated Learning and Correlated Equilibrium," *Games and Economic Behavior*, 21, 40–55.

—— (1998), "Asymptotic Calibration," *Biometrika*, 85, 379–390.

Fudenberg, D. and D. K. Levine (1998), *Theory of Learning in Games*, Cambridge, MA: MIT Press.

—— (1999), "Conditional Universal Consistency," *Games and Economic Behavior*, 29, 104–130.

Hannan, J. (1957), "Approximation to Bayes Risk in Repeated Play," in *Contributions to the Theory of Games*, Vol. III *(Annals of Mathematics Studies 39)*, ed. by M. Dresher, A. W. Tucker and P. Wolfe, Princeton: Princeton University Press, pp. 97–139.

Hart, S. and A. Mas-Colell (2000), "A Simple Adaptive Procedure Leading to Correlated Equilibrium," *Econometrica*, 68, 1127–1150. [Chapter 2]

—— (2001), A General Class of Adaptive Strategies, *Journal of Economic Theory*, 98, 26–54. [Chapter 3]

Loève, M. (1978), *Probability Theory*, Vol. II, 4th Edition, Amsterdam: Springer.

Megiddo, N. (1980), "On Repeated Games with Incomplete Information Played by Non-Bayesian Players," *International Journal of Game Theory*, 9, 157–167.

Roth, A. E. and I. Erev (1995), "Learning in Extensive-Form Games: Experimental Data and Simple Dynamic Models in the Intermediate Term," *Games and Economic Behavior*, 8, 164–212.

Chapter 5

REGRET-BASED CONTINUOUS-TIME DYNAMICS*

Sergiu Hart and Andreu Mas-Colell

Regret-based dynamics have been introduced and studied in the context of discrete-time repeated play. Here we carry out the corresponding analysis in continuous time. We observe that, in contrast to (smooth) fictitious play or to evolutionary models, the appropriate state space for this analysis is the space of distributions on the product of the players' pure action spaces (rather than the product of their mixed action spaces). We obtain relatively simple proofs for some results known in the discrete case (related to "no-regret" and correlated equilibria), and also a new result on two-person potential games (for this result we also provide a discrete-time proof).

1. Introduction

"Regret matching" as a strategy of play in long-run interactions has been introduced and studied in a number of earlier papers (Hart and Mas-Colell 2000, 2001a, 2001b). We have shown that, under general conditions, regret matching leads to distributions of play that are related to the concept of correlated equilibrium. The purpose of the current paper is to reexamine the dynamics of regret matching from the standpoint of differential dynamics in continuous time. It is well known that this approach often leads to a

---

Originally published in *Games and Economic Behavior*, 45 (2003), 375–394.

*Previous version: August 2001.
JEL classification: C7; D7; C6
Research is partially supported by grants of the Israel Academy of Sciences and Humanities, the Spanish Ministry of Education, the Generalitat de Catalunya, and the EU-TMR Research Network.
We thank Drew Fudenberg, Josef Hofbauer, Gil Kalai, David Levine, Abraham Neyman, Yosef Rinott, William Sandholm, and Benjamin Weiss for their comments and suggestions.

simplified and streamlined treatment of the dynamics, to new insights and also to new results—and this will indeed happen here.

An important insight comes already in the task of formulating the differential setup. The appropriate state space for regret matching is not the product of the mixed action spaces of the players but a larger set: the distributions on the product of the pure action spaces of the players. Of course the players play independently at every point in time—but this in no way implies that the state variable evolves over time as a product distribution.

In Section 2 we present the model and specify the general setup of the dynamics we consider. In Section 3 we analyze general regret-based dynamics, the continuous-time analog to Hart and Mas-Colell (2001a), to which we refer for extensive discussion and motivation. In Section 4 we establish that for some particularly well-behaved classes of two-person games—zero-sum games, and potential games—the dynamics in fact single out the Nash equilibria of the game. The result for potential games is new and so we present a discrete-time version in Appendix A. In Section 5 we move to the analysis of conditional regret dynamics and prove convergence to the set of correlated equilibria. Finally, Section 6 offers some remarks. Appendix B provides a technical result, and Appendix C deals with the continuous-time version of the approachability theorems *à la* Blackwell (1956), which are basic mathematical tools for this area of research.

We dedicate this paper to the memory of Bob Rosenthal. It is not really necessary to justify this by exhibiting a connection between our topics of interest here and some particular paper of his. The broadness of his intellectual gaze guarantees that he would have been engaged and that, as usual, he would have contributed the insightful comments that were a trademark of his. At any rate, we mention that one of the cases that we examine with some attention is that of potential games and that Bob Rosenthal was the first to identify the remarkable properties of this class of games (Rosenthal 1973).

2. Model

2.1. *Preliminaries*

An *N-person game* Γ *in strategic form* is given by a finite set N of *players*, and, for each player $i \in N$, by a finite set S^i of *actions* and a *payoff function* $u^i : S \to \mathbb{R}$, where $S := \prod_{i \in N} S^i$ is the set of N-tuples of actions (we call

the elements of S^i "actions" rather than strategies, a term we will use for
the repeated game). We write $S^{-i} := \prod_{j \in N, j \neq i} S^j$ for the set of action
profiles of all players except player i, and also $s = (s^i, s^{-i})$. Let M be a
bound on payoffs: $|u^i(s)| \leq M$ for all $i \in N$ and all $s \in S$.

A randomized (mixed) action x^i of player i is a probability distribution
over i's pure actions, i.e.,[1] $x^i \in \Delta(S^i)$. A *randomized joint action* (or *joint
distribution*) z is a probability distribution over the set of N-tuples of pure
actions S, i.e., $z \in \Delta(S)$. Given such z, we write $z^i \in \Delta(S^i)$ and $z^{-i} \in
\Delta(S^{-i})$ for the marginals of z, i.e., $z^i(s^i) = \sum_{s^{-i} \in S^{-i}} z(s^i, s^{-i})$ for all
$s^i \in S^i$, and $z^{-i}(s^{-i}) = \sum_{s^i \in S^i} z(s^i, s^{-i})$ for all $s^{-i} \in S^{-i}$. When the
joint action is the result of independent randomizations by the players, we
have $z(s) = \prod_{i \in N} z^i(s^i)$ for all $s \in S$; we will say in this case that z is
independent, or that it is a *product measure*.[2]

2.2. Dynamics

We consider continuous-time dynamics on $\Delta(S)$ of the form

$$\dot{z}(t) = \frac{1}{t} \left(q(t) - z(t) \right), \tag{2.1}$$

where $q(t) \in \Delta(S)$ is[3] the joint play at time t and $z(t)$ is the "time-average
joint play." Assume one starts at $t = 1$ with some[4] $z(1) \in \Delta(S)$.

To justify (2.1), recall the discrete-time model: Time is $t = 1, 2, \ldots$;
player i at period t plays $s_t^i \in S^i$, and the time-average joint play at time
t is $z_t \in \Delta(S)$, given inductively by[5] $z_t = (1/t)(1_{s_t} + (t-1)z_{t-1})$, or

$$z_t - z_{t-1} = \frac{1}{t} \left(1_{s_t} - z_{t-1} \right).$$

Taking the expectation over s_t—whose distribution is q_t—leads to (2.1).

---

[1] For a finite set A, we write $|A|$ for the number of elements of A, and $\Delta(A)$ for the
set of probability distributions on A, i.e., $\Delta(A) := \{x \in \mathbb{R}_+^A : \sum_{a \in A} x(a) = 1\}$ (the
$(|A| - 1)$-dimensional unit simplex).
[2] We thus view $\prod_{i \in N} \Delta(S^i)$ as the subset of independent distributions in $\Delta(S)$.
[3] If in fact the players play independently then $q(t) \in \prod_{i \in N} \Delta(S^i) \subset \Delta(S)$.
[4] Note that if $z(t)$ is on the boundary of $\Delta(S)$, i.e., if $(z(s))(t) = 0$ for some $s \in S$, then
(2.1) implies $(\dot{z}(s))(t) \geq 0$, and thus $z(t)$ can never leave $\Delta(S)$.
[5] We write 1_s for the unit vector in $\Delta(S)$ corresponding to the pure $s \in S$.

3. Regret-Based Strategies

3.1. *Regrets and the Hannan set*

Given a joint distribution $z \in \Delta(S)$, the *regrets* of player i are defined by[6]

$$D_k^i(z) := u^i(k, z^{-i}) - u^i(z), \quad \text{for each } k \in S^i;$$

put $D^i(z) := (D_k^i(z))_{k \in S^i}$ for the *vector* of regrets.

It is useful to introduce the concept of the *Hannan set* H (of a given game Γ) as the set of all $z \in \Delta(S)$ satisfying

$$u^i(z) \geq \max_{k \in S^i} u^i(k, z^{-i}) \quad \text{for all } i \in N$$

(recall that z^{-i} denotes the marginal of z on S^{-i}); i.e., $z \in H$ if all regrets of all players are non-positive: $D^i(z) \leq 0$ for all $i \in N$. Thus, a joint distribution of actions lies in the Hannan set if the payoff of each player is no less than his best-reply payoff against the joint distribution of actions of the other players (in the context of a repeated game, this is the Hannan (1957) condition).

We note that:

- The Hannan set H is a convex set (in fact a convex polytope).
- The Hannan set H contains all correlated equilibria,[7] and thus *a fortiori* all Nash equilibria.
- If z is independent over the players, then z is in the Hannan set H if and only if z is a Nash equilibrium.

3.2. *Potential functions*

General regret-based strategies make use of potential functions, introduced in Hart and Mas-Colell (2001a). A *potential function* on $\mathbb{R}^m$ is a function

---

[6]It is convenient to extend multilinearly the payoff functions u^i from S to $\Delta(S)$, in fact to all $\mathbb{R}^S$; i.e., $u(z) := \sum_{s \in S} z(s) u(s)$ for all $z \in \mathbb{R}^S$. We slightly abuse notation and write expressions of the form (k, z^{-i}) or $k \times z^{-i}$ instead of $e_k^i \times z^{-i}$, where $k \in S^i$ and $e_k^i \in \Delta(S^i)$ is the k-unit vector.

[7]Consider the setup where players get "recommendations" before the play of the game. Correlated equilibria are those outcomes where no player can unilaterally gain by deviating from some recommendation. If only *constant* deviations (i.e., playing a fixed action regardless of the recommendation) are allowed, this yields the Hannan set. Note that if every player has 2 strategies, then the Hannan set coincides with the set of correlated equilibria. See Section 5.

$P : \mathbb{R}^m \to \mathbb{R}$ satisfying:

(P1) P is a C^1 function; $P(x) > 0$ for all $x \notin \mathbb{R}^m_-$, and $P(x) = 0$ for all $x \in \mathbb{R}^m_-$;

(P2) $\nabla P(x) \geq 0$ and $\nabla P(x) \cdot x > 0$ for all $x \notin \mathbb{R}^m_-$;

(P3) $P(x) = P([x]_+)$ for[8] all x; and

(P4) there exist $0 < \rho_1 \leq \rho_2 < \infty$ such that $\rho_1 P(x) \leq \nabla P(x) \cdot x \leq \rho_2 P(x)$ for all $x \notin \mathbb{R}^m_-$.

Note that (P1), (P2), and (P3) correspond to (R1), (R2), and (R3) of Hart and Mas-Colell (2001a) for[9,10] $C = \mathbb{R}^m_-$. Condition (P4) is technical.[11]

The potential function P may be viewed as a generalized distance to $\mathbb{R}^m_-$; for example, take $P(x) = \min\{(\|x - y\|_p)^p : y \in \mathbb{R}^m_-\} = (\|[x]_+\|_p)^p$ where $\|\cdot\|_p$ is the l^p-norm on $\mathbb{R}^m$ and $1 < p < \infty$.

From now on we will always assume (P1)–(P4). By (P2), the gradient of P at $x \notin \mathbb{R}^m_-$ is a non-negative and non-zero vector; we introduce the notation[12]

$$\widehat{\nabla} P(x) := \frac{1}{\|\nabla P(x)\|} \nabla P(x) \in \Delta(m) \tag{3.1}$$

for the *normalized gradient* of P at x; thus $\widehat{\nabla}_k P(x) := \nabla_k P(x) / (\sum_{\ell \in S^i} \nabla_\ell P(x))$ for each $k = 1, \ldots, m$.

3.3. *Regret-based strategies*

"Regret matching" is a repeated game strategy where the probabilities of play are proportional to the positive part of the regrets (i.e., to $[D^i(z)]_+$). This is a special case of what we will call regret-based strategies.

---

[8]We write $[\xi]_+$ for the positive part of the real ξ, i.e., $[\xi]_+ = \max\{\xi, 0\}$; for a vector $x = (x_1, \ldots, x_m)$, we write $[x]_+$ for $([x_1]_+, \ldots, [x_m]_+)$.

[9]The second part of (P1) is without loss of generality—see Lemma 2.3 (c1) and the construction of P_1 in the Proof of Theorem 2.1 of Hart and Mas-Colell (2001a).

[10]The "better play" condition (R3) is "If $x_k < 0$ then $\nabla_k P(x) = 0$," which indeed implies that $P(x) = P([x]_+)$.

[11]$\nabla P(x) \cdot x / P(x) = dP(\tau x)/d\tau$ evaluated at $\tau = 1$; therefore it may be interpreted as the "local returns to scale of P at x." Condition (P4) thus says that the local returns to scale are uniformly bounded from above and from below (away from 0). If P is homogeneous of degree α then one can take $\rho_1 = \rho_2 = \alpha$.

[12]It will be convenient to use throughout the l^1-norm $\|x\| = \sum_k |x_k|$. The partial derivative $\partial P(x)/\partial x_k$ of $P(x)$ with respect to x_k is denoted $\nabla_k P(x)$ (it is the k-coordinate of the gradient vector $\nabla P(x)$). Finally, we write $\Delta(m)$ for the unit simplex of $\mathbb{R}^m$.

We say that player i uses a *regret-based strategy* if there exists a potential function $P^i : \mathbb{R}^{S^i} \to \mathbb{R}$ (satisfying (P1)–(P4)) such that at each time t where some regret of player i is positive, the mixed play $q^i(t) \in \Delta(S^i)$ of i is proportional to the gradient of the potential evaluated at the current regret vector; that is,

$$q^i(t) = \widehat{\nabla} P^i(D^i(z(t))) \quad \text{when } D^i(z(t)) \notin \mathbb{R}^{S^i}_-. \tag{3.2}$$

Note that there are no conditions when the regret vector is non-positive. Such a strategy is called a P^i-*strategy* for short.

Condition (3.2) is the counterpart of the discrete-time P^i-strategy of Hart and Mas-Colell (2001a):

$$q^i_k(T+1) \equiv \Pr[s^i_{T+1} = k \,|\, h_T] = \widehat{\nabla} P^i_k(D^i(z_T)) \quad \text{when } D^i(z_T) \notin \mathbb{R}^{S^i}_-.$$

Remark. The class of regret-based strategies of a player i is invariant to transformations of i's utility function which preserve i's mixed-action best-reply correspondence (i.e., replacing u^i with $\tilde{u}^i$ given by $\tilde{u}^i(s) := \alpha u^i(s) + v(s^{-i})$ for some $\alpha > 0$; indeed, $v(\cdot)$ does not affect the regrets, and α changes the scale, which requires a corresponding change in P^i).

The main property of regret-based strategies (see Hart and Mas-Colell 2001a, Theorem 3.3, for the discrete-time analog) is:

Theorem 3.1. *Let $z(t)$ be a solution of (2.1) and (3.2). Then $\varlimsup_{t\to\infty} D^i_k(z(t)) \leq 0$ for every $k \in S^i$.*

Remark. This result holds for *any* strategies of the other players q^{-i}; in fact, one may allow correlation between the players in $N\backslash\{i\}$ (but, of course, q^{-i} must be independent of q^i—thus $q(t) = q^i(t) \times q^{-i}(t)$).

Proof. For simplicity rescale the time t so that (2.1) becomes[13] $\dot{z} = q - z$. Assume $D^i(z) \notin \mathbb{R}^{S^i}_-$, so $P^i(D^i(z)) > 0$. We have (recall Footnote 6)

$$\begin{aligned}
\dot{D}^i_k(z) &= u^i(k \times \dot{z}^{-i} - \dot{z}) \\
&= u^i(k \times (q^{-i} - z^{-i}) - q^i \times q^{-i} + z) \\
&= u^i(k \times q^{-i} - q^i \times q^{-i} - k \times z^{-i} + z) \\
&= u^i(k, q^{-i}) - u^i(q^i, q^{-i}) - D^i_k(z).
\end{aligned}$$

---

[13] Take $\tilde{t} = \exp(t)$.

Multiplying by q_k^i and summing over $k \in S^i$ yields

$$q^i \cdot \dot{D}^i(z) = -q^i \cdot D^i(z). \qquad (3.3)$$

Define $\pi^i(z) := P^i(D^i(z))$; then (recall (3.2))

$$\dot{\pi}^i(z) = \nabla P^i(D^i(z)) \cdot \dot{D}^i(z) = \left\| \nabla P^i(D^i(z)) \right\| q^i \cdot \dot{D}^i(z)$$
$$= -\left\| \nabla P^i(D^i(z)) \right\| q^i \cdot D^i(z) = -\nabla P^i(D^i(z)) \cdot D^i(z). \qquad (3.4)$$

Using condition (P2) implies that $\dot{\pi}^i < 0$ when $D^i(z) \notin \mathbb{R}_-^{S^i}$—thus π^i is a strict Lyapunov function for the dynamical system. It follows that[14] $\pi^i(z) \to 0$. $\qquad \square$

Corollary 3.2. *If all players play regret-based strategies, then $z(t)$ converges as $t \to \infty$ to the Hannan set H.*

One should note that here (as in all the other results of this paper), the convergence is to the *set* H, and not to a specific point in that set. That is, the distance between $z(t)$ and the set H converges to 0; or, equivalently, the limit of any convergent subsequence lies in the set.

We end this section with a technical result: Once there is some positive regret, then a regret-based strategy will maintain this forever (of course, the regrets go to zero by Theorem 3.1).

Lemma 3.3. *If $D^i(z(t_0)) \notin \mathbb{R}_-^{S^i}$ then $D^i(z(t)) \notin \mathbb{R}_-^{S^i}$ for all $t \geq t_0$.*

Proof. Let $\pi^i := P^i(D^i(z))$. Then $\pi^i(t_0) > 0$, and (3.4) together with (P3) implies that $\dot{\pi}^i \geq -\rho_2 \pi^i$ and thus $\pi^i(t) \geq e^{-\rho_2(t-t_0)} \pi^i(t_0) > 0$ for all $t > t_0$. $\qquad \square$

4. Nash Equilibria

In this section we consider two-person games, and show that in some special classes of games regret-based strategies by both players do in fact lead to the set of Nash equilibria (not just to the Hannan set, which is in general a strictly larger set).

---

[14]Note that only (P1) and (P2) were used in this proof.

If z belongs to the Hannan set H, then $u^i(z) \geq u^i(k^i, z^j)$ for all $k^i \in S^i$ and $i \neq j$. Averaging according to z^i yields

$$u^i(z) \geq u^i(z^1, z^2) \quad \text{for } i = 1, 2. \tag{4.1}$$

Lemma 4.1. *In a two-person game, if z belongs to the Hannan set and the payoff of z is the same as the payoff of the product of its marginals, i.e., if*

$$u^i(z) = u^i(z^1, z^2) \quad \text{for } i = 1, 2, \tag{4.2}$$

then (z^1, z^2) is a Nash equilibrium.

Proof. If $z \in H$ then $u^i(k^i, z^j) \leq u^i(z) = u^i(z^1, z^2)$ for all $k^i \in S^i$. □

4.1. *Two-person zero-sum games*

Consider a two-person zero-sum game Γ, i.e., $u^1 = u$ and $u^2 = -u$. Let v denote the minimax *value* of Γ. A pair of (mixed) strategies (y^1, y^2) is a Nash equilibrium if and only if y^i is an *optimal* strategy of player i (i.e., if it guarantees the value v).

Theorem 4.2. *Let Γ be a two-person zero-sum game. If both players play regret-based strategies, then $(z^1(t), z^2(t))$ converges to the set of Nash equilibria of Γ, and $u(z(t))$ and $u(z^1(t), z^2(t))$ both converge as $t \to \infty$ to the minimax value v of Γ.*

Proof. The inequalities (4.1) for both players imply the equalities (4.2), and the result follows from Theorem 3.1 and Lemma 4.1. □

See Corollary 4.5 in Hart and Mas-Colell (2001a) for the discrete-time analog.

4.2. *Two-person potential games*

Consider a two-person *potential game* Γ: Without loss of generality the two players have identical payoff functions[15] $u^1 = u^2 = u : S \to \mathbb{R}$.

We will show first that if initially[16] each player has some positive regret, then both players using regret-based strategies leads to the set of Nash

---

[15] Recall the remark preceding Theorem 3.1.
[16] I.e., at $t = 1$—or, in fact, at any $t = t_0$.

equilibria. Regret-based strategies allow a player to behave arbitrarily when all his regrets are non-positive—in particular, inside the Hannan set (which is larger than the set of Nash equilibria). In order to extend our result and always guarantee convergence to the set of Nash equilibria, the strategies need to be appropriately defined in the case of non-positive regrets; we do so at the end of this subsection.

Before proceeding we need a technical lemma.

Lemma 4.3. *Let P be a potential function (satisfying* (P1)–(P4)*). Then for every $K > 0$ there exists a constant $c > 0$ such that*

$$\max_k x_k \leq c(P(x))^{1/\rho_2} \quad \text{for all } x \in [-K, K]^m.$$

Proof. Since replacing P with P^{1/ρ_2} does not affect (P1)–(P4), we can assume without loss of generality that $\rho_2 = 1$ in (P4). Take a non-negative $x \in [0, K]^m$, and let $f(\tau) := P(\tau x)$ for $\tau \geq 0$. Then $f'(\tau) = \nabla P(\tau x) \cdot x \leq P(\tau x)/\tau = f(\tau)/\tau$ for all $\tau > 0$; hence $(f(\tau)/\tau)' \leq 0$, which implies that $f(\tau)/\tau \geq f(1)$ for all $\tau \leq 1$. Thus $P(\tau x) \geq \tau P(x)$ for all $x \geq 0$ and all $0 \leq \tau \leq 1$. Let $a := \min\{P(x) : x \geq 0, \|x\| = K\}$, then $a > 0$ since the minimum is attained. Hence

$$P(x) = P(x_1, \ldots, x_m) \geq P(x_1, 0, \ldots, 0) \geq \frac{x_1}{K} P(K, 0, \ldots, 0) \geq \frac{x_1}{K} a$$

(the first inequality since $\nabla P \geq 0$). Altogether we get $x_1 \leq cP(x)$ where $c = K/a$; the same applies to the other coordinates. For $x \in [-K, K]^m$ which is not non-negative, use (P3):

$$\max_k x_k \leq \max_k [x_k]_+ \leq cP([x]_+) = cP(x).$$

This completes the proof.   □

By replacing P with cP^{1/ρ_2} for an appropriate $c > 0$—which does not affect the normalized gradient—we will assume from now on without loss of generality that the potential P^i for each player i is chosen so that

$$\max_{k \in S^i} x_k \leq P^i(x) \quad \text{for all } x \in [-2M, 2M]^{S^i}. \tag{4.3}$$

We deal first with the case where initially, at $t = 1$, both players have some positive regret.

Theorem 4.4. *Let Γ be a two-person potential game. Assume that initially both players have some positive regret, i.e., $D^i(z(1)) \notin \mathbb{R}_-^{S^i}$ for $i = 1, 2$.*

If both players use regret-based strategies, then the pair of marginal distributions $(z^1(t), z^2(t)) \in \Delta(S^1) \times \Delta(S^2)$ converges as $t \to \infty$ to the set of Nash equilibria of the game. Moreover, there exists a number $\bar{v}$ such that $(z^1(t), z^2(t))$ converges to the set of Nash equilibria with payoff $\bar{v}$ (to both players), and the average payoff $u(z(t))$ also converges to $\bar{v}$.

Proof. We again rescale t so that $\dot{z} = q - z$. Lemma 3.3 implies that $\pi^i(t) := P^i(D^i(z^i(t))) > 0$ for all t. We have

$$\dot{u}(z^1, z^2) = \dot{u}(z^1 \times z^2) = u(\dot{z}^1 \times z^2 + z^1 \times \dot{z}^2)$$
$$= u((q^1 - z^1) \times z^2 + z^1 \times (q^2 - z^2))$$
$$= u(q^1, z^2) + u(z^1, q^2) - 2u(z^1, z^2).$$

Now

$$u(q^1, z^2) = \sum_{k \in S^1} q_k^1 u(k, z^2) = u(z) + \sum_{k \in S^1} q_k^1 D_k^1(z)$$
$$= u(z) + q^1 \cdot D^1 > u(z)$$

(by (P2) since q^1 is proportional to $\nabla P^1(D^1)$). Thus

$$\dot{u}(z^1, z^2) > 2u(z) - 2u(z^1, z^2). \tag{4.4}$$

Next, (4.3) implies

$$u(k, z^2) - u(z) = D_k^1(z) \le P^1(D^1(z)) = \pi^1$$

for all $k \in S^1$, and therefore

$$u(z^1, z^2) - u(z) \le \pi^1. \tag{4.5}$$

Similarly for player 2, and thus from (4.4) we get

$$\dot{u}(z^1, z^2) > -\pi^1 - \pi^2. \tag{4.6}$$

Now

$$\dot{\pi}^i = -\nabla P^i(D^i(z)) \cdot D^i(z) \le -\rho P^i(D^i(z)) = -\rho \pi^i \tag{4.7}$$

(we have used (3.4) and (P3), with ρ the minimum of ρ_1^i of (P4) for $i = 1, 2$). Define $v := u(z^1, z^2) - \pi^1/\rho - \pi^2/\rho$; from (4.6) we get

$$\dot{v} = \dot{u}(z^1, z^2) - \dot{\pi}^1/\rho - \dot{\pi}^2/\rho \ge \dot{u}(z^1, z^2) + \pi^1 + \pi^2 > 0. \tag{4.8}$$

Therefore v increases; since it is bounded, it converges; let $\bar{v}$ be its limit. Theorem 3.1 implies that $\pi^i \to 0$, so $u(z^1, z^2) \to \bar{v}$.

By Lemma 4.1, it remains to show that $u(z) - u(z^1, z^2) \to 0$. We use the following

Lemma 4.5. *Let* $f : \mathbb{R}_+ \to \mathbb{R}_+$ *be a non-negative, uniformly Lipschitz function such that* $\int_0^\infty f(t)\, dt < \infty$. *Then* $f(t) \to 0$ *as* $t \to \infty$.

Proof. Let L be such that $|f(t) - f(T)| \le L\,|t - T|$ for all t, T. If $f(T) \ge 2\varepsilon > 0$, then $f(t) \ge \varepsilon$ for all $T \le t \le T + \varepsilon/L$, so $\int_T^{T+\varepsilon/L} f(t)\, dt \ge \varepsilon^2/L$. Since the integral is bounded, it follows that there can be at most finitely many such occurrences, so $f(T) < 2\varepsilon$ for all T large enough. $\square$

To get back to the proof of Theorem 4.4: define $f := 2u(z) - 2u(z^1, z^2) + \pi^1 + \pi^2$; then f is non-negative (by (4.5)) and uniformly Lipschitz, and $\int_0^\infty f(t)\, dt$ is finite (it is bounded by $\bar{v}$, since $f < \dot{u}(z^1, z^2) + \pi^1 + \pi^2 \le \dot{v}$ by (4.4) and (4.8)). Lemma 4.5 implies that $f \to 0$; thus $u(z) - u(z^1, z^2) \to 0$ (since $\pi^i \to 0$).

We handle now the case where at the initial condition $z(1)$ all the regrets of a player i are non-positive. We define the strategy of i as follows: i plays an arbitrary *fixed* mixed strategy $\bar{y}^i \in \Delta(S^i)$, up to such time T^i when some regret is at least $1/T^i$ (i.e., T^i is the first $t > 1$ such that[17] $\max_{k \in S^i} D_k^i(z(t)) \ge 1/t$); of course, if this never happens (i.e., if $T^i = \infty$), then i always plays $\bar{y}^i$. After time T^i player i plays P^i-regret matching (recall Lemma 3.3). That is,

$$q^i(t) := \begin{cases} \bar{y}^i, & \text{for } t \le T^i, \\[2mm] \widehat{\nabla} P^i(D^i(z(t))) & \text{for } t > T^i. \end{cases} \tag{4.9}$$

Corollary 4.6. *The result of Theorem 4.4 holds for any initial* $z(1)$ *when the strategies are given by* (4.9).

Proof. If there is some time T after which both players play $q^i = \widehat{\nabla} P^i(D^i(z))$, then we apply Theorem 4.4 starting at T. Otherwise, for a player i that plays $\bar{y}^i$ forever, we have $\max_{k \in S^i} D_k^i(z(t)) < 1/t$ for all t, so $D^i(z(t)) \to \mathbb{R}_-^{S^i}$. Moreover $z^i(t)$ converges to the constant $\bar{y}^i$ and so $z(t)$ becomes independent in the limit (i.e., $z(t) - z^i(t) \times z^{-i}(t) \to 0$); the convergence to the set of Nash equilibria follows from Lemma 4.1. Finally, the payoff $\bar{v}$ is just the best-reply payoff against $\bar{y}^i$. $\square$

---

[17] We use $1/t$ rather than 0 in order to avoid difficulties at the boundary of $\mathbb{R}_-^{S^i}$; any positive function of t converging to 0 as $t \to \infty$ will do.

The analog of this result for discrete-time—which is a new result—is stated and proved in Appendix A.

4.3. *Other classes of games*

Smooth fictitious play—which may be viewed as (approximately) a limiting case of regret-based strategies—has been shown to converge to the set of (approximate) Nash equilibria for additional classes of two-person games, namely, games with a unique interior ESS, and supermodular games (see Hofbauer and Sandholm 2002). It turns out that general regret-based strategies converge to Nash equilibria for the first class (Hofbauer, personal communication 2002); we do not know about the second class.

5. Correlated Equilibria

Given a joint distribution $z \in \Delta(S)$, the regret of player i for action k may be rewritten as follows:

$$D_k^i(z) = \sum_{s \in S} \left[u^i(k, s^{-i}) - u^i(s) \right] z(s)$$

$$= \sum_{j \in S^i} \sum_{s^{-i} \in S^{-i}} \left[u^i(k, s^{-i}) - u^i(j, s^{-i}) \right] z(j, s^{-i}).$$

We now define the *conditional regret of player i from action j to action k* (for $j, k \in S^i$ with $j \neq k$) as follows:

$$C_{jk}^i(z) := \sum_{s^{-i} \in S^{-i}} \left[u^i(k, s^{-i}) - u^i(j, s^{-i}) \right] z(j, s^{-i}). \tag{5.1}$$

This is the change in the payoff of i if action j had always been replaced by action k. Denote $L := \{(j, k) \in S^i \times S^i : j \neq k\}$ and let $C^i(z) := (C_{jk}^i(z))_{(j,k) \in L}$ be the vector of conditional regrets. A distribution $z \in \Delta(S)$ is a *correlated equilibrium* if and only if $C^i(z) \leq 0$ for all $i \in N$ (see Hart and Mas-Colell 2000).[18]

Conditional regret-based strategies for a player i will define the action of i by the way it changes with time—i.e., by a differential equation. This requires us to add $q^i(t) \in \Delta(S^i)$ as a state variable—in addition

---

[18]Note that $C_{jk}^i(z) \leq 0$ for all $j \neq k$ implies $D_k^i(z) = \sum_{j \neq k} C_{jk}^i(z) \leq 0$; this shows that the Hannan set contains the set of correlated equilibria (recall Section 3.1 and footnote 7).

to $z(t) \in \Delta(S)$, which changes according to (2.1). Specifically, we say that player i plays a *conditional regret-based strategy* if there exists a potential function $P^i : \mathbb{R}^L \to \mathbb{R}$ (satisfying (P1)–(P4)), such that, when $C^i(z(t)) \notin \mathbb{R}^L_-$,

$$\dot{q}^i_j(t) = \sum_{k \neq j} \nabla_{(k,j)} P^i(C^i(z(t))) q^i_k(t) - \sum_{k \neq j} \nabla_{(j,k)} P^i(C^i(z(t))) q^i_j(t) \quad (5.2)$$

for all $j \in S^i$, where $\nabla_{(k,j)}$ denotes the derivative with respect to the (k,j)-coordinate;[19] again, there are no conditions when all conditional regrets are non-positive, i.e., when $C^i(z(t)) \in \mathbb{R}^L_-$.

To see where (5.2) comes from, recall the discrete-time strategy of Hart and Mas-Colell (2000, (2.2)):

$$q^i_j(t+1) = \mathbf{1}_{s^i_t = j} \left[1 - \frac{1}{\mu} \sum_{k \neq j} R^i_{jk}(t) \right] + \sum_{k \neq j} \mathbf{1}_{s^i_t = k} \frac{1}{\mu} R^i_{kj}(t),$$

which, when taking expectations, yields

$$q^i_j(t+1) = q^i_j(t) \left[1 - \frac{1}{\mu} \sum_{k \neq j} R^i_{jk}(t) \right] + \sum_{k \neq j} q^i_k(t) \frac{1}{\mu} R^i_{kj}(t)$$

$$= q^i_j(t) + \frac{1}{\mu} \sum_{k \neq j} \left[R^i_{kj}(t) q^i_k(t) - R^i_{jk}(t) q^i_j(t) \right].$$

Replacing the positive part of the regrets $R^i_{kj} = [C^i_{kj}]_+$ with their generalizations $\nabla_{(k,j)} P^i(C^i)$ leads to (5.2) (see Hart and Mas-Colell 2001a, Section 5.1).

Remarks.

(1) The "speeds of adjustment" of q and z (a constant for q, and $1/t$ for z) are different.

(2) We have $\sum_j \dot{q}^i_j = 0$ and $\dot{q}^i_j \geq 0$ when $q^i_j = 0$; therefore q^i never leaves the simplex $\Delta(S^i)$ if we start there (i.e., if $q^i(1) \in \Delta(S^i)$).

---

[19](5.2) may be viewed as the differential equation for the expected probability of a continuous-time Markov process.

Theorem 5.1. *If player i plays a conditional regret-based strategy, then*

$$\overline{\lim_{t \to \infty}} \max_{j,k} C^i_{jk}(z(t)) \leq 0$$

for any play $q^{-i}(t)$ of the other players.

Corollary 5.2. *If all players use conditional regret-based strategies, then $z(t)$ converges as $t \to \infty$ to the set of correlated equilibria of the game Γ.*

Remark. Unlike the discrete-time case (see the discussion in Hart and Mas-Colell 2000, Section 4(d)), the result for continuous time applies to each player separately; that is, no assumption is needed on q^{-i} in Theorem 5.1. The reason is that, in the "limit"—as the time periods become infinitesimal—the condition of Cahn (2004) is essentially satisfied by any continuous solution.[20] Thus continuous-time conditional regret-based strategies are "universally conditionally consistent" or "universally calibrated" (cf. Fudenberg and Levine 1998, 1999).

Proof of Theorem 5.1. Assume without loss of generality that in (P4) we have $\rho_1 = 1$ (replace P^i with $(P^i)^{1/\rho_1}$). Throughout this proof, j and k will always be elements of S^i; we have:

$$\dot{C}^i_{jk}(z) = \sum_{s^{-i} \in S^{-i}} \left[u^i(k, s^{-i}) - u^i(j, s^{-i}) \right] \dot{z}(j, s^{-i})$$

$$= \frac{1}{t} \sum_{s^{-i} \in S^{-i}} \left[u^i(k, s^{-i}) - u^i(j, s^{-i}) \right] \left(-z(j, s^{-i}) + q^i_j q^{-i}_{s^{-i}} \right)$$

$$= \frac{1}{t} \left(-C^i_{jk}(z) + \sum_{s^{-i} \in S^{-i}} \left[u^i(k, s^{-i}) - u^i(j, s^{-i}) \right] q^i_j q^{-i}_{s^{-i}} \right). \quad (5.3)$$

Denote $\pi(t) := P^i(C^i(z(t)))$ and $G(t) = \nabla P^i(C^i(z(t)))$. Then

$$\dot{\pi} = G \cdot \dot{C}^i(z)$$

$$\leq -\frac{1}{t} \pi + \frac{1}{t} \sum_{s^{-i} \in S^{-i}} q^{-i}_{s^{-i}} \left\{ \sum_{j,k} G_{jk} \left[u^i(k, s^{-i}) - u^i(j, s^{-i}) \right] q^i_j \right\},$$

$$(5.4)$$

---

[20]The Cahn condition is that the effect of the choice of player i at time t on the choice of another player j at some future time goes to zero as t goes to infinity. More precisely, if the histories h_{t+w-1} and h'_{t+w-1} differ *only* in their s^i_t-coordinate, then for all $j \neq i$ we have $|\Pr[s^j_{t+w} = s^j | h_{t+w-1}] - \Pr[s^j_{t+w} = s^j | h'_{t+w-1}]| \leq f(w)/g(t)$ for some functions f and g such that $g(t) \to \infty$ as $t \to \infty$.

where we have used (P4) (recall that $\rho_1 = 1$). Denote by E the right-hand sum over s^{-i}, and by $E(s^{-i})$ the expression in the curly brackets $\{\ldots\}$ (thus E is a weighted average of the $E(s^{-i})$). Rearranging terms yields

$$E(s^{-i}) = \sum_j u(j, s^{-i}) \left[\sum_k G_{kj} q_k^i - \sum_k G_{jk} q_j^i \right] = \sum_j u(j, s^{-i}) \dot{q}_j^i.$$

We now claim that $\dot{q}_j^i \to 0$ as $t \to \infty$ for all $j \in S^i$.

Indeed, let $m := |S^i|$; then $|C_{jk}^i| \leq 2M$ and so $\|C^i\| \leq 2Mm(m-1) =: M_1$; also $|\dot{C}_{jk}^i| \leq 2M|S^{-i}|/t$ (since $|\dot{z}(s)| \leq 1/t$ for all s by (2.1)) and thus

$$\|\dot{C}^i\| \leq 2M|S^{-i}|m(m-1)/t =: M_2/t.$$

Let K be a Lipschitz bound for $\nabla P^i(x)$ over $\|x\| \leq M_1$; then for all $t_2 \geq t_1 \geq 1$ we have

$$\|G(t_2) - G(t_1)\| \leq K \left\| C^i(z(t_2)) - C^i(z(t_1)) \right\| \leq K \left\| \dot{C}^i(z(\tau)) \right\| (t_2 - t_1)$$

$$\leq K M_2 \frac{t_2 - t_1}{t_1} \tag{5.5}$$

($\tau \in [t_1, t_2]$ is some intermediate point).

Let $M_3 := \max_{\|x\| \leq M_1} \|\nabla P^i(x)\|$, and define

$$A_{jk}(t) := \frac{1}{M_3} G_{jk}(t), \quad \text{for } j \neq k, \quad \text{and}$$

$$A_{jj}(t) := 1 - \frac{1}{M_3} \sum_{k \neq j} G_{kj}(t).$$

Then $A(t)$ is a stochastic matrix,[21] and (5.2) can be rewritten as[22]

$$\dot{q}^i(t) = M_3 \, q^i(t)(A(t) - I). \tag{5.6}$$

Finally, (5.5) yields[23]

$$\|A(t_2) - A(t_1)\| \leq \frac{m}{M_3} \|G(t_2) - G(t_1)\| \leq \frac{mKM_2}{M_3} \frac{t_2 - t_1}{t_1}$$

for all $t_2 \geq t_1 \geq 1$.

---

[21] I.e., its elements are non-negative and the sum of each row is 1.

[22] Vectors (like q) are viewed as row vectors; I denotes the identity matrix.

[23] The norm $\|A\|$ of a matrix A is taken to be $\max\{\|xA\| : \|x\| = 1\}$, so that always $\|xA\| \leq \|x\| \|A\|$. Note that if $A = (A_{jk})$ is an $m \times m$ matrix then $\max_{j,k} |A_{jk}| \leq \|A\| \leq m \max_{j,k} |A_{jk}|$; if moreover A is a stochastic matrix, then $\|A\| = 1$.

Applying Proposition B.1 (see Appendix B; the constant M_3 in (5.6) does not matter—replace t by $M_3 t$) implies that indeed $\dot{q}^i \to 0$ as $t \to \infty$.

Therefore $E(s^{-i}) \to 0$ and so (recall (5.4)) $t\dot{\pi}(t) + \pi(t) \leq E(t) \to 0$ as $t \to \infty$, from which it follows that $\pi(t) \to 0$ (indeed, for each $\varepsilon > 0$ let $t_0 \equiv t_0(\varepsilon)$ be such that $|E(t)| \leq \varepsilon$ for all $t \geq t_0$; then $d(t\pi(t))/dt \leq \varepsilon$ for all $t \geq t_0$, which yields $t\pi(t) \leq t_0\pi(t_0) + \varepsilon(t - t_0)$ and thus $\overline{\lim}_{t\to\infty}\pi(t) \leq \varepsilon$).

6. Remarks

(a) It is worthwhile to emphasize, once again, that the appropriate state space for our analysis is not the product of the mixed action spaces of the players $\prod_i \Delta(S^i)$, but the space of joint distributions on the product of their pure action sets $\Delta(\prod_i S^i)$. This is so because, as we pointed out in Hart and Mas-Colell (2001a, Section 4), with the exception of the limiting case constituted by fictitious play, the dynamics of regret matching depend on $u(z)$, the time-average of the realized payoffs, and therefore on the joint distribution z. It is interesting to contrast this with, for example, Hofbauer (2000) and Sandholm (2002), where, in an evolutionary context, dynamics similar to regret matching are considered but where, nonetheless, the context dictates that the appropriate state space is the product of the mixed action spaces. This family of evolutionary dynamics is named by Hofbauer (2000) "Brown–von Neumann–Nash dynamics."

(b) The fact that the state space variable is the time-average distribution of play $z(t)$ does not impose on players informational requirements additional to those familiar from, say, fictitious play. It only asks that players record also their own play at each period (i.e., i keeps track of the frequency of each s, and not only of s^{-i}).

(c) One could ask to what extent the discrete-time analog of the results in this paper can be obtained by appealing to stochastic approximation techniques (see Benaïm 1999 or Benaïm and Weibull 2003). We have not investigated this matter in detail. However, it seems to us that for the results of Section 3 and Appendix C it should be a relatively simple matter, but for those of Sections 4 (Nash equilibria) and 5 (correlated equilibria) there may be a real challenge.

Appendix A. Discrete-Time Dynamics for Potential Games

In this Appendix we deal with discrete-time dynamics for two-person potential games (see Section 4.2). We assume that the potential function P^i of each player satisfies (P1)–(P4) and, in addition,

(P5) P is a C^2 function.

A *discrete-time regret-based strategy* of player i is defined as follows: If $D^i(z_{t-1}) \notin \mathbb{R}^{S^i}_-$ (i.e., if there is some positive regret), then the play probabilities are proportional to the gradient of the potential $\nabla P^i(D^i(z_{t-1}))$. If $D^i(z_{t-1}) \in \mathbb{R}^{S^i}_-$ (i.e., if there is no positive regret),[24] then we assume that i uses the empirical distribution of his past choices z^i_{t-1}. One simple way to implement this is to choose at random a past period $r = 1, 2, \ldots, t-1$ (with equal probabilities of $1/(t-1)$ each) and play at time t the same action that was played at time r (i.e., $s^i_t = s^i_r$).[25] To summarize: At time t the action of player i is chosen according to the probability distribution $q^i_t \in \Delta(S^i)$ given by

$$q^i_t(k) = \Pr[s^i_t = k \mid h_{t-1}] := \begin{cases} \widehat{\nabla}_k P^i(D^i(z_{t-1})), & \text{if } D^i(z_{t-1}) \notin \mathbb{R}^{S^i}_-, \\ \\ z^i_{t-1}(k) & \text{if } D^i(z_{t-1}) \in \mathbb{R}^{S^i}_-, \end{cases}$$

(A.1)

for each $k \in S^i$ (starting at $t = 1$ with an arbitrary $q^i_1 \in \Delta(S^i)$).

Theorem A.1. *Let Γ be a two-person potential game. If both players use regret-based strategies (A.1), then, with probability 1, the pair of empirical marginal distributions (z^1_t, z^2_t) converges as $t \to \infty$ to the set of Nash equilibria of the game, and the average realized payoff $u(z_t)$ (and $u(z^1_t, z^2_t)$) converges to the set of Nash equilibrium payoffs.*

Proof. Without loss of generality assume that (4.3) holds for both players (thus $\rho^i_2 = 1$ in (P4)), and let $\rho > 0$ be the minimum of the ρ^i_1 in (P4). Put $d^i_t(k) := D^i_k(z_t)$ for the k-regret and $d^i_t := D^i(z_t)$ for the vector of regrets, and $\pi^i_t := P^i(D^i(z_t)) = P^i(d^i_t)$. For clarity, we divide the proof into five steps.

• **Step 1.** $\pi^i_t \to 0$ as $t \to \infty$ a.s., and there exists a constant M_1 such that

$$\mathrm{E}[\pi^i_t \mid h_{t-1}] \le (1 - \frac{\rho}{t})\pi^i_{t-1} + \frac{M_1}{t^2}.$$

(A.2)

$\square$

---

[24]Unlike the continuous-time case (recall Lemma 3.3), here the regret vector may enter and exit the negative orthant infinitely often—which requires a more delicate analysis.
[25]In short: There is no change when there is no regret. Other definitions are possible in this case of "no regret"—for example, the result of Theorem A.1 can be shown to hold also if a player plays optimally against the empirical distribution of the other player (i.e., "fictitious play") when all his regrets are non-positive.

Proof. [26] Consider player 1; for each $k \in S^1$ we have

$$
E[d_t^1(k) - d_{t-1}^1(k) \mid h_{t-1}] = \frac{t-1}{t} u(k, z_{t-1}^2) + \frac{1}{t} u(k, q_t^2) - \frac{t-1}{t} u(z_{t-1})
$$
$$
- \frac{1}{t} u(q_t^1, q_t^2) - u(k, z_{t-1}^2) + u(z_{t-1})
$$
$$
= \frac{1}{t} \left(u(k, q_t^2) - u(q_t^1, q_t^2) \right) - \frac{1}{t} d_{t-1}^1(k).
$$

The first term vanishes when averaging according to q_t^1, so

$$
E[q_t^1 \cdot (d_t^1 - d_{t-1}^1) \mid h_{t-1}] = -\frac{1}{t} q_t^1 \cdot d_{t-1}^1
$$

(compare with (3.3)). If $d_{t-1}^1 \notin \mathbb{R}_-^{S^1}$ then q_t^1 is proportional to $\nabla P^1(d_{t-1}^1)$; hence

$$
E[\nabla P^1(d_{t-1}^1) \cdot (d_t^1 - d_{t-1}^1) \mid h_{t-1}] = -\frac{1}{t} \nabla P^1(d_{t-1}^1) \cdot d_{t-1}^1
$$
$$
\leq -\frac{\rho}{t} P^1(d_{t-1}^1)
$$

by (P4). This also holds when $d_{t-1}^1 \in \mathbb{R}_-^{S^1}$ (since then both P^1 and ∇P^1 vanish). Therefore, by (P5), there exists some constant M_1 such that

$$
E[P^1(d_t^1) - P^1(d_{t-1}^1) \mid h_{t-1}] \leq -\frac{\rho}{t} P^1(d_{t-1}^1) + \frac{M_1}{t^2},
$$

which is (A.2). Finally, $\pi_t^i \to 0$ follows from Theorem 3.3 in Hart and Mas-Colell (2001a) (or use (A.2) directly).

- **Step 2.** Let[27] $\alpha_{t-1}^i := u(q_t^i, z_{t-1}^j) - u(z_{t-1}^1, z_{t-1}^2) + \pi_{t-1}^i$. Then $\alpha_{t-1}^i \geq 0$ and moreover:

If $\pi_{t-1}^i > 0$ then $\alpha_{t-1}^i > u(z_{t-1}) - u(z_{t-1}^1, z_{t-1}^2) + \pi_{t-1}^i \geq 0.$ (A.3)

Proof. Take $i = 1$. We have

$$
u(k, z_{t-1}^2) - u(z_{t-1}) = d_{t-1}^1(k) \leq P^1(d_{t-1}^1) = \pi_{t-1}^1   \quad (A.4)
$$

for all $k \in S^1$ by (4.3). Averaging over k according to z_{t-1}^1 yields

$$
u(z_{t-1}^1, z_{t-1}^2) - u(z_{t-1}) \leq \pi_{t-1}^1.
$$

---

[26] Compare with (4.7) and with the computation of Lemma 2.2 in Hart and Mas-Colell (2001a).
[27] We use j for the other player (i.e., $j = 3 - i$).

If $\pi_{t-1}^1 = 0$ then $q_t^1 = z_{t-1}^1$ and so $\alpha_{t-1}^1 = 0$. If $\pi_{t-1}^1 > 0$ then $q_t^1 \cdot d_{t-1}^1 = \widehat{\nabla} P^1(d_{t-1}^1) \cdot d_{t-1}^1 > 0$ by (P2); thus averaging the equality in (A.4) according to q_{t-1}^1 implies that

$$u(q_t^1, z_{t-1}^2) - u(z_{t-1}) > 0.$$

Adding the last two displayed inequalities completes the proof. $\qquad\square$

• **Step 3.** Let $\pi_t := \pi_t^1 + \pi_t^2$ and $\alpha_t := \alpha_t^1 + \alpha_t^2$, and define

$$v_t := u(z_t^1, z_t^2) - \frac{1}{\rho}\pi_t - \sum_{r=t+1}^{\infty} \frac{M_2}{r^2},$$

where $M_2 := 2M + 2M_1/\rho$. Then

$$E[v_t \mid h_{t-1}] \geq v_{t-1} + \frac{t-1}{t^2}\alpha_{t-1} \geq v_{t-1}, \qquad (A.5)$$

and there exists a bounded random variable v such that $u(z_t^1, z_t^2) \to v$ as $t \to \infty$ a.s.

Proof. We have

$$E[t^2 u(z_t^1, z_t^2) \mid h_{t-1}] = (t-1)^2 u(z_{t-1}^1, z_{t-1}^2) + (t-1)u(q_t^1, z_{t-1}^2)$$
$$+ (t-1)u(z_{t-1}^1, q_t^2) + u(q_t^1, q_t^2)$$

and thus (recall the definition of α_{t-1}^i, and $|u(\cdot)| \leq M$)

$$E[u(z_t^1, z_t^2) \mid h_{t-1}] \geq u(z_{t-1}^1, z_{t-1}^2) - \frac{t-1}{t^2}\pi_{t-1} + \frac{t-1}{t^2}\alpha_{t-1} - \frac{2M}{t^2}. \quad (A.6)$$

Using the inequality (A.2) of Step 1, $\pi_{t-1} \geq 0$, and $\alpha_{t-1} \geq 0$, we get

$$E[v_t \mid h_{t-1}] \geq u(z_{t-1}^1, z_{t-1}^2) - \frac{t-1}{t^2}\pi_{t-1} + \frac{t-1}{t^2}\alpha_{t-1} - \frac{2M}{t^2}$$
$$- \frac{1}{\rho}\left(1 - \frac{\rho}{t}\right)\pi_{t-1} - \frac{2M_1}{\rho t^2} - \sum_{r=t+1}^{\infty} \frac{M_2}{r^2}$$
$$\geq u(z_{t-1}^1, z_{t-1}^2) - \frac{1}{\rho}\pi_{t-1} - \sum_{r=t}^{\infty} \frac{M_2}{r^2} + \frac{t-1}{t^2}\alpha_{t-1}$$
$$= v_{t-1} + \frac{t-1}{t^2}\alpha_{t-1} \geq v_{t-1}.$$

Therefore $(v_t)_{t=1,2,\ldots}$ is a bounded submartingale, which implies that there exists a bounded random variable v such that $v_t \to v$ a.s., and so $u(z_t^1, z_t^2) \to v$ (since $\pi_t^i \to 0$ by Step 1).

• **Step 4.**

$$1_{\pi_t^1 > 0}(u(z_t) - u(z_t^1, z_t^2)) \to 0 \text{ as } t \to \infty \text{ a.s.}$$

Proof. From (A.5) we get

$$w_T := \sum_{t=1}^{T} (\mathrm{E}[v_{t+1} \mid h_t] - v_t) \geq \sum_{t=1}^{T} \frac{\beta_t}{t},$$

where

$$\beta_{t-1} := \frac{(t-1)^2}{t^2} \alpha_{t-1} \geq 0.$$

Thus $(w_T)_{T=1,2,\dots}$ is a non-negative non-decreasing sequence, with $\sup_T \mathrm{E}(w_T) = \sup_T \mathrm{E}(v_{T+1}) - \mathrm{E}(v_1) < \infty$ (the sequence v_t is bounded). Therefore a.s. $\lim w_T$ exists and is finite, which implies that

$$\sum_{t=1}^{\infty} \frac{\beta_t}{t} < \infty. \tag{A.7}$$

In addition, $|z_t(s) - z_{t-1}(s)| \leq 1/t$ for all $s \in S$, and therefore

$$|\beta_t - \beta_{t-1}| \leq \frac{M_3}{t} \tag{A.8}$$

for some constant M_3.

Lemma A.2. *Let $(\beta_t)_{t=1,2,\dots}$ be a non-negative real sequence satisfying (A.7) and (A.8). Then $\beta_t \to 0$ as $t \to \infty$.*

Proof. [28] Without loss of generality take $M_3 = 1$. Let $0 < \varepsilon \leq 1$, and assume that $\beta_t \geq 2\varepsilon$ for some t. Then (A.8) yields, for all $t \leq r \leq t + \varepsilon t$,

$$\beta_r \geq \beta_t - \frac{1}{t+1} - \cdots - \frac{1}{r} \geq 2\varepsilon - \frac{r-t}{t} \geq 2\varepsilon - \varepsilon = \varepsilon,$$

and thus

$$\frac{\beta_r}{r} \geq \frac{\varepsilon}{(1+\varepsilon)t} \geq \frac{\varepsilon}{2t}.$$

Therefore

$$\sum_{t \leq r \leq t + \varepsilon t} \frac{\beta_r}{r} \geq \varepsilon t \frac{\varepsilon}{2t} = \frac{\varepsilon^2}{2} > 0.$$

---

[28] (A.7) implies that the Cesaro averages of the β_t converge to 0 (this is Kronecker's Lemma); together with (A.8), we obtain that the β_t themselves converge to 0.

By (A.7), this implies that there can be at most finitely many t such that $\beta_t \geq 2\varepsilon$, so indeed $\beta_t \to 0$. $\qquad\square$

Using Lemma A.2 shows that a.s. $\beta_t \to 0$ and so $\alpha_t \to 0$, which together with $\pi_t \to 0$ proves Step 4 (recall (A.3)).

• **Step 5.**

$$1_{\pi_t^1=0}(u(z_t) - u(z_t^1, z_t^2)) \to 0 \text{ as } t \to \infty \text{ a.s.}$$

Proof. Let $\gamma_t := 1_{\pi_t^1=0}$ be the indicator of the event $\pi_t^1 = 0$ and define $X_t := t(u(z_t) - u(z_t^1, z_t^2))$. Then $|X_t - X_{t-1}| \leq 4M$, and

$$\begin{aligned}
E[X_t \mid h_{t-1}, \gamma_{t-1} = 1] &= (t-1)u(z_{t-1}) + u(z_{t-1}^1, q_{t-1}^2) \\
&\quad -(t-1)u(z_{t-1}^1, z_{t-1}^2) - u(z_{t-1}^1, q_{t-1}^2) \\
&= X_{t-1}
\end{aligned}$$

(since $q_t^1 = z_{t-1}^1$ when $\gamma_{t-1} = 1$). Let $Y_t := \gamma_{t-1}(X_t - X_{t-1})$; then the Y_t are uniformly bounded martingale differences. Azuma's inequality[29] yields, for each $\varepsilon > 0$ and $r < t$,

$$\Pr\left[\sum_{\tau=r+1}^{t} Y_\tau > t\varepsilon\right] < \exp\left(-\frac{(t\varepsilon)^2}{2(4M)^2(t-r)}\right) \leq \exp(-\delta t)$$

where $\delta := \varepsilon^2/32M^2 > 0$, and thus

$$\Pr\left[\sum_{\tau=r+1}^{t} Y_\tau > t\varepsilon \text{ for some } r < t\right] < t\exp(-\delta t).$$

For each $t \geq 1$ define $R \equiv R(t)$ to be the maximal index $r < t$ such that $\gamma_r = 0$; if there is no such r, put $R(t) = 0$ and, for convenience, take $\gamma_0 \equiv 0$ and $X_0 \equiv 0$. Thus $\gamma_\tau = 1$ for $R + 1 \leq \tau \leq t - 1$ and $\gamma_R = 0$. Therefore $\gamma_{t-1}X_t - X_{R(t)+1} = \sum_{\tau=R(t)+2}^{t} Y_\tau$, and so

$$\Pr[\gamma_{t-1}X_t - X_{R(t)+1} > t\varepsilon] < t\exp(-\delta t). \tag{A.9}$$

The series $\sum_t t\exp(-\delta t)$ converges; therefore, by the Borel–Cantelli Lemma, the event of (A.9) happening for infinitely many t has probability

---

[29] Azuma's inequality is: $\Pr[\sum_{i=1}^{m} Y_i > \lambda] < \exp(-\lambda^2/(2K^2 m))$, where the Y_i are martingale differences with $|Y_i| \leq K$; see Alon and Spencer (2000, Theorem 7.2.1).

0. Thus a.s. $\gamma_{t-1}X_{t-1} - X_{R(t)} \leq \gamma_{t-1}X_t - X_{R(t)+1} + 8M \leq t\varepsilon + 8M$ for all t large enough (recall that $|X_t - X_{t-1}| \leq 4M$), which implies that

$$\overline{\lim_{t\to\infty}} \frac{1}{t}\gamma_t X_t \leq \overline{\lim_{t\to\infty}} \frac{1}{t}X_{R(t)} + \varepsilon.$$

Now either $R(t) \to \infty$, in which case $(1/t)X_{R(t)} \leq (1/R(t))X_{R(t)} \to 0$ by Step 4 since $\pi^1_{R(t)} > 0$; or $R(t) = r_0$ for all $t \geq r_0$, in which case $(1/t)X_{R(t)} \leq (1/t)(4M)r_0 \to 0$. Thus $\mathbf{1}_{\pi^1_t=0}(u(z_t)-u(z_t^1, z_t^2)) = \gamma_t X_t/t \to 0$ a.s., as claimed. □

Proof of Theorem A.1. Steps 4 and 5 show that $u(z_t)$ converges (a.s.) to the same (random) limit v of $u(z_t^1, z_t^2)$ (recall Step 3), which proves that any limit point of the sequence (z_t^1, z_t^2) is indeed a Nash equilibrium (see Lemma 4.1). □

Remark. The proof shows that in fact, with probability one, all limit points are Nash equilibria with the same payoff; that is, for almost every realization (i.e., infinite history) there exists an equilibrium payoff v such that $u(z_t^1, z_t^2)$—and also $u(z_t)$—converges to v.

Appendix B. Continuous-Time Markov Processes

In this Appendix we prove a result on continuous-time Markov processes that we need in Section 5.

Proposition B.1. *For each $t \geq 1$, let $A(t)$ be a stochastic $m \times m$ matrix, and assume that there exists K such that*

$$\|A(t_2) - A(t_1)\| \leq K\frac{t_2 - t_1}{t_1} \quad \text{for all } t_2 \geq t_1 \geq 1.$$

Consider the differential system

$$\dot{x}(t) = x(t)\,(A(t) - I)$$

starting with some[30] $x(1) \in \Delta(m)$. Then

$$\dot{x}(t) \to 0.$$

The proof consists of considering first the case where $A(t) = A$ is independent of t (Proposition B.2), and then estimating the difference in the general case (Lemma B.3).

---

[30]Recall that $\Delta(m)$ is the $(m-1)$-dimensional unit simplex in $\mathbb{R}^m$. Note that $x(1) \in \Delta(m)$ implies that $x(t) \in \Delta(m)$ for all $t \geq 1$.

Proposition B.2. *There exists a universal constant c such that*

$$\left\| e^{t(A-I)}(A - I) \right\| \leq \frac{c}{\sqrt{t}}$$

for any stochastic matrix[31] *A and any* $t \geq 1$.

Proof. We have $e^{t(A-I)} = e^{-tI} e^{tA} = e^{-t} e^{tA}$ and

$$e^{tA}(A - I) = \sum_{n=0}^{\infty} \frac{t^n}{n!}(A^{n+1} - A^n) = \sum_{n=0}^{\infty} \alpha_n A^n,$$

where

$$\alpha_n := \frac{t^{n-1}}{(n-1)!} - \frac{t^n}{n!}$$

(put $t^{-1}/(-1)! = 0$). The matrix A^n is a stochastic matrix for all n; therefore $\|A^n\| = 1$, and thus

$$\left\| e^{tA}(A - I) \right\| \leq \sum_{n=0}^{\infty} |\alpha_n|.$$

Now $\alpha_n > 0$ for $n > t$ and $\alpha_n \leq 0$ for $n \leq t$, so $\sum_n |\alpha_n| = \sum_{n>t} \alpha_n - \sum_{n \leq t} \alpha_n$. Each one of the two sums is telescopic and reduces to[32] $t^r/r!$, where $r := \lfloor t \rfloor$ denotes the largest integer that is $\leq t$. Using Stirling's formula[33] $r! \sim \sqrt{2\pi r} \, r^r e^{-r}$ together with $t/r \to 1$ and $(t/r)^r \to e^{t-r}$ yields

$$\frac{t^r}{r!} \sim \frac{t^r e^r}{\sqrt{2\pi r} \, r^r} \sim \frac{e^t}{\sqrt{2\pi t}}.$$

Therefore

$$\overline{\lim_{t \to \infty}} \left\| e^{t(A-I)}(A - I) \right\| \sqrt{t} \leq \sqrt{\frac{2}{\pi}},$$

from which the result follows.[34] $\qquad \square$

Remark. For each stochastic matrix A it can be shown that[35] $\left\| e^{t(A-I)}(A - I) \right\| = O(e^{\mu t})$, where $\mu < 0$ is given by[36] $\mu := \max\{\mathrm{Re}\lambda : \lambda \neq$

---

[31] Of arbitrary size $m \times m$.

[32] They are equal since $\sum_n \alpha_n = 0$.

[33] $f(t) \sim g(t)$ means that $f(t)/g(t) \to 1$ as $t \to \infty$.

[34] Note that all estimates are uniform: They depend neither on A nor on the dimension m.

[35] $f(t) = O(g(t))$ means that there exists a constant c such that $|f(t)| \leq c|g(t)|$ for all t large enough.

[36] λ is an eigenvalue of $A - I$ if and only if $\lambda + 1$ is an eigenvalue of A. Thus $|\lambda + 1| \leq 1$, which implies that either $\lambda = 0$ or $\mathrm{Re}\lambda < 0$.

0 is an eigenvalue of $A - I$}. However, this estimate—unlike the $O(t^{-1/2})$ of Proposition B.2—is *not* uniform in A and thus does not suffice.

Lemma B.3. *For each $t \geq 1$, let $A(t), B(t)$ be stochastic $m \times m$ matrices, where the mappings $t \to A(t)$ and $t \to B(t)$ are continuous. Let $x(t)$ and $y(t)$ be, respectively, the solutions of the differential systems*

$$\dot{x}(t) = x(t)\,(A(t) - I) \quad and$$
$$\dot{y}(t) = y(t)\,(B(t) - I),$$

starting with some $x(1), y(1) \in \Delta(m)$. Then for all $t \geq 1$

$$\|x(t) - y(t)\| \leq \|x(1) - y(1)\| + \int_1^t \|A(\tau) - B(\tau)\|\,d\tau.$$

Proof. Let $z(t) := e^{t-1} x(t)$ and $w(t) := e^{t-1} y(t)$; then $\dot{z}(t) = z(t) A(t)$ and $\dot{w}(t) = w(t) B(t)$. We have $\|\dot{w}(t)\| \leq \|w(t)\|\,\|B(t)\| = \|w(t)\|$, which implies that $\|w(t)\| \leq e^{t-1}\|w(1)\| = e^{t-1}$. Put $v(t) := z(t) - w(t)$; then

$$\|\dot{v}(t)\| \leq \|(z(t) - w(t))\,A(t)\| + \|w(t)\,(A(t) - B(t))\|$$
$$\leq \|z(t) - w(t)\|\,\|A(t)\| + \|w(t)\|\,\|A(t) - B(t)\| \leq \|v(t)\| + e^{t-1}\delta(t),$$

where $\delta(t) := \|A(t) - B(t)\|$. The solution of $\dot{\eta}(t) = \eta(t) + e^{t-1}\delta(t)$ is $\eta(t) = e^{t-1}\left(\eta(1) + \int_1^t \delta(\tau)\,d\tau\right)$, so $\|v(t)\| \leq e^{t-1}\left(\|v(1)\| + \int_1^t \delta(\tau)\,d\tau\right)$, which, after dividing by e^{t-1}, is precisely our inequality.   □

We can now prove our result.

Proof of Proposition B.1. Let $\alpha = 2/5$. Given T, put $T_0 := T - T^\alpha$. Let $y(T_0) = x(T_0)$ and $\dot{y}(t) = y(t)(A(T_0) - I)$ for $t \in [T_0, T]$. By Proposition B.2,

$$\|\dot{y}(T)\| \leq O((T - T_0)^{-1/2}) = O(T^{-\alpha/2}).$$

Now $\|A(t) - A(T_0)\| \leq K(t - T_0)/T_0 \leq KT^\alpha/(T - T^\alpha) = O(T^{\alpha-1})$ for all $t \in [T_0, T]$, and thus, by Lemma B.3, we get $\|x(T) - y(T)\| \leq (T - T_0)O(T^{\alpha-1}) = O(T^{2\alpha-1})$. Therefore

$$\|\dot{x}(T) - \dot{y}(T)\| = \|x(T)\,(A(T) - I) - y(T)\,(A(T_0) - I)\|$$
$$\leq \|x(T)\|\,\|A(T) - A(T_0)\| + \|x(T) - y(T)\|\,\|A(T_0) - I\|$$
$$\leq O(T^{\alpha-1}) + O(T^{2\alpha-1}) = O(T^{2\alpha-1}).$$

Adding the two estimates yields

$$\|\dot{x}(T)\| \leq O(T^{-\alpha/2}) + O(T^{2\alpha-1}) = O(T^{-1/5})$$

(recall that $\alpha = 2/5$). □

Appendix C. Continuous-Time Approachability

We state and prove here the continuous-time analog of the Blackwell (1956) Approachability Theorem and its generalization in Hart and Mas-Colell (2001a, Section 2); all the notations follow the latter paper. The vector-payoff function is $A : S^i \times S^{-i} \to \mathbb{R}^m$, and we are given a convex closed set $C \subset \mathbb{R}^m$, which is *approachable*, i.e., for every $\lambda \in \mathbb{R}^m$ there exists $\sigma^i \in \Delta(S^i)$ such that

$$\lambda \cdot A(\sigma^i, s^{-i}) \leq w(\lambda) := \sup\{\lambda \cdot y : y \in C\} \quad \text{for all } s^{-i} \in S^{-i} \quad \text{(C.1)}$$

(see (2.1) there).

Let $P : \mathbb{R}^m \to \mathbb{R}$ be a C^1 function satisfying

$$\nabla P(x) \cdot x > w(\nabla P(x)) \quad \text{for all } x \notin C, \quad \text{(C.2)}$$

and also, without loss of generality,[37] $P(x) > 0$ for all $x \notin C$ and $P(x) = 0$ for all $x \in C$. We say that player i plays a *generalized approachability strategy* if the play $q^i(t) \in \Delta(S^i)$ of i at time t satisfies

$$\lambda(t) \cdot A(q(t), s^{-i}) \leq w(\lambda(t)) \quad \text{for all } s^{-i} \in S^{-i}, \quad \text{(C.3)}$$

where

$$\lambda(t) = \nabla P(A(z(t))) \quad \text{(C.4)}$$

(such a $q^i(t)$ exists since C is approachable—see (C.1)). Note that the original Blackwell strategy corresponds to $P(x)$ being the squared Euclidean distance from x to the set C.

Theorem C.1. *Let $z(t)$ be a solution of (2.1), (C.3), and (C.4). Then $A(z(t)) \to C$ as $t \to \infty$.*

---

[37] Cf. the Proof of Theorem 2.1 of Hart and Mas-Colell (2001a).

Proof. Rescale t so that $\dot{z} = q - z$. Denote $\pi(t) := P(A(z(t)))$. If $z(t) \notin C$, then

$$\dot{\pi} = \nabla P \cdot A(\dot{z}) = \lambda \cdot A(q^i, q^{-i}) - \lambda \cdot A(z) < w(\lambda) - w(\lambda) = 0$$

(we have used (C.4), (C.3), and (C.2)). Thus π is a strict Lyapunov function, and so $\pi \to 0$ as $t \to \infty$. $\qquad\qquad\square$

References

Alon, N. and J. H. Spencer (2000), *The Probabilistic Method*, 2nd Edition. New York: Wiley.

Benaïm, M. (1999), "Dynamics of Stochastic Approximation Algorithms," in *Seminaire de Probabilites XXXIII (Lecture Notes in Mathematics 1709)*, ed. by J. Azema *et al.* New York: Springer, pp. 1–68.

Benaïm, M. and J. Weibull (2003), "Deterministic Approximation of Stochastic Evolution in Games," *Econometrica*, 71, 873–903.

Blackwell, D. (1956), "An Analog of the Minmax Theorem for Vector Payoffs," *Pacific Journal of Mathematics*, 6, 1–8.

Cahn, A. (2004), "General Procedures Leading to Correlated Equilibria," *International Journal of Game Theory*, 33, 21–40. [Chapter 6]

Fudenberg, D. and D. K. Levine (1998), *Theory of Learning in Games*. Cambridge, MA: MIT Press.

———— (1999), "Conditional Universal Consistency," *Games and Economic Behavior*, 29, 104–130.

Hannan, J. (1957), "Approximation to Bayes Risk in Repeated Play," in *Contributions to the Theory of Games*, Vol. III, Annals of Mathematics Studies 39, ed. by M. Dresher, A. W. Tucker, and P. Wolfe. Princeton: Princeton University Press, pp. 97–139.

Hart, S. and A. Mas-Colell (2000), "A Simple Adaptive Procedure Leading to Correlated Equilibrium," *Econometrica*, 68, 1127–1150. [Chapter 2]

———— (2001a), "A General Class of Adaptive Strategies," *Journal of Economic Theory*, 98, 26–54. [Chapter 3]

———— (2001b), "A Reinforcement Procedure Leading to Correlated Equilibrium," in *Economic Essays: A Festschrift for Werner Hildenbrand*, ed. by G. Debreu, W. Neuefeind, and W. Trockel. New York: Springer, pp. 181–200.

Hofbauer, J. (2000), "From Nash and Brown to Maynard Smith: Equilibria, Dynamics and ESS," *Selection* 1, 81–88.

Hofbauer, J. and W. H. Sandholm (2002), "On the Global Convergence of Stochastic Fictitious Play," *Econometrica*, 70, 2265–2294.

Rosenthal, R. W. (1973), "A Class of Games Possessing Pure Strategy Nash Equilibria," *International Journal of Game Theory*, 2, 65–67.

Sandholm, W. H. (2002), "Potential Dynamics and Stable Games," University of Wisconsin (mimeo).

Chapter 6

GENERAL PROCEDURES LEADING
TO CORRELATED EQUILIBRIA*

Amotz Cahn

Hart and Mas-Colell (2000) show that if all players play "regret matching" strategies, i.e., they play with probabilities proportional to the regrets, then the empirical distribution of play converges to the set of correlated equilibria, and the regrets of every player converge to zero. Here we show that if only one player, say player i, plays with these probabilities, while the other players are "not too sophisticated," then the result that player i's regrets converge to zero continues to hold. The condition of "not too sophisticated" essentially says that the effect of one change of action of player i on the future actions of the other players decreases to zero as the horizon goes to infinity. Furthermore, we generalize all these results to a whole class of "regret-based" strategies introduced in Hart and Mas-Colell (2001). In particular, these simplify the "conditional smooth fictitious play" of Fudenberg and Levine (1999).

1. Introduction

A game G of N players is a triplet $\langle N, \{S^i\}_{i \in N}, \{u^i\}_{i \in N} \rangle$, where N is the set of players, S^i is the set of strategies of player i, and $u^i : \prod_{j \in N} S^j \to \mathbb{R}$ is the payoff function of player i. All sets of players and strategies are finite. Denote by $S := \prod_{i \in N} S^i$ the set of strategies of all players, and by $S^{-i} := \prod_{j \neq i, j \in N} S^j$ the set of strategies of all players different from i. Denote by ΔS^i the set of probabilities on S^i (similarly ΔS is the set of

Originally published in *International Journal of Game Theory*, 33 (2004), 21–40.

*Based on the author's M.Sc. thesis (2000), written under the supervision of Sergiu Hart.

Keywords: adaptive procedures; correlated equilibrium; regret matching; adaptive heuristics; fictitious play.

The author thanks Professor Sergiu Hart for his help and guidance, and the Associate Editor and an anonymous referee for their comments. I am grateful to my parents and wife for everything.

probabilities on S). A strategy $s^i \in S^i$ is a pure strategy of player i, and the elements of ΔS^i are the mixed strategies of player i. When dealing with mixed strategies one often is interested not in the actual payoff received, but rather in the expected payoff using those mixed strategies.

One can also consider a situation where the game G is repeated over and over again. In this situation the strategy the players play at time t is denoted s_t (and the strategy of player i is correspondingly s_t^i). In this paper we study repeated games, and their relations to the solution concepts of the one-shot game.

1.1. *Correlated equilibria*

In a two-player zero-sum game there exists the value of the game, which is the payoff that the players can ensure they will get, and optimal strategies, which ensure this payoff. In a non-zero-sum game and in a game with more than two players, we cannot talk about a value of the game since such a value does not exist. Instead of value and optimal strategies one considers equilibrium. Equilibrium is a vector of strategies, such that no player will increase his payoff by unilaterally changing his strategy. The leading non-cooperative equilibrium notion for N-person games in strategic (normal) form is Nash equilibrium, which is an N-tuple of probabilities on S^i, such that no player can increase his payoff by unilaterally changing his action.

The notion of Nash equilibrium has been generalized by Aumann (1974), who introduced the concept of *correlated equilibrium*. Assume that, before the game is played, every player receives a private signal (which does not affect the payoffs). The player may (but need not) choose his action in the game depending on this signal. A correlated equilibrium of the original game is just a Nash equilibrium of the game with the signals. If the signals are (stochastically) independent across the players, this is just a Nash equilibrium (in mixed or pure strategies) of the original game. But the signals could well be correlated, in which case new equilibria may obtain.

Equivalently, a correlated equilibrium is a probability distribution on N-tuples of actions, which can be interpreted as the distribution-of-play instructions given to the players by some "device" or "referee." Every player is given—privately—instructions for his play only; the joint distribution is known to all of them. Also, for every possible instruction that a player receives, the player realizes that the instruction provides a best response to the random estimated play of the other players—assuming they all follow their instructions.

Finally, one can think of the set of correlated equilibria as the following subset of ΔS, the set of probability distributions on N-tuples of actions: $x \in \Delta S$ is a correlated equilibrium if for any random variable $Y = (Y^i)_{i \in N}$ (Y^i with values in S^i) such that[1] $Y \sim x$, the following holds for all $i \in N$, and for all $s^i \in S^i$ such that $\Pr(Y^i = s^i) > 0$:

$$E[u^i(s^i, Y^{-i}) \mid Y^i = s^i] = \max_{j \in S^i} E[u^i(j, Y^{-i}) \mid Y^i = s^i]. \qquad (1.1)$$

1.2. *Regrets*

A player may be sorry because he played one way instead of another. We can quantify how sorry he is by the difference between the payoff he would have gotten had he played differently, and the payoff he actually received. This difference between payoffs is called the *regret*.

The regret we consider is obtained by comparing the actual payoff received to the payoff one would have gotten by playing another pure strategy. In a one-shot game this regret does not make much sense when dealing with mixed strategies. However, when dealing with a repeated game the picture is different. In a repeated game this regret can be described for player i as the difference between the average payoff of playing k instead of j every time player i played strategy j, and the average payoff of the actually played strategies. We denote this regret by $D_t^i(j, k)$—the *regret at time t of player i from j to*[2] k. This regret is of course a function of how often every strategy $s \in S$ was played, or, put differently, a function of the proportion of play of every strategy $s \in S$. This proportion is called the *empirical distribution of play*, and it depends on the time t of the game. It is denoted by[3] z_t. Notice that z_t is a probability distribution on S, the set of all strategies.[4]

---

[1] The notation $Y \sim x$ means that Y and x have the same probability distribution.
[2] A formula for this is

$$D_t^i(j, k) = \frac{1}{t} \sum_{\tau \leq t; s_\tau^i = j} [u^i(k, s_\tau^{-i}) - u^i(j, s_\tau^{-i})].$$

[3] $z_t(s)$ equals the number of times s was actually played in the first t periods, divided by t.
[4] A way of describing the regret of player i from j to k in terms of z_t is

$$D_t^i(j, k) = \sum_{s \in S: s^i = j} z_t(s)[u^i(k, s^{-i}) - u^i(s)].$$

Hart and Mas-Colell (2000) found an interesting connection (see Section 3 there) between regrets and the set of correlated equilibria (which can be described as a subset of the set of probability distributions on S). They prove: Given any $\varepsilon \geq 0$, let $\{s_t\}_{t=1,2,\ldots}$ be a sequence of plays such that the *limsup* of the regret for every player and every strategy is less than or equal to ε. Then the sequence of empirical distributions z_t converges to the set of correlated ε-equilibria. Furthermore, Hart and Mas-Colell (2000, Theorem A) use Blackwell's (1956) Approachability Theorem to prove that the set of all nonpositive regret vectors is an approachable set for every player. This means that every player has an adaptive strategy such that, no matter what the other players do, all his regrets converge to the nonpositive orthant. However, this strategy is quite complicated, and one must calculate eigenvectors of a different matrix in every period t in order to evaluate it. Therefore Hart and Mas-Colell (2000, Section 2) construct a simple adaptive procedure in which the transition probabilities are linearly proportional to the regrets. This procedure has the property that if all players follow it, then the regrets of every player converge to the nonpositive orthant. (Hence the empirical distribution z_t converges to the set of correlated equilibria.) Nevertheless, as they state, this property holds only if all players follow this procedure. Here (in Section 3) we give weak conditions on the other players' play that suffice for the regrets of player i to converge to the nonpositive orthant.[5] Furthermore, following Hart and Mas-Colell (2001), we generalize (in Section 4) the Hart–Mas-Colell strategy to a larger class of strategies and we prove that the same convergence theorems hold. In particular, in Section 4.5, we strengthen a theorem of Fudenberg and Levine dealing with conditional smooth fictitious play.

2. Preliminaries

2.1. *The Hart–Mas-Colell simple procedure*

As mentioned in the introduction Hart and Mas-Colell develop a simple method that ensures that the empirical distribution of play will converge with probability one to the set of correlated equilibria.

---

[5]Note that a procedure that is totally correlated between players is not of major interest. Players can decide on any Nash equilibrium, and this would be a strategy that leads to equilibria. Hart and Mas-Colell's procedure is not such a procedure, since it is not based on the game played. Yet, our result gives added importance to this procedure.

We describe here the procedure that they developed.

Let G be a game with a finite number of players. Suppose that G is played repeatedly through time: $t = 1, 2, 3, \ldots$. Let s_t^i be the strategy that player i played at time t (and s_t^{-i} the strategy other players played, and s_t the strategy combination of all players at time t). Let $h_t := (s_\tau)_{\tau \leq t}$ be the history of the game until time t.

Let

$$A_t^i(j, k) := 1_{\{s_t^i = j\}}[u^i(k, s_t^{-i}) - u^i(s_t)]$$

be the regret at the specific time t from j to k. The regret $D_t^i(j, k)$ is the average of $A^i(j, k)$, i.e.,

$$D_t^i(j, k) = \frac{1}{t} \sum_{\tau=1}^{t} A_\tau^i(j, k).$$

The positive part of the regret, denoted $R_t^i(j, k)$, is $R_t^i(j, k) := [D_t^i(j, k)]_+$. As we mentioned, the transition probabilities $\pi_t^i(j, k)$ of the *Hart–Mas-Colell strategy* (henceforth HMS) are proportional to the positive part of the regret. Let μ be sufficiently large,[6] let $\pi_t^i(j, k) := (1/\mu) \cdot R_t^i(j, k)$ for $k \neq j \in S^i$, and $\pi_t^i(j, j) := 1 - \sum_{k \in S^i : k \neq j} \pi_t^i(j, k)$.

In the HMS player i plays at time $t + 1$ according to the probabilities

$$\Pr(s_{t+1}^i = s^i \mid h_t) = \pi_t^i(s_t^i, s^i); \tag{2.1}$$

that is, the transition probabilities from one period to the next are linearly proportional to the regrets. Hart and Mas-Colell prove (2000, Section 2) that if all players follow this strategy then with probability one the empirical distribution of play converges to the set of correlated equilibria. The method used in order to prove this is to show that the positive part of the regret converges to zero almost surely for every player and every strategy. (Henceforth, whenever we use the term "regret converges to zero," we mean the positive part, i.e., $R_t^i(j, k) \to_{t \to \infty} 0$ for all $k \neq j \in S^i$.)

However, if other players do not follow HMS then the regret of player i need not converge to zero. In Section 3 we show that with some slight conditions on the other players' play we can still get this convergence.

---

[6]Specifically, μ should be large enough to ensure $\pi_t^i(j, j) > 0$ for every $i \in N$ and every $j \in S^i$.

2.2. *Approachable sets*

Consider a game in strategic form played by a player i against an opponent $-i$ (which can be nature, or another player, or many other players). The action sets are the finite sets S^i for i and S^{-i} for $-i$. The payoff functions are vectors in some Euclidean space. Let a_t be the payoff to player i at time t, and $\bar{a}_t$ be the average payoff to player i up to time t. A set C is called an approachable set for player i if player i can guarantee, no matter what player $-i$ does, that the Euclidean distance, $\text{dist}(\bar{a}_t, C)$, tends to zero almost surely as $t \to \infty$.

Given a game with a scalar payoff u^i, we can look at the vector of regrets of player i in the one-shot game A^i (defined in the previous subsection) as a vector payoff. Now we can consider this vector payoff and ask which sets are approachable. Hart and Mas-Colell (2000, Section 3) prove that the nonpositive orthant (denoted $\mathbb{R}^{m_i}_-$ where $m_i = |S^i|$) is approachable for every player i. (Obviously, other sets are also approachable, e.g., any set that includes $\mathbb{R}^m_-$, which may correspond to correlated ε-equilibria.)

Consider a convex closed set C such that $\mathbb{R}^m_- \subseteq C$ and a mapping $\Lambda \colon \mathbb{R}^m \backslash C \to \mathbb{R}^m$ such that Λ is continuous, integrable, and, for every $x \in \mathbb{R}^m \backslash C$, the vector $\Lambda(x)$ represents a direction from C to x, in the following sense: $\Lambda(x) \cdot x > \Lambda(x) \cdot y$ for all $y \in C$. Hart and Mas-Colell (2001) prove that if a player uses a strategy that guarantees that the one-shot payoffs lie in the C-side of the half-space generated by $\Lambda(x)$ (and not in the x-side), then the average payoff will converge to C almost surely (2001, Theorem 2.1). In Section 4 we prove, similarly to what Hart and Mas-Colell do in their simple adaptive procedure, that we can use a strategy linearly proportional to the direction pointed out by operating Λ on the regret (that is, $\Lambda(D^i_t)$), to get, if all players follow this strategy, a similar result. Furthermore, we show that with some slight conditions on other players' play, we can get the convergence of the regret for player i to this set C, no matter what the other players do.

3. Consistency of the Hart–Mas-Colell Strategy

3.1. *Introduction*

In their paper (2000), Hart and Mas-Colell show a simple adaptive procedure leading to correlated equilibrium, as we described in Section 2.1. However, as they point out in Section 4(d), this procedure is not "conditionally universally consistent."[7] In particular, if only player i follows the procedure we cannot conclude that all his regrets go to zero. In this section we give sufficient conditions on the behavior of the other players, which imply that all regrets of player i will necessarily converge to zero.

3.2. *Main results of this section*

Let G be a game with a finite number of players. When dealing with player i we can always look at G as a two-player game between i and $-i$, where $-i$ is $N \setminus \{i\}$.[8] For any strategy used by $-i$, we can ask: what is the effect of the action actually used by player i at stage t of the game, on the action player $-i$ uses at step $t + w$? We show that if this effect is as small as $f(w)/g(t)$, for some functions f, g such that $g(t) \to_{t \to \infty} \infty$, and player i uses the HMS (2.1), then, no matter what strategy player $-i$ uses, the regrets of player i will converge to zero as time goes to infinity.

Formally, assume that for all $t, w > 0$, given two histories h_{t+w-1} and h'_{t+w-1} such that for every $\tau < t + w$, $\tau \neq t$ we have $s_\tau = s'_\tau$ and for $\tau = t$ we have $s_\tau^{-i} = s'^{-i}_\tau$, $s_\tau^i \neq s'^i_\tau$ (that is, the two histories h_{t+w-1} and h'_{t+w-1} differ only in player i's action at time t), then for all s^{-i} in S^{-i}

$$\left| \Pr(s^{-i}_{t+w} = s^{-i} \mid h_{t+w-1}) - \Pr(s^{-i}_{t+w} = s^{-i} \mid h'_{t+w-1}) \right| \leq \frac{f(w)}{g(t)} \qquad (3.1)$$

for some functions f, g such that $g(t) \to_{t \to \infty} \infty$. (Note that there are no conditions on f; its role is to get a uniform bound for every w.)

What this condition says is that the effect of one change in the action of i on the action of the other players converges to zero as the horizon goes to infinity. (For example, in the HMS this is so since the effect is of the order $1/t$.)

Remarks.

(1) In fact, we need this condition only for w such that $w = o(t)$.
(2) The interdependence between the strategies of the players other than i is irrelevant.
(3) If all players follow HMS as in (2.1), then by Step M3 of the Appendix of Hart and Mas-Colell (2000) $R^i_{t+w}(j, k) - R^i_t(j, k) = O(w/t)$ (and the

---

[7] The result is not guaranteed for a player unless all players play according to this strategy.
[8] We allow other players to be correlated among themselves; hence we may refer to them as one player.

same holds for the corresponding transition probabilities); hence (3.1) holds with f a linear function of w (specifically, $f(w) = 2cw$ where c is the constant for $O(w/t)$ and $g(t) = t$).

Theorem 3.1. *If player i uses the Hart–Mas-Colell simple strategy of (2.1), then the regrets of player i are guaranteed to converge to zero a.s. as $t \to \infty$, for any strategies of the other players that satisfy (3.1).*

3.3. *Proof of theorem*

We shall make use of the following remarks.

3.3.1. *Remarks*

In (3.1) we may assume without loss of generality that:

(*1) $f(w) \geq w$; and f is monotone increasing. (Otherwise, take $F(w) := w + \max_{w' \leq w} f(w')$ instead of f.) We extend f to the entire positive real line in a manner that will make it one-to-one.

(*2) $g(t) \leq t$, $(g(t) \geq 1)$ and g is monotone nondecreasing. (Otherwise, define $G(t) := \min\{t, \min_{x \geq t} g(x)\}$, and take G instead of g.)

(*3) $g(t) - g(t-1) \leq 1/t$. (Otherwise, one can define $\hat{G}(t)$ by $\hat{G}(1) := g(1)$ and $\hat{G}(t+1) := \min\{g(t+1), \hat{G}(t)+1/(t+1)\}$, satisfying $\hat{G}(t) \to_{t \to \infty} \infty$ and $\hat{G}(t) - \hat{G}(t-1) \leq 1/t$, and take $\hat{G}(t)$ instead of g. $[\hat{G}$ satisfies (*1) and (*2) if g does.])

Notice that we require that (3.1) hold for every history. Thus, assume that we have two histories h_{t+w-1} and h'_{t+w-1} such that for every $\tau < t$ we have $s_\tau = s'_\tau$ and for $\tau \geq t$ we possibly have $s^i_\tau \neq s'^i_\tau$, $s^{-i}_\tau = s'^{-i}_\tau$. Then we can define a sequence of histories $h^0_{t+w-1}, h^1_{t+w-1}, \ldots, h^w_{t+w-1}$, such that $h^0_{t+w-1} = h_{t+w-1}$, and h^l_{t+w-1} differs from h^{l+1}_{t+w-1} only in that at stage $t+l$, we have $h^{l+1}_{t+w-1}(s_{t+l}) = s'_{t+l}$. Hence $h^w_{t+w-1} = h'_{t+w-1}$, and

$$\left| \Pr\left(s^{-i}_{t+w} = s^{-i} \mid h_{t+w-1}\right) - \Pr\left(s^{-i}_{t+w} = s^{-i} \mid h'_{t+w-1}\right) \right|$$

$$\leq \sum_{l=0}^{w-1} \left| \Pr\left(s^{-i}_{t+w} = s^{-i} \mid h^l_{t+w-1}\right) \right.$$

$$\left. - \Pr\left(s^{-i}_{t+w} = s^{-i} \mid h^{l+1}_{t+w-1}\right) \right| \leq \sum_{l=0}^{w-1} \frac{f(w-l)}{g(t+l)}.$$

Now since we have assumed that f is monotone increasing, and g is monotone nondecreasing, it follows that:

$$\left|\Pr(s_{t+w}^{-i} = s^{-i} \mid h_{t+w-1}) - \Pr(s_{t+w}^{-i} = s^{-i} \mid h'_{t+w-1})\right| \leq \frac{wf(w)}{g(t)}.$$

If we define $f^*(w) := wf(w)$ instead of $f(w)$ we get the following: there exists f, g such that $g \to \infty$, and for every h_{t+w-1} and h'_{t+w-1} such that for every $\tau < t$ we have $s_\tau = s'_\tau$ and for $\tau \geq t$ we possibly have $s_\tau^i \neq s_\tau'^i$, $s_\tau^{-i} = s_\tau'^{-i}$

$$\left|\Pr(s_{t+w}^{-i} = s^{-i} \mid h_{t+w-1}) - \Pr(s_{t+w}^{-i} = s^{-i} \mid h'_{t+w-1})\right| \leq \frac{f(w)}{g(t)} \qquad (3.2)$$

holds. Henceforth, we assume that the strategy of $-i$ satisfies (3.2), with f and g satisfying (*1)−(*3).

An analogous way of looking at this situation is as follows. If all players follow HMS, then there exists a matrix of transition probabilities $\Pi_t^{i'}$, for every player i' and stage t. For every player i' we have $|\Pi_t^{i'} - \Pi_{t+w}^{i'}| = O(w/t)$; hence

$$\begin{aligned}
&\left|\Pr(s_{t+w}^{-i} = s^{-i} \mid h_{t+w-1})\right. \\
&\quad \left. - \Pr(s_{t+w}^{-i} = s^{-i} \mid h_t, s_{t+1}^{-i}, \ldots, s_{t+w-1}^{-i})\right| = O\left(\frac{w}{t}\right).
\end{aligned}$$

Consider a given process s such that player i follows HMS and the other players do not. Suppose that

$$\begin{aligned}
&\left|\Pr(s_{t+w}^{-i} = s^{-i} \mid h_{t+w-1})\right. \\
&\quad \left. - \Pr(s_{t+w}^{-i} = s^{-i} \mid h_t, s_{t+1}^{-i}, \ldots, s_{t+w-1}^{-i})\right| = O\left(\frac{f(w)}{g(t)}\right)
\end{aligned}$$

for some functions f, g such that $g(t) \to_{t \to \infty} \infty$. We prove that the regrets of i converge to zero almost surely.

3.3.2. Lemma

Before proving the theorem we state a simple lemma.

Lemma 3.2. *Assume that the players $-i$ are using strategies that are independent of player i's moves, that is,*

$$\Pr(s_t^{-i} = s^{-i} \mid h_{t-1}) = \Pr(s_t^{-i} = s^{-i} \mid s_1^{-i}, \ldots, s_{t-1}^{-i}),$$

and that player i uses HMS as in (2.1). *Then the regrets of player i converge to zero a.s. as $t \to \infty$.*

Proof.   The proof is exactly as in Hart and Mas-Colell (2000), except that in Step M4 here we define $\hat{s}_{t+w}$ differently. $\hat{s}_{t+w}$ is defined by $\hat{s}_t := s_t$, and the transition probabilities are:

$$\Pr(\hat{s}_{t+w} = s \mid \hat{s}_t, \dots, \hat{s}_{t+w-1}) = \Pi_t^i(\hat{s}_{t+w-1}^i, s^i)$$
$$\cdot \Pr(s_{t+w}^{-i} = s^{-i} \mid h_t, \hat{s}_{t+1}^{-i}, \dots, \hat{s}_{t+w-1}^{-i}).$$

One can verify that Step M4 is still true with the same proof given in Hart and Mas-Colell (2000). (Given h_t, player $-i$ plays with the same probabilities for $\hat{s}$ and s; therefore

$$\left| \Pr(\hat{s}_{t+w}^{-i} = s^{-i} \mid h_{t+w-1}) - \Pr(s_{t+w}^{-i} = s^{-i} \mid h_{t+w-1}) \right| = 0;$$

and for player i we still have

$$\left| \Pr(\hat{s}_{t+w}^i = s^i \mid h_{t+w-1}) - \Pr(s_{t+w}^i = s^i \mid h_{t+w-1}) \right| = \mathrm{O}\left(\frac{w}{t}\right);$$

hence one can use the same proof.) A similar statement is true also for Step M5 of Hart and Mas-Colell (2000). Step M6 is also true, and the proof remains unchanged. (Notice that in the proof of M6 we are using only the independence of i and $-i$, which still holds, and the fact that player i uses Π_t^i as transition probabilities.) The last step involving the $\hat{s}$ strategies is M7, which involves only the stochastic matrix Π_t^i, and it is obviously still true; therefore the continuation of the proof as in Hart and Mas-Colell (2000) is still valid. $\square$

We now can prove our theorem.

3.3.3. *Proof of Theorem 3.1*

This proof follows Hart and Mas-Colell (2000), and we use the same steps. We use lowercase letters to distinguish steps in our proof from those in theirs.

- Steps M1, M2, M3 of Hart and Mas-Colell (2000) are generally true, independently of the strategies used.

Define $\hat{s}_{t+w}$ as in the proof of Lemma 3.2.

- **Step m4.** $\left| \Pr(\hat{s}_{t+w} = s \mid h_t) - \Pr(s_{t+w} = s \mid h_t) \right| = \mathrm{O}\left(\frac{w f(w)}{g(t)}\right).$

We use Hart and Mas-Colell's lemma in the Proof of Step M4 (2000). Since for every player, the transition probability for the $\hat{s}$ process differs from the corresponding one for the s process by at most $O(f(w)/g(t))$ (for player i it differs by $O(w/t)$ which is $\leq O(f(w)/g(t))$ by (*1) and (*2)), it follows that

$$|\Pr(\hat{s}_{t+w} = s \mid h_t) - \Pr(s_{t+w} = s \mid h_t)|$$

$$= \sum_{w' \leq w} O\left(\frac{f(w')}{g(t)}\right) = O\left(\frac{wf(w)}{g(t)}\right).$$

(The last equality follows since f is increasing.)

- **Step m5.** $\left|\alpha_{t,w}(j, s^{-i}) - \widehat{\alpha}_{t,w}(j, s^{-i})\right| = O\left(\frac{wf(w)}{g(t)}\right).$

This is immediate by Step m4.

- **Step m6.** $\widehat{\alpha}_{t,w}(j, s^{-i}) = \Pr(\hat{s}_{t+w}^{-i} = s^{-i} \mid h_t)[\Pi_t^{w+1} - \Pi_t^w](s_t^i, j).$

The proof of Hart and Mas-Colell (2000, Step M6) holds, since, by definition of the $\hat{s}$ process, the transitions of i and $-i$ are independent.

- **Step m7.** $\widehat{\alpha}_{t,w}(j, s^{-i}) = O(w^{-1/2}).$

The proof is the same as in Hart and Mas-Colell (2000, Step M7).

- **Step m8.** $\mathrm{E}\left[(t+v)^2 \rho_{t+v} \mid h_t\right] \leq t^2 \rho_t + O\left(\frac{tv^2 f(v)}{g(t)} + tv^{1/2}\right).$

By Steps m5, m7, and M2, it follows that

$$\sum_{w=1}^{v} R_t \cdot \mathrm{E}[A_{t+w} \mid h_t] = \sum_{w=1}^{v} O\left(\frac{wf(w)}{g(t)} + w^{-1/2}\right)$$

$$= O\left(\frac{v^2 f(v)}{g(t)} + v^{1/2}\right).$$

Substituting into M1(i) yields the result. (Note that the term $O(v^2)$ is not needed since $tv^2 f(v)/g(t) \geq v^2$.)

Now let t_n be an increasing sequence of positive integers (to be defined later), and let $v_n := t_{n+1} - t_n$. Then

- **Step m9.1.** $\mathrm{E}[t_{n+1}^2 \rho_{t_{n+1}} \mid h_{t_n}] \leq t_n^2 \rho_{t_n} + O\left(\frac{t_n v_n^2 f(v_n)}{g(t_n)} + t_n v_n^{1/2}\right).$

Step m9.1 follows immediately from m8.

Let $\widetilde{f}(w) := w^2 f(w)$. (Notice that $\widetilde{f}$ is a continuous strictly increasing function, and thus $\widetilde{f}$ has an inverse function, denoted by $\widetilde{f}^{-1}$.)

Let $a_n := \frac{1}{2}\widetilde{f}^{-1}(g(n))$, and let $t_n := \lceil na_n \rceil$; $(v_n = t_{n+1} - t_n)$.

- **Step m9.2:**

  (*i*) a_n is a nondecreasing sequence, and $a_n \to_{n\to\infty} \infty$.

  (*ii*) $\frac{v_n}{t_n} = O(n^{-1})$.

  (*iii*) $\widetilde{f}(v_n)/g(t_n) = O(1)$.

  (*iv*) $\frac{t_n v_n^2 f(v_n)}{g(t_n)} + t_n v_n^{1/2} = O(na_n^{3/2})$.

Proof. (*i*) is immediate since all the functions involved are increasing and go to infinity.

(*ii*) By (*3) we have $g(t) - g(t-1) \leq 1/t$; also, for $x \geq y \geq 1$ we have $\widetilde{f}^{-1}(x) - \widetilde{f}^{-1}(y) \leq x - y$ since f is increasing and greater than or equal to 1 by (*1). Hence

$$2\left(a_n - a_{n-1}\right) \leq g\left(n\right) - g\left(n-1\right) \leq \frac{1}{n}.$$

Thus

$$\frac{v_n}{t_n} \leq \frac{1 + (n+1)\,a_{n+1} - na_n}{na_n}$$

$$\leq \frac{1 + (n+1)\left(a_n + \frac{1}{2(n+1)}\right) - na_n}{na_n} = \frac{a_n + \frac{3}{2}}{na_n} = O\left(\frac{1}{n}\right),$$

as claimed.

(*iii*) By (*i*) there exists an n_0 such that $a_n \geq 3/2$ for all $n > n_0$. First, we have $v_n \leq 2a_n$ for all $n > n_0$. Indeed,

$$v_n - 2a_n \leq 1 + (n+1)a_{n+1} - na_n - 2a_n$$

$$\leq 1 + (n+1)\left(a_n + \frac{1}{2(n+1)}\right) - (n+2)a_n = 1.5 - a_n \leq 0.$$

Thus $\widetilde{f}(v_n) \leq \widetilde{f}(2a_n) = g(n) \leq g(t_n)$ since $t_n \geq n \cdot 1.5$, which yields the result.

(*iv*) By (*iii*), the first term is $O(t_n)$; thus, in total we have $O(t_n v_n^{1/2})$, which by (*ii*) is $O(t_n(t_n/n)^{1/2}) = O(na_n^{3/2})$.

- **Step m10.** $\lim_{n\to\infty} \rho_{t_n} = 0$ a.s.

Proof. Define $b_n := t_n^2 \approx n^2 a_n^2$ and $X_n := b_n \rho_{t_n} - b_{n-1}\rho_{t_{n-1}} = t_n^2 \rho_{t_n} - t_{n-1}^2 \rho_{t_{n-1}}$.

By Step M1 (ii) it follows that $|X_n| \leq O(t_n v_n + v_n^2)$, which by m9.2$(ii)$ equals $O(t_n^2/n)$. Thus

$$\sum_n \frac{1}{b_n^2} \mathrm{Var}(X_n) = \sum_n O\left(\frac{1}{n^2}\right) < \infty.$$

Next, Steps m9.1 and m9.2 imply that

$$\frac{1}{b_n} \sum_{\nu \leq n} \mathrm{E}[X_\nu \mid X_1, ..., X_{\nu-1}] = O(n^{-2} a_n^{-2}) \cdot \sum_{\nu \leq n} O\left(\frac{t_\nu v_\nu^2 f(v_\nu)}{g(t_\nu)} + t_\nu v_\nu^{1/2}\right)$$

$$= O(n^{-2} a_n^{-2}) \cdot \sum_{\nu \leq n} O(\nu a_\nu^{1.5})$$

$$= O(a_n^{-0.5}) \to_{n \to \infty} 0.$$

(For the first equality we use m9.1, for the second equality we use m9.2(iv), and for the third we use $a_\nu \leq a_n$ for $\nu \leq n$ which follows by m9.2(i), and $\sum_{\nu \leq n} \nu = O(n^2)$.)

Applying the Strong Law of Large Numbers for Dependent Random Variables yields that ρ_{t_n}, which is nonnegative and equals $(1/b_n) \sum_{\nu \leq n} X_\nu$, must converge to 0 a.s. (Here and in the sequel, whenever we mention the Strong Law of Large Numbers for Dependent Random Variables, we refer to Theorem 32.1.E of Loève (1978), also quoted in Step M10 of Hart and Mas-Colell 2000.)

- **Step m11.** $\lim_{t \to \infty} R_t(j, k) = 0$ a.s.

The proof in Hart and Mas-Colell (2000) applies.

3.4. *Counterexample*

We exhibit an example where the conditions fail and the regrets do not converge to zero.

Let $M > 0$ be as large as desired, and $1 > \rho > 0$ be as small as desired. We show that even if we demand that player i's action have no effect on the strategy of player $-i$ for the following $M - 1$ periods after it is played, and that after M periods the effect be no larger than ρ, yet we can construct an example where the regrets of player i will not converge to zero almost surely.

Example 3.3. Consider a two-player game in which the payoff matrix of player 1 is:

	L	R
T	1	0
B	0	1

Let $M > 0$ be as large as desired, and $1 > \rho > 0$ be as small as desired. Let player 1 use HMS (2.1), and let player 2 use the following strategy:

$$\Pr\left[s_t^2 = R \mid h_t\right] = \begin{cases} 1, & \text{if } s_{t-1}^2 = R \text{ and } s_{t-M}^1 = T, \\ 1 - \rho, & \text{if } s_{t-1}^2 = R \text{ and } s_{t-M}^1 = B, \\ \rho, & \text{if } s_{t-1}^2 = L \text{ and } s_{t-M}^1 = T, \\ 0, & \text{if } s_{t-1}^2 = L \text{ and } s_{t-M}^1 = B, \end{cases}$$

for all $t > M$ (and arbitrary for $t \le M$). That is, player 2 changes his action with probability ρ to the worst reply (from the point of view of player 1) to player 1's action M periods ago.

Note that in our example for all $t, w > 0$

$$\left|\Pr(s_{t+w}^{-i} = s^{-i} \mid h_{t+w-1}) - \Pr(s_{t+w}^{-i} = s^{-i} \mid h'_{t+w-1})\right| = \begin{cases} 0, & \text{if } w \ne M, \\ \rho, & \text{if } w = M, \end{cases}$$

for any two histories h_{t+w-1} and h'_{t+w-1} such that for every $\tau < t + w$, $\tau \ne t$ we have $s_\tau = s'_\tau$ and for $\tau = t$ we have $s_\tau^{-i} = s_\tau^{-i}$, $s_\tau^i \ne s_\tau^i$ (that is, the two histories h_{t+w-1} and h'_{t+w-1} differ only in i's action at time t).

However, $O(\frac{f(M)}{g(t)}) \to 0$ for any g such that $g \to \infty$; hence our condition (3.1) is not fulfilled.

Claim: *The regrets of player 1 do not converge to zero with probability one.*

Proof. Suppose that the regrets of player 1 do converge to zero with probability one; we shall show that this leads to a contradiction.

For all positive integers $t, k > 0$ we have

$$\Pr[s_{t+k+M}^2 = R \mid s_{t+M}^2 = R \text{ and } s_{t+i}^1 = T \text{ for } i = 1, \dots, k] = 1,$$

and

$$\Pr[s_{t+k+M}^2 = R \mid s_{t+M}^2 = L \text{ and } s_{t+i}^1 = T \text{ for } i = 1, \dots, k] = 1 - (1 - \rho)^k$$

(since, once a switch from L to R occurs, s^2 remains at R; the probability of no switch, thus always L, is $(1 - \rho)^k$). Hence

$$\Pr[s_{t+k+M}^2 = R \mid s_{t+i}^1 = T \text{ for } i = 1, \dots, k] \ge 1 - (1 - \rho)^k. \qquad (3.3)$$

The same argument implies that for any positive integer $c > k$

$$\Pr[s_{t+M+j}^2 = R \text{ for } j = k, \dots, c \mid s_{t+i}^1 = T \text{ for } i = 1, \dots, c] \ge 1 - (1 - \rho)^k, \qquad (3.4)$$

since once player 2 is at R he will not switch to L. The same argument obviously holds for the pair of actions (B, L), replacing T by B and R by L; hence

$$\Pr[s_{t+M+j}^2 = L \text{ for } j = k, \ldots, c \mid s_{t+i}^1 = B \text{ for } i = 1, \ldots, c] \geq 1 - (1-\rho)^k. \tag{3.5}$$

Let $\varepsilon > 0$, let k be such that $(1-\rho)^k < \varepsilon$, and let c be such that $2(M+k)/c < \varepsilon$. We divide time into blocks of length c. Let

$$H_1 := \{1, 2, \ldots, c\}, \; H_2 := \{c+1, c+2, \ldots, 2c\}, \ldots.$$

Let

$$X_i := \frac{1}{c} \sum_{v \in H_i : s_v^1 = T} [u^1(B, s_v^2) - u^1(s_v^1, s_v^2)], \quad i = 1, 2, 3, \ldots$$

(note that $|X_i| \leq 1$).[9] Similarly, define

$$Y_i := \frac{1}{c} \sum_{v \in H_i : s_v^1 = B} [u^1(T, s_v^2) - u^1(s_v^1, s_v^2)], \quad i = 1, 2, 3, \ldots.$$

Suppose no change has occurred in player 1's play in block H_i (denote this event by $\widetilde{H_i}$); then by (3.4) and (3.5) we get

$$E[X_i + Y_i \mid \widetilde{H_i}] > (1-\varepsilon)^2 - \varepsilon > 1 - 3\varepsilon. \tag{3.6}$$

Indeed, if the action of player 1 had been T (denote this event by $\widetilde{H_{i,T}}$), then by (3.4) $\Pr[s_{t+M+j}^2 = R \text{ for } j = k, \ldots, c] \geq 1 - (1-\rho)^k > 1 - \varepsilon$. Therefore

$$E[X_i + Y_i \mid \widetilde{H_{i,T}}] > \left(\frac{c - (M+k)}{c} - \frac{M+k}{c} \right)$$
$$\times \Pr[s_{t+M+j}^2 = R \quad \text{for } j = k, \ldots, c \mid \widetilde{H_{i,T}}]$$
$$- (1 - \Pr[s_{t+M+j}^2 = R \quad \text{for } j = k, \ldots, c \mid \widetilde{H_{i,T}}])$$
$$> (1-\varepsilon)^2 - \varepsilon > 1 - 3\varepsilon.$$

We can now check the frequency of no change in player 1's action. The probability of a change from t to $t+1$ is either $(1/\mu)R_t^1(T, B)$ or

---

[9]Since there are only two strategies for player 1 we can conclude that $X_i = (1/c) \sum_{v \in H_i^1} [u(B, s_{t_i+v}^2) - u^1(s_{t_i+v}^1, s_{t_i+v}^2)]$, which is the ith-block Hannan (1957) regret, which in this case coincides with the Hart–Mas-Colell regret.

$(1/\mu)R_t^1(B,T)$, hence less than $(1/\mu)R_t^1(T,B) + (1/\mu)R_t^1(B,T)$. Therefore in a block of length c the probability of a change is no more than

$$c \cdot \frac{1}{\mu} \max_{t \in H_i} \{R_t^1(T,B) + R_t^1(B,T)\}.$$

Now since $R_t^1(T,B)$ and $R_t^1(B,T)$ converge to zero a.s. there exists i_0 such that

$$\Pr\left[\frac{1}{\mu} \max_{t \in H_i} \{R_t^1(T,B) + R_t^1(B,T)\} < \frac{\varepsilon}{c}\right] > 1 - \varepsilon$$

for all $i > i_0$. Hence the probability of a change in player 1's action in block H_i is less than $c \cdot \varepsilon/c \cdot (1 - \varepsilon) + \varepsilon < 2\varepsilon$. Therefore

$$E[X_i + Y_i] \geq E[X_i + Y_i \mid \widetilde{H_i}]\Pr(\widetilde{H_i}) - (1 - \Pr(\widetilde{H_i}))$$
$$> (1 - 2\varepsilon)(1 - 3\varepsilon) - 2\varepsilon > 1 - 7\varepsilon$$

for all $i > i_0$. We can easily choose ε such that $1 - 7\varepsilon > 0.9$; hence the average satisfies

$$\liminf_{n \to \infty} E[\overline{X_n + Y_n}] > 0.9.$$

But $\liminf_{n \to \infty} E[\overline{X_n + Y_n}]$ is less than or equal to $\lim_{t \to \infty} R_t^1(T,B) + R_t^1(B,T)$, which contradicts our assumption. □

- An interesting question is, what happens if our $g(t)$ does not converge to infinity but has a subsequence that converges to infinity. Can one still get convergence of the regrets to zero? (The answer probably depends on how dense this subsequence is.)
- Another interesting question is, what happens if we do not have f,g, as in (3.1) but only require that for any w:

$$\left|\Pr(s^{-i} \mid h_{t+w-1}) - \Pr(s^{-i} \mid h_t, s_{t+1}^{-i}, \dots, s_{t+w-1}^{-i})\right| \to_{t \to \infty} 0.$$

Can one still get convergence of the regrets to zero?

4. A General Class of Simple Adaptive Procedures

4.1. *Introduction*

As explained in Section 2.2 above, Hart and Mas-Colell (2001) exhibit a class of adaptive strategies that have a convergence property. Specifically, they define a function Λ defined on all $\mathbb{R}^m$ except for a closed and convex approachable set C, which defines "directions" from this set C. Using a strategy which follows Λ in some sense causes the average payoffs to converge to the set C. Applying this to the setup of conditional regrets—see Hart and Mas-Colell (2001, Section 5.1)—yields strategies that require the computation of an eigenvector at every step. The question raised in Hart and Mas-Colell (2001) is whether we can find a simple adaptive procedure with reference to Λ, in the same way as is done in Hart and Mas-Colell (2000) (which corresponds to the special case of the l_2-potential). In this section we show that the answer to this question is in the affirmative.

4.2. *The model*

Consider a game in strategic form played by a finite set of players N, each having a finite set of strategies S^i. Fix a player i, let $m^i := |S^i|$ and $L^i := \{(j,k) : j \neq k \text{ and } j,k \in S^i\}$. (From now on we omit i whenever it is obvious that we are dealing with player i.) Let K' be a closed bounded set in $\mathbb{R}^L$ containing in its interior the set of all possible payoffs of player i; we can without loss of generality assume that $0 \in K'$ (as in Section 2.2, we view the vector of regrets of player i in the one-shot game as his payoff vector). Let $K := K' + (-K')$; note that K is compact. Let $C \subseteq \mathbb{R}^L$ be a closed convex set[10] such that $C \supseteq \mathbb{R}^L_-$. Let $w : \mathbb{R}^L \to \mathbb{R}$ be defined by $w(x) := \sup_{y \in C}\{x \cdot y\}$. Notice that since $C \supseteq \mathbb{R}^L_-$,

$$w(x) \geq 0 \quad \text{if } x(j,k) \geq 0 \text{ for all } j \neq k, \quad \text{and } w(x) = \infty \quad \text{otherwise.} \quad (4.1)$$

Let Λ and P be as in Hart and Mas-Colell (2001, Section 2); i.e., $\Lambda : \mathbb{R}^L \backslash C \to \mathbb{R}^L$, and $P : \mathbb{R}^L \to \mathbb{R}$ but with the following slightly stronger conditions:[11]

(D1) Λ is Lipschitz on $K \backslash C$.

(D2) P is differentiable; ∇P is Lipschitz on K; and $\nabla P(x) = \phi(x)\Lambda(x)$ for almost every $x \notin C$, where $\phi : \mathbb{R}^L \backslash C \to \mathbb{R}_{++}$ is a continuous positive function.

(D3) $\Lambda(x) \cdot x > w(\Lambda(x))$ for all $x \notin C$.

(D4) Λ can be extended to a Lipschitz function on[12] K.

---

[10]Using a general set C, rather than $\mathbb{R}^L_-$, allows us to handle strategies like Fudenberg and Levine's (1995, 1998, 1999) smooth fictitious play. This will be discussed later in Section 4.5.

[11]The change is that in both (D1) and (D2) we require Λ and ∇P to be Lipschitz rather than just continuous, and we have added (D4).

[12]This added condition is not very strong. In most cases we have a trivial extension to Λ, since ∇P is proportional to Λ (notice that ∇P is Lipschitz on K). This condition is needed since, unlike Hart and Mas-Colell (2001), we want a simple procedure in which probabilities are proportional to Λ; therefore Λ should be defined globally.

Notice that (D1) and (D2) imply $\nabla P(x) = \phi(x)\Lambda(x)$ for every $x \notin C$. By (4.1), (D2), and (D3), for every $x \notin C$ we have $\Lambda(x)(j,k) \geq 0$ and $\nabla P(x)(j,k) \geq 0$ for all $j \neq k$; therefore we can w.l.o.g. also assume in (D4) $\Lambda(x)(j,k) \geq 0$ on[13] K.

Define $a_t^i, d_t^i, \lambda_t^i \in \mathbb{R}^L$ by

$$a_t^i(j,k) := \begin{cases} 0, & \text{if } s_t^i \neq j, \\ u^i(k, s_t^{-i}) - u^i(j, s_t^{-i}), & \text{if } s_t^i = j, \end{cases}$$

$$d_t^i := \frac{1}{t}\sum_{v=1}^{t} a_v^i,$$

$$\lambda_t^i := \Lambda(d_t^i).$$

Notice that $\lambda_t^i(j,k) \geq 0$ for all j, k. Since a_t^i, d_t^i lie in the compact set K^L and Λ is continuous, it follows that λ_t^i is bounded.

Let $\mu > 0$ be large enough for the following π_t^i to be a probability function such that[14] $\pi_t^i(j,j) > 0$. For every $j \in S^i$, let

$$\pi_t^i(j,k) := \frac{1}{\mu}\lambda_t^i(j,k) \quad \text{for } k \neq j \text{ and } \pi_t^i(j,j) := 1 - \sum_{k \in S^i : k \neq j} \pi_t^i(j,k)$$

$$(4.2)$$

be the transition probabilities from stage t to $t+1$ (we can let the probabilities π_0^i of the first move s_1^i be arbitrary). Notice that these probabilities, the set C, and the functions Λ and P are defined separately for every different player; for convenience, we drop the superscript i when it is clear.

Now assume that for every player i' there is a set $C^{i'}$ and functions $P^{i'}$ and $\Lambda^{i'}$. We show that if all players use these strategies (which are similar to HMS) as in Hart and Mas-Colell (2000), then the regrets of every player i' will converge to his set $C^{i'}$. Furthermore, as in Section 3, we show that even if the other players do not follow this strategy, but change their actions with only slight connection to player i's actions, then player i's regrets converge to C^i.

Theorem 4.1. *If every player i uses the strategy given in (4.2), then $d_t^i \rightarrow_{t\to\infty} C^i$ a.s. for every player i.*

Suppose now that $-i$ does not follow the above procedure but, as in the previous section, $-i$ plays in a way that given two histories h_t and h_t'

---

[13] One can take $\Lambda^*(j,k) := \max\{0, \Lambda(j,k)\}$ instead of Λ.

[14] Any μ greater than $\max_{x \in K}\sum_{k \neq j}\Lambda_{(k,j)}(x)$ will suffice.

such that for every $\tau \neq t - w$ we have $s_\tau = s'_\tau$ and for $\tau = t - w$ we have $s_\tau^{-i} = s_\tau'^{-i}$ and possibly $s_\tau^i \neq s_\tau'^i$, then

$$\left| \Pr(s_{t+1}^{-i} = s^{-i} \mid h_t) - \Pr(s_{t+1}^{-i} = s^{-i} \mid h_t') \right| \leq \frac{f(w)}{g(t)} \qquad (4.3)$$

for some functions f, g such that $g(t) \to_{t \to \infty} \infty$.

Theorem 4.2. *If player i uses strategy (4.2), i.e., $\Pr(s_{t+1}^i = s^i \mid h_t) = \pi_t^i(s_t^i, s^i)$, then it is guaranteed that $d_t^i \to_{t \to \infty} C^i$ a.s. for any strategies of the other players satisfying (4.3).*

4.3. *Proof of Theorem 4.1*

Lemma 2.3 of Hart and Mas-Colell (2001) shows that there exists a constant c such that

$$P(x) = c, \quad \text{if } x \in \partial C \ (= \text{the boundary of } C),$$
$$P(x) > c, \quad \text{if } x \notin C.$$

Choose $\varepsilon > 0$. We shall show that $\Pr(\limsup P(d_t) \leq c + \varepsilon) = 1$, and since this is true for every ε, it follows that $\Pr(\limsup P(d_t) \leq c) = 1$, which is equivalent to $d_t \to_{t \to \infty} C$ a.s. since P is continuous.

Let P_1 be as in Hart and Mas-Colell (2001, Section 2.2), namely, $P_1(x) := [P(x) - c]^2$ for every $x \notin C$ and $P_1(x) := 0$ for $x \in C$. Notice that

$$\nabla P_1(x) = 2 \nabla P(x)[P(x) - c].$$

Let $Q : \mathbb{R}^L \to \mathbb{R}$ be as in Hart and Mas-Colell (2001, Section 2.2, Proof of Theorem 2.1), i.e.,

$$Q(x) \geq 0 \quad \text{for all } x \in \mathbb{R}^L, \quad \text{and} \quad Q(x) = 0 \quad \text{if and only if} \quad P_1(x) \leq \varepsilon; \qquad (4.4)$$

$$\nabla Q(x) \cdot x - w(\nabla Q(x)) \geq Q(x); \text{ and} \qquad (4.5)$$

$$\nabla Q(x) = \begin{cases} 0, & \text{if } Q(x) = 0, \\ r(P_1(x) - \varepsilon)^{r-1} 2 \nabla P(x)[P(x) - c], & \text{otherwise.} \end{cases} \qquad (4.6)$$

Let $y(x) := r[P_1(x) - \varepsilon]_+^{r-1} 2[P(x) - c]$; now by (4.4) we can write

$$\nabla Q(x) = y(x) \cdot \nabla P(x). \qquad (4.7)$$

Notice that by (4.1) it follows that (4.5) can be written as:

$$\nabla Q(x)(j, k) \geq 0 \quad \text{for all } j \neq k, \quad \text{and} \quad \nabla Q(x) \cdot x \geq Q(x). \qquad (4.8)$$

Let $q_t := \nabla Q(d_t)$. Notice that by (D2) and the fact that $q_t = 0, \lambda_t = 0$ for all $x \in C$ it follows that

$$q_t = \lambda_t \cdot y(d_t). \tag{4.9}$$

In the sequel, the present proof will be divided into steps similar to those of the Proof of the Main Theorem in Hart and Mas-Colell (2000, Appendix).

4.3.1. *Steps of the Proof*

- **Step N1.**
  (i) $\mathrm{E}[(t+v)Q(d_{t+v}) \mid h_t] \le tQ(d_t) + \sum_{w=1}^{v} \mathrm{E}[a_{t+w} \mid h_t] \cdot q_t + \mathrm{O}\left(\frac{v^2}{t}\right)$.
  (ii) $(t+v)Q(d_{t+v}) - tQ(d_t) = \mathrm{O}(v)$.

Proof.

$$Q(d_{t+v}) = Q\left(\frac{t}{t+v}d_t + \frac{1}{t+v}\sum_{w=1}^{v} a_{t+w}\right)$$

$$= Q(d_t) + \left(\frac{t}{t+v}d_t + \frac{1}{t+v}\sum_{w=1}^{v} a_{t+w} - d_t\right) \cdot \nabla Q(d_t)$$

$$+ \mathrm{O}\left(\frac{t}{t+v}d_t + \frac{1}{t+v}\sum_{w=1}^{v} a_{t+w} - d_t\right)^2.$$

The second equality follows since ∇Q is Lipschitz[15] on K and hence there exists $0 \le t \le 1$ such that

$$Q(x+y) = Q(x) + y \cdot \nabla Q(x+ty)$$
$$\le Q(x) + y \cdot \nabla Q(x) + \|y\|k\|ty\| = Q(x) + y \cdot \nabla Q(x) + \mathrm{O}(\|y\|^2).$$

By (4.8) $\nabla Q(d_t) \cdot d_t \ge Q(d_t)$, hence

$$(t+v)Q(d_{t+v}) = (t+v)Q(d_t) + \left(\sum_{w=1}^{v}(a_{t+w} - d_t)\right) \cdot \nabla Q(d_t) + \mathrm{O}\left(\frac{v^2}{(t+v)}\right)$$

$$\le tQ(d_t) + \sum_{w=1}^{v} a_{t+w} \cdot q_t + \mathrm{O}\left(\frac{v^2}{t+v}\right).$$

Taking expectation yields (i), and since $\mathrm{O}(\frac{v^2}{(t+v)}) \le \mathrm{O}(v)$ and a_{t+w}, q_t are bounded we get (ii). $\square$

---

[15]The Lipschitz constant depends on ε, but this does not affect the proof.

Define

$$\alpha_{t,w}(j, s^{-i}) := \sum_{k \in S^i} \pi_t^i(j, k) \Pr[s_{t+w} = (k, s^{-i}) \mid h_t] - \Pr[s_{t+w} = (j, s^{-i}) \mid h_t].$$

- **Step N2.**

$$\mathrm{E}[a_{t+w} \mid h_t] \cdot q_t = \mu y(d_t) \cdot \sum_{(j,s^{-i}) \in S^i \times S^{-i}} \alpha_{t,w}(j, s^{-i}) u^i(j, s_t^{-i}).$$

Proof. By the same method as in Step M2 of Hart and Mas-Colell (2000) we get:

$$\mathrm{E}[a_{t+w} \mid h_t] \cdot \lambda_t = \sum_{(j,s^{-i}) \in S^i \times S^{-i}} \alpha_{t,w}(j, s^{-i}) u^i(j, s_t^{-i})$$

and (4.9) yields the result. □

- **Step N3.**

  (i) $d_{t+v}(j, k) - d_t(j, k) = \mathrm{O}\left(\frac{v}{t}\right).$
  (ii) $q_{t+v}(j, k) - q_t(j, k) = \mathrm{O}\left(\frac{v}{t}\right).$
  (iii) $\lambda_{t+v}(j, k) - \lambda_t(j, k) = \mathrm{O}\left(\frac{v}{t}\right).$
  (iv) $\pi_{t+v}(j, k) - \pi_t(j, k) = \mathrm{O}\left(\frac{v}{t}\right).$

Proof. Since $d_{t+v} = d_t + \frac{1}{t+v} \sum_{w=1}^{v}(a_{t+w} - d_t)$ and $(a_{t+w} - d_t)$ is bounded, (i) is true. Since by (4.6) ∇Q is Lipschitz on the compact set K, (ii) follows. The fact that Λ is Lipschitz yields (iii) and (iv). □

- **Steps N4–N7** are exactly the same as Steps M4–M7 in Hart and Mas-Colell (2000) based on N3 *(iv)*.

- **Step N8.** $\mathrm{E}[(t + v)Q(d_{t+v}) \mid h_t] \leq tQ(d_t) + \mathrm{O}\left(\frac{v^3}{t} + v^{\frac{1}{2}}\right).$

Proof. Steps N5 and N7 imply that $\alpha_{t,w}(j, s^{-i}) = \mathrm{O}(\frac{w^2}{t} + w^{-0.5})$. The fact that K is compact and that y is continuous yields that y is bounded on K. Therefore the formula of Step N2 yields $\mathrm{E}[a_{t+w} \mid h_t] \cdot q_t = \mathrm{O}(\frac{w^2}{t} + w^{-0.5})$. Summing over w and Step N1 *(i)* yield the result.
  Let $t_n := \lfloor n^{5/3} \rfloor$, and let $v_n := t_{n+1} - t_n = \mathrm{O}(n^{2/3})$.

- **Step N9.** $\mathrm{E}[t_{n+1}Q(d_{t_{n+1}}) \mid h_{t_n}] \leq t_n Q(d_{t_n}) + \mathrm{O}(n^{1/3}).$

Proof. Immediate by Step N8.

- **Step N10.** $\lim_{n\to\infty} Q(d_{t_n}) \to 0$ a.s.

Proof. Define $b_n := t_n \approx n^{5/3}$ and $X_n := b_n Q(d_{t_n}) - b_{n-1} Q(d_{t_{n-1}})$.
By Step N1(ii) we have $|X_n| \leq O(v_n) = O(t_n/n)$, thus

$$\sum_n \frac{1}{b_n^2} \operatorname{Var}(X_n) = \sum_n O(\frac{1}{n^2}) < \infty.$$

Next, by Step N9 we have

$$\frac{1}{b_n} \sum_{\nu \leq n} \mathrm{E}[X_\nu \mid X_1, \ldots, X_{\nu-1}] \leq O(n^{-5/3} \sum_{\nu \leq n} \nu^{1/3}) = O(n^{-1/3}) \to 0.$$

Applying the Strong Law of Large Numbers for Dependent Random Variables yields

$$\frac{1}{b_n} \sum_{\nu \leq n} X_\nu = Q(d_{t_n}) \to_{n\to\infty} 0$$

a.s.

- **Step N11.**

  (i) $\lim_{t\to\infty} Q(d_t) \to 0$ a.s.
  (ii) $\Pr(\limsup P(d_t) \leq c + \varepsilon) = 1$.

Proof.   Since for $t_n < t \leq t_{n+1}$ we have $\frac{t - t_n}{t_n} \leq \frac{v_n}{t_n} = O(n^{-1})$, by N3 we get $Q(d_t) \to_{t\to\infty} 0$ a.s., and by (4.4) and the fact that Q and P are continuous it follows that $\Pr(\limsup P(d_t) \leq c + \varepsilon) = 1$.   □

4.4. *Proof of Theorem 4.2*

It is easy to see that all the claims in Section 3.3.1 about f, g and the histories h, h' hold also in this case. The Proof of Theorem 4.1 can be used for the Proof of Theorem 4.2; the only changes necessary are precisely those that were needed in the Proof of Theorem 3.1 of Section 3. (We use lowercase letters for the steps in this proof.) Namely, in Step n4 define $\hat{s}_{t+w}$ differently. Here $\hat{s}_{t+w}$ will be defined by $\hat{s}_t := s_t$, and the transition probabilities are:

$$\Pr(\hat{s}_{t+w} = s \mid \hat{s}_t, \ldots, \hat{s}_{t+w-1})$$
$$= \pi_t^i(\hat{s}_{t+w-1}^i, s^i) \cdot \Pr(s_{t+w}^{-i} = s^{-i} \mid h_t, \hat{s}_{t+1}^{-i}, \ldots, \hat{s}_{t+w-1}^{-i})$$

and we get results similar to m4–m7 of Section 3. In Step n8 we get

$$\mathrm{E}[(t+v)Q(d_{t+v}) \mid h_t] \le tQ(d_t) + \mathrm{O}\left(\frac{f(v)v^2}{g(t)} + v^{\frac{1}{2}}\right),$$

similar to m8.

We now need Steps n9.1 and n9.2 similar to m9.1 and m9.2. Specifically:

- **Step n9.1.** $\mathrm{E}[t_{n+1}Q(d_{t_{n+1}}) \mid h_{t_n}] \le t_n Q(d_{t_n}) + \mathrm{O}\left(\frac{v_n^2 f(v_n)}{g(t_n)} + v_n^{1/2}\right).$

This follows immediately from n8.

Define $\widetilde{f}(w) := w^2 f(w)$. (Notice that $\widetilde{f}$ has an inverse function, denoted by $\widetilde{f}^{-1}$.) Let $a_n := \frac{1}{2}\widetilde{f}^{-1}(g(n))$, and let $t_n := \lceil na_n \rceil$; $(v_n = t_{n+1} - t_n)$.

- **Step n9.2.**

  (i) a_n is a nondecreasing sequence, and $a_n \to_{n\to\infty} \infty$.
  (ii) $\frac{v_n}{t_n} = \mathrm{O}(n^{-1})$.
  (iii) $\widetilde{f}(v_n)/g(t_n) = \mathrm{O}(1)$.
  (iv) $\frac{v_n^2 f(v_n)}{g(t_n)} + v_n^{1/2} = \mathrm{O}(a_n^{1/2})$.

The only thing different from Step m9.2 is *(iv)*, which follows immediately from *(iii)* and the fact that $v = \mathrm{O}(a_n)$ (proved in Step m9.2).

The rest of the proof is the same[16] as in Theorem 4.1.

---

[16]In Step n10 we get (with $b_n = t_n$)

$$\sum_n \mathrm{Var}(X_n)/b_n^2 = \sum_n \mathrm{O}(n^{-2}) < \infty$$

by Steps n1*(ii)* and n9.2*(ii)*. The following inequality

$$b_n^{-1} \sum_{\nu \le n} \mathrm{E}[X_\nu \mid X_1, \dots, X_{\nu-1}] \le \mathrm{O}\left(n^{-1} a_n^{-1} \sum_{\nu \le n} a_\nu^{0.5}\right)$$

we get by Steps n9.1 and n9.2*(iv)*, and since a_n is increasing

$$\sum_{\nu \le n} a_\nu^{0.5} \le na_n^{0.5};$$

hence

$$b_n^{-1} \sum_{\nu \le n} \mathrm{E}[X_\nu \mid X_1, \dots, X_{\nu-1}] \to 0.$$

4.5. *Conditional smooth fictitious play*

Fictitious play is a strategy where a player plays a best reply to the empirical distribution of play z_t. It may be viewed as a regret-based strategy, corresponding to the l_∞-potential (cf. Hart and Mas-Colell 2001, Section 4.1). However, this strategy does not satisfy the above conditions, specifically condition (D1): Λ is not continuous. Fudenberg and Levine (1995, 1998) present a smoothing of fictitious play. Let $\sigma_{t+1}^i \in \Delta\left(S^i\right)$ denote the (possibly mixed) choice of player i at time $t+1$. Fictitious play requires

$$\sigma_{t+1}^i \in \operatorname*{argmax}_{\sigma^i \in \Delta(S^i)} \{u^i(\sigma^i, z_t^{-i})\}$$

(note that the maximizer is not necessarily unique). In *smooth fictitious play*, we have instead

$$\sigma_{t+1}^i = \operatorname*{argmax}_{\sigma^i \in \Delta(S^i)} \{u^i(\sigma^i, z_t^{-i}) + \nu(\sigma^i)\}$$

where $\nu \equiv \nu^i$ is a smooth strictly differentiably concave function with gradient vector approaching infinite length as one approaches the boundary of $\Delta(S^i)$; hence, the maximizer is unique.

In the conditional case (Fudenberg and Levine 1998, 1999), instead of z_t^{-i} one considers for every $j \in S^i$ the distribution of play $z_t^i(j)$ only in those periods where i played j, i.e.,

$$z_t^{-i}(j)(s^{-i}) := \frac{1}{t}|\{\tau \le t : s_\tau = (j, s^{-i})\}|.$$

Since u^i is linear,

$$\operatorname*{argmax}_{\sigma^i \in \Delta(S^i)} \{u^i(\sigma^i, z_t^{-i}(j)) + \nu(\sigma^i)\} = \operatorname*{argmax}_{\sigma^i \in \Delta(S^i)} \{\sigma^i \cdot D_t^i(j, \cdot) + \nu(\sigma^i)\}$$

(we write $\sigma^i \cdot D_t^i(j, \cdot)$ for $\sum_{k \ne j} \sigma^i(k) \cdot D_t^i(j, k)$).

Therefore we define *conditional smooth fictitious play* as the Hart–Mas-Colell conditional-regret-based strategy (4.2), with $\Lambda(x) := \nabla P(x)$, where

$$P(x) := \sum_{j \in S^i} \max_{\sigma_j^i \in \Delta(S^i)} \left\{ \sum_{k \in S^i, k \ne j} \sigma_j^i(k) \cdot x(j, k) + \nu(\sigma_j^i) \right\}, \quad \text{for all } x \in \mathbb{R}^L$$

$$\tag{4.10}$$

is the corresponding potential. It follows immediately from the definition of a Λ-strategy in Hart and Mas-Colell (2001, Section 2.1) that the argmax is a Λ-strategy. Notice that $\mathrm{argmax}_{\sigma_j^i \in \Delta(S^i)} \{\sigma_j^i \cdot x(j, \cdot) + \nu(\sigma_j^i)\}$ is a smooth (and in particular twice differentiable) function, as shown by Fudenberg and Levine (1999, Section 3); thus P and Λ satisfy properties (D1)–(D4) for[17] $C := \{x \mid P(x) \le m\|\nu\|\}$ (again, see Hart and Mas-Colell 2001, Section 4.1).

Now, by our results we get that if player i plays conditional smooth fictitious play as defined above, and for the other players (4.3) holds, then all the conditional regrets of player i will in the limit be at most $m\|\nu\|$ (a.s.). Formally, this can be written, according to Fudenberg and Levine's notations, as:

Proposition 4.3. *The strategy (4.2), where π_t^i are transition probabilities given by the conditional smooth fictitious play potential (4.10), is $m^i\|\nu^i\|$-calibrated for any strategies of the other players satisfying (4.3). Moreover, if all players play this way, then the empirical distribution of play z_t converges a.s. to the set of correlated ε-equilibria, where $\varepsilon = \max_{i \in N} m^i\|\nu^i\|$.*

The difference between our strategy and that of Fudenberg and Levine is that we do not have to evaluate eigenvectors, as do Fudenberg and Levine, but our probabilities are just proportional to Λ.

References

Aumann, R. J. (1974), "Subjectivity and Correlation in Randomized Strategies," *Journal of Mathematical Economics*, 1, 67–96.

Blackwell, D. (1956), "An Analog of the Minmax Theorem for Vector Payoffs," *Pacific Journal of Mathematics*, 6, 1–8.

Fudenberg, D. and D. K. Levine (1995), "Universal Consistency and Cautious Fictitious Play," *Journal of Economic Dynamics and Control*, 19, 1065–1090.

—— (1998), *Theory of Learning in Games*. Boston, MA: MIT Press.

—— (1999), "Conditional Universal Consistency," *Games and Economic Behavior*, 29, 104–130.

Hannan, J. (1957), "Approximation to Bayes Risk in Repeated Play," in *Contributions to the Theory of Games*, Vol. III, Annals of Mathematics Studies

---

[17]Recall that $m \equiv m^i := |S^i|$.

39, ed. by M. Dresher, A. W. Tucker, and P. Wolfe. Princeton: Princeton University Press, pp. 97–139.

Hart, S. and A. Mas-Colell (2000), "A Simple Adaptive Procedure Leading to Correlated Equilibrium," *Econometrica*, 68, 1127–1150. [Chapter 2]

—— (2001), "A General Class of Adaptive Strategies," *Journal of Economic Theory*, 98, 26–54. [Chapter 3]

Loève, M. (1978), *Probability Theory*, Vol. II, 4th edn. New York: Springer.

Part III

Uncoupled Dynamics

Chapter 7

UNCOUPLED DYNAMICS DO NOT LEAD TO NASH EQUILIBRIUM*

Sergiu Hart and Andreu Mas-Colell

It is notoriously difficult to formulate sensible adaptive dynamics that guarantee convergence to Nash equilibrium. In fact, short of variants of exhaustive search (deterministic or stochastic), there are no general results; of course, there are many important, interesting, and well-studied particular cases. See the books of Weibull (1995), Vega-Redondo (1996), Samuelson (1997), Fudenberg and Levine (1998), Hofbauer and Sigmund (1998), Young (1998), and the discussion in Section 4 below.

Here we provide a simple answer to the question: Why is that so? Our answer is that the lack of a general result is an intrinsic consequence of the natural requirement that dynamics of play be "uncoupled" among the players, that is, the adjustment of a player's strategy does not depend on the payoff functions (or utility functions) of the other players (it may depend on the other players' strategies, as well as on the payoff function of the player himself). This is a basic informational condition for dynamics of the "adaptive" or "behavioral" type.

It is important to emphasize that, unlike the existing literature (see Section 4), we make no "rationality" assumptions: our dynamics are *not* best-reply dynamics, or better-reply, or payoff-improving, or monotonic,

Originally published in *American Economic Review*, 93 (2003), 1830–1836.

*The research is partially supported by grants of the Israel Academy of Sciences and Humanities, the Spanish Ministry of Education, the Generalitat de Catalunya, and the EU-TMR Research Network.

We thank Robert J. Aumann, Yaacov Bergman, Vincent Crawford, Josef Hofbauer, Piero La Mura, Eric Maskin, Motty Perry, Alexander Vasin, Bob Wilson, the referees, and the co-editor for their useful comments.

and so on. What we show is that the impossibility result is due only to an "informational" requirement—that the dynamics be uncoupled.

1. The Model

The setting is that of games in strategic (or normal) form. Such a game Γ is given by a finite set of players N, and, for each player $i \in N$, a strategy set S^i (not necessarily finite) and a payoff function[1] $u^i : \prod_{j \in N} S^j \to \mathbb{R}$.

We examine differential dynamical systems defined on a convex domain X, which will be either $\prod_{i \in N} S^i$ or[2] $\prod_{i \in N} \Delta(S^i)$, and are of the form

$$\dot{x}(t) = F(x(t); \Gamma),$$

or $\dot{x} = F(x; \Gamma)$ for short. We also write this as $\dot{x}^i = F^i(x; \Gamma)$ for all i, where $x = (x^i)_{i \in N}$ and[3] $F = (F^i)_{i \in N}$.

From now on we keep N and $(S^i)_{i \in N}$ fixed, and identify a game Γ with its N-tuple of payoff functions $(u^i)_{i \in N}$, and a family of games with a set $\mathcal{U}$ of such N-tuples; the dynamics are thus

$$\dot{x}^i = F^i(x; (u^j)_{j \in N}) \quad \text{for all } i \in N. \tag{1}$$

We consider families of games $\mathcal{U}$ where every game $\Gamma \in \mathcal{U}$ has a single Nash equilibrium $\bar{x}(\Gamma)$. Such families are the most likely to allow for well-behaved dynamics. For example, the dynamic $\dot{x} = \bar{x}(\Gamma) - x$ will guarantee convergence to the Nash equilibrium starting from any initial condition.[4] Note, however, that in this dynamic $\dot{x}^i$ depends on $\bar{x}^i(\Gamma)$, which, in turn, depends on all the components of the game Γ, in particular on u^j for $j \neq i$. This motivates our next definition.

We call a dynamical system $F(x; \Gamma)$ (defined for Γ in a family of games $\mathcal{U}$) *uncoupled* if, for every player $i \in N$, the function F^i does *not* depend

---

[1] $\mathbb{R}$ denotes the real line.

[2] We write $\Delta(A)$ for the set of probability measures over A.

[3] For a well-studied example (see for instance Hofbauer and Sigmund 1998), consider the class of "fictitious play"-like dynamics: the strategy $q^i(t)$ played by i at time t is some sort of "good reply" to the past play of the other players j, i.e., to the time average $x^j(t)$ of $q^j(\tau)$ for $\tau \leq t$; then (after rescaling the time axis) $\dot{x}^i = q^i - x^i \equiv G^i(x; \Gamma) - x^i \equiv F^i(x; \Gamma)$.

[4] The same applies to various generalized Newton methods and fixed-point-convergent dynamics.

on u^j for $j \neq i$; i.e.,

$$\dot{x}^i = F^i(x; u^i) \quad \text{for all } i \in N \tag{2}$$

(compare with (1)). Thus the change in player i's strategy can be a function of the current N-tuple of strategies x and i's payoff function u^i only.[5] In other words, if the payoff function of player i is identical in two games in the family, then at each x his strategy x^i will change in the same way.[6]

If, given a family $\mathcal{U}$ with the single-Nash-equilibrium property, the dynamical system always converges to the unique Nash equilibrium of the game for any game $\Gamma \in \mathcal{U}$—i.e., if $F(\bar{x}(\Gamma); \Gamma) = 0$ and $\lim_{t \to \infty} x(t) = \bar{x}(\Gamma)$ for any solution $x(t)$ (with any initial condition)—then we will call F a *Nash-convergent* dynamic for $\mathcal{U}$. To facilitate the analysis, we always restrict ourselves to C^1 functions F with the additional property that at the (unique) rest point $\bar{x}(\Gamma)$ the Jacobian matrix J of $F(\cdot; \Gamma)$ is hyperbolic and (asymptotically) stable—i.e., all eigenvalues of J have negative real parts.

We will show that:

There exist no uncoupled dynamics that guarantee Nash convergence.

Indeed, in the next two sections we present two simple families of games (each game having a single Nash equilibrium), for which uncoupledness and Nash convergence are mutually incompatible.

More precisely, in each of the two cases we exhibit a game Γ_0 and show that:[7]

Theorem 1. *Let $\mathcal{U}$ be a family of games containing a neighborhood of the game Γ_0. Then every uncoupled dynamic for $\mathcal{U}$ is not Nash-convergent.*

Thus an arbitrarily small neighborhood of Γ_0 is sufficient for the impossibility result (of course, nonexistence for a family $\mathcal{U}$ implies nonexistence for any larger family $\mathcal{U}' \supset \mathcal{U}$).

---

[5]It may depend on the *function* $u^i(\cdot)$, not just on the current payoffs $u^i(x)$.

[6]What the other players *do* (i.e., x^{-i}) is much easier to observe than *why* they do it (i.e., their utility functions u^{-i}).

[7]An ε-*neighborhood* of a game $\Gamma_0 = (u_0^i)_{i \in N}$ consists of all games $\Gamma = (u^i)_{i \in N}$ satisfying $|u^i(s) - u_0^i(s)| < \varepsilon$ for all $s \in \prod_{i \in N} S^i$ and all $i \in N$.

2. An Example with a Continuum of Strategies

Take $N = \{1, 2\}$ and $S^1 = S^2 = D$, where $D := \{z = (z_1, z_2) \in \mathbb{R}^2 : \|z\| \leq 1\}$ is the unit disk. Let $\phi : D \to D$ be a continuous function that satisfies:

- $\phi(z) = 2z$ for z in a neighborhood of 0; and
- $\phi(\phi(z)) \neq z$ for all $z \neq 0$.

Such a function clearly exists; for instance, let us put $\phi(z) = 2z$ for all $\|z\| \leq 1/3$, define ϕ on the circle $\|z\| = 1$ to be a rotation by, say, $\pi/4$, and interpolate linearly on rays between $\|z\| = 1/3$ and $\|z\| = 1$.

Define the game Γ_0 with payoff functions u_0^1 and u_0^2 given by[8]

$$u_0^i(x^i, x^j) := -\|x^i - \phi(x^j)\|^2 \quad \text{for all } x^i,\ x^j \in D.$$

Γ_0 has a unique Nash equilibrium[9] $\bar{x} = (0, 0)$.

We embed Γ_0 in the family $\mathcal{U}_0$ consisting of all games $\Gamma \equiv (u^1, u^2)$ where, for each $i = 1, 2$, we have $u^i(x^i, x^j) = -\|x^i - \xi^i(x^j)\|^2$, with $\xi^i : D \to D$ a continuous function, such that the equation $\xi^i(\xi^j(x^i)) = x^i$ has a unique solution $\bar{x}^i$. Then $\bar{x} = (\bar{x}^1, \bar{x}^2)$ is the unique Nash equilibrium of the game[10] Γ.

We will now prove that every uncoupled dynamic for $\mathcal{U}_0$ is not Nash-convergent. This proof contains the essence of our argument, and the technical modifications needed for obtaining Theorem 1 are relegated to the Appendix. Let F thus be, by contradiction, a dynamic for the family $\mathcal{U}_0$ which is uncoupled and Nash-convergent. The dynamic can thus be written: $\dot{x}^i = F^i(x^i, x^j; u^i)$ for $i = 1, 2$.

The following key lemma uses uncoupledness repeatedly.

Lemma 2. *Assume that y^i is the unique u^i-best-reply of i to a given y^j, i.e., $u^i(y^i, y^j) > u^i(x^i, y^j)$ for all $x^i \neq y^i$. Then $F^i(y^i, y^j; u^i) = 0$, and the eigenvalues of the 2×2 Jacobian matrix[11] $J^i = (\partial F_k^i(y^i, y^j; u^i)/\partial x_\ell^i)_{k,\ell=1,2}$ have negative real parts.*

---

[8]We use $j := 3 - i$ throughout this section. In the game Γ_0, each player i wants to choose x^i so as to match as closely as possible a function of the other player's choice, namely, $\phi(x^j)$.

[9]$\bar{x}$ is a pure Nash equilibrium if and only if $\bar{x}^1 = \phi(\bar{x}^2)$ and $\bar{x}^2 = \phi(\bar{x}^1)$, or $\bar{x}^i = \phi(\phi(\bar{x}^i))$ for $i = 1, 2$. There are no mixed-strategy equilibria since the best reply of i to any mixed strategy of j is always unique and pure.

[10]Moreover $\bar{x}$ is a strict equilibrium.

[11]Subscripts denote coordinates: $x^i = (x_1^i, x_2^i)$ and $F^i = (F_1^i, F_2^i)$.

Proof. Let Γ_1 be the game (u^i, u^j) with $u^j(x^i, x^j) := -\|x^j - y^j\|^2$ (i.e., ξ^j is the constant function $\xi^j(z) \equiv y^j$); then (y^i, y^j) is its unique Nash equilibrium, and thus $F^i(y^i, y^j; u^i) = 0$. Apply this to player j, to get $F^j(x^i, y^j; u^j) = 0$ for all x^i (since y^j is the unique u^j-best-reply to any x^i). Hence $\partial F_k^j(x^i, y^j; u^j)/\partial x_\ell^i = 0$ for $k, \ell = 1, 2$. The 4×4 Jacobian matrix J of $F(\cdot, \cdot; \Gamma_1)$ at (y^i, y^j) is therefore of the form

$$J = \begin{bmatrix} J^i & K \\ 0 & L \end{bmatrix}.$$

The eigenvalues of J—which all have negative real parts by assumption—consist of the eigenvalues of J^i together with the eigenvalues of L, and the result follows. $\qquad\square$

Put $f^i(x) := F^i(x; u_0^i)$; Lemma 2 implies that the eigenvalues of the 2×2 Jacobian matrix $J^i := (\partial f_k^i(0, 0)/\partial x_\ell^i)_{k,\ell=1,2}$ have negative real parts. Again by Lemma 2, $f^i(\phi(x^j), x^j) = 0$ for all x^j, and therefore in particular $f^i(2x^j, x^j) = 0$ for all x^j in a neighborhood of 0. Differentiating and then evaluating at $\bar{x} = (0, 0)$ gives

$$2\partial f_k^i(0, 0)/\partial x_\ell^i + \partial f_k^i(0, 0)/\partial x_\ell^j = 0 \quad \text{for all } k, \ell = 1, 2.$$

Therefore the 4×4 Jacobian matrix J of the system (f^1, f^2) at $\bar{x} = (0, 0)$ is

$$J = \begin{bmatrix} J^1 & -2J^1 \\ -2J^2 & J^2 \end{bmatrix}.$$

Lemma 3. *If the eigenvalues of J^1 and J^2 have negative real parts, then J has at least one eigenvalue with positive real part.*

Proof. The coefficient a_3 of λ in the characteristic polynomial $\det(J - \lambda I)$ of J equals the negative of the sum of the four 3×3 principal minors; a straightforward computation shows that

$$a_3 = 3 \det(J^1)\text{trace}(J^2) + 3 \det(J^2)\text{trace}(J^1).$$

But $\det(J^i) > 0$ and $\text{trace}(J^i) < 0$ (since the eigenvalues of J^i have negative real parts), so that $a_3 < 0$.

Let $\lambda_1, \lambda_2, \lambda_3, \lambda_4$ be the eigenvalues of J. Then

$$\lambda_1\lambda_2\lambda_3 + \lambda_1\lambda_2\lambda_4 + \lambda_1\lambda_3\lambda_4 + \lambda_2\lambda_3\lambda_4 = -a_3 > 0$$

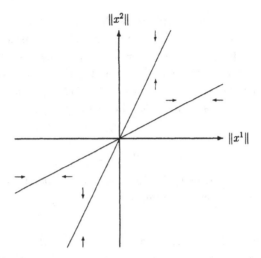

Fig. 1.   The dynamic for the game Γ_0 of Section 2 around $(0,0)$.

from which it follows that at least one λ_r must have positive real part (consider the three cases: four real eigenvalues; two real eigenvalues and one conjugate pair; two conjugate pairs). □

This shows that the unique Nash equilibrium $\bar{x} = (0,0)$ is unstable for $F(\cdot;\Gamma_0)$—a contradiction that establishes our claim.

For a suggestive illustration,[12] see Figure 1, which is drawn for x in a neighborhood of $(0,0)$ where $\phi(x^i) = 2x^i$. In the region $\|x^2\|/2 < \|x^1\| < 2\|x^2\|$ the dynamic leads "away" from $(0,0)$ (the arrows show that, for x^j fixed, the dynamic on x^i must converge to $x^i = 2x^j$—see Lemma 2).

3. An Example with Finitely Many Strategies

If the games have a finite number of strategies (i.e., if the S^i are finite), then the state space for the dynamics is the space of N-tuples of mixed strategies $\prod_{i \in N} \Delta(S^i)$.

Consider a family $\mathcal{U}_0$ of three-player games where each player has two strategies, and the payoffs are:

0, 0, 0	a^1, 1, 0
1, 0, a^3	0, 1, a^3

0, a^2, 1	a^1, 0, 1
1, a^2, 0	0, 0, 0

---

[12]The actual dynamic is 4-dimensional and may be quite complex.

where all the a^i are close to 1 (say, $1 - \varepsilon < a^i < 1 + \varepsilon$ for some small $\varepsilon > 0$), and, as usual, player 1 chooses the row, player 2 the column, and player 3 the matrix.[13] Let Γ_0 be the game with $a^i = 1$ for all i; this game has been introduced by Jordan (1993).

Denote by $x^1, x^2, x^3 \in [0, 1]$ the probability of the top row, left column, and left matrix, respectively. For every game in the family $\mathcal{U}_0$ there is a unique Nash equilibrium[14]: $\bar{x}^i(\Gamma) = a^{i-1}/(a^{i-1} + 1)$. In particular, for Jordan's game $\bar{x}_0 \equiv \bar{x}(\Gamma_0) = (1/2, 1/2, 1/2)$.

Let F be, by way of contradiction, an uncoupled Nash-convergent dynamic $\mathcal{U}_0$. For the game $\Gamma_0 \equiv (u_0^i)_{i=1,2,3}$ we denote $f^i(x^1, x^2, x^3) := F^i(x^1, x^2, x^3; u_0^i)$; let J be the 3×3 Jacobian matrix of f at $\bar{x}_0$.

For any y^1 (close to 1/2), the unique equilibrium of the game $\Gamma_1 = (u_0^1, u_0^2, u^3)$ in $\mathcal{U}_0$ with u^3 given by $a^3 = y^1/(1 - y^1)$ is $(y^1, 1/2, 1/2)$, and so $F^1(y^1, 1/2, 1/2; \Gamma_1) = 0$. This holds therefore also for Γ_0 since the dynamic is uncoupled: $f^1(y^1, 1/2, 1/2) = 0$ for all y^1 close to 1/2. Hence $\partial f^1(\bar{x}_0)/\partial x^1 = 0$. The same applies to the other two players, and we conclude that the diagonal—and thus the trace—of the Jacobian matrix J vanishes. Together with hyperbolicity (in fact, $\det(J) \neq 0$ suffices here), this implies the existence of an eigenvalue with positive real part,[15] thus establishing our contradiction—which proves Theorem 1 in this case.

We put on record that the uncoupledness of the dynamic implies additional structure on J. Indeed, we have $f^1(x^1, 1/2, x^3) = 0$ for all x^1 and x^3 close to 1/2 (since $(x^1, 1/2, x^3)$ is the unique Nash equilibrium when $a^1 = 1$—as in Γ_0—and $a^2 = x^3/(1 - x^3)$, $a^3 = x^1/(1 - x^1)$). Therefore $\partial f^1(\bar{x}_0)/\partial x^3 = 0$ too, and so J is of the form

$$J = \begin{bmatrix} 0 & c & 0 \\ 0 & 0 & d \\ e & 0 & 0 \end{bmatrix}$$

for some real[16] c, d, e.

---

[13]Each player i wants to mismatch the next player $i + 1$, regardless of what player $i - 1$ does. (Of course, $i \pm 1$ is always taken modulo 3.)

[14]In equilibrium: if i plays pure, then $i - 1$ plays pure, so all players play pure—but there is no pure equilibrium; if i plays completely mixed, then $a^i(1 - x^{i+1}) = x^{i+1}$.

[15]Indeed: otherwise the real parts of all eigenvalues are 0. The dimension being odd implies that there must be a real eigenvalue. Therefore 0 is an eigenvalue—and the determinant vanishes.

[16]If $cde \neq 0$ there is an eigenvalue with positive part, and if $cde = 0$ then 0 is the only eigenvalue.

We conclude by observing that the specificities of the example have played very little role in the discussion. In particular, the property that the trace of the Jacobian matrix is null, or that f^i vanishes over a linear subspace of co-dimension 1, which is determined from the payoff function of player i only, will be true for any uncoupled dynamics at the equilibrium of a game with a completely mixed Nash equilibrium—provided, of course, that the game is embedded in an appropriate family of games.

4. Discussion

(a) There exist uncoupled dynamics converging to *correlated equilibria*[17]— see Foster and Vohra (1997), Fudenberg and Levine (1999), Hart and Mas-Colell (2000),[18] and Hart and Mas-Colell (2003). It is thus interesting that Nash equilibrium, a notion that does not predicate coordinated behavior, cannot be guaranteed to be reached in an uncoupled way, while correlated equilibrium, a notion based on coordination, can.[19]

(b) In a general economic equilibrium framework, the parallel of Nash equilibrium is Walrasian (competitive) equilibrium. It is again well known that there are no dynamics that guarantee the general convergence of prices to equilibrium prices if the dynamic has to satisfy natural uncoupledness-like conditions, for example, the nondependence of the adjustment of the price of one commodity on the conditions of the markets for other commodities (see Saari and Simon 1978).

(c) In a mechanism-design framework, the counterpart of the uncoupledness condition is Hurwicz's "privacy-preserving" or "decentralized" condition—see Hurwicz (1986).

(d) There are various results in the literature, starting with Shapley (1964, Section 5), showing that certain classes of dynamics cannot be Nash-convergent. These dynamics assume that the players adjust to the current state $x(t)$ in a way that is, roughly speaking, payoff-improving; this includes fictitious play, best-reply dynamics, better-reply dynamics,

---

[17] Of course, these dynamics are defined on the appropriate state space of joint distributions $\Delta(\prod_{i \in N} S^i)$, i.e., probability vectors on N-tuples of (pure) strategies.

[18] In fact, the notion of "decoupling" appears in Section 4 (i) there.

[19] *Cum grano salis* this may be called the "Coordination Conservation Law": there must be some coordination either in the equilibrium concept or in the dynamic.

monotonic dynamics, adjustment dynamics, replicator dynamics, and so on; see Crawford (1985), Jordan (1993), Gaunesdorfer and Hofbauer (1995), Foster and Young (1998, 2001), and Hofbauer and Sigmund (1998, Theorem 8.6.1). All these dynamics are necessarily uncoupled (since a player's "good reply" to $x(t)$ depends only on his own payoff function). Our result shows that what underlies such impossibility results is not necessarily the rationality-type assumptions on the behavior of the players—but rather the informational requirement of uncoupledness.

(e) In a two-population evolutionary context, Vasin (1999) shows that dynamics that depend only on the vector of payoffs of each pure strategy against the current state—a special class of uncoupled dynamics—cannot be Nash-convergent.

(f) There exist uncoupled dynamics that are guaranteed to converge to (the set of) Nash equilibria for *special families* of games, like two-person zero-sum games, two-person potential games, dominance-solvable games, and others;[20] for some recent work see Hofbauer and Sandholm (2002) and Hart and Mas-Colell (2003).

(g) There exist uncoupled dynamics that are most of the time close to Nash equilibria, but are not Nash-convergent (they exit infinitely often any neighborhood of Nash equilibria); see Foster and Young (2003).

(h) Sufficient epistemic conditions for Nash equilibrium—see Aumann and Brandenburger (1995, Preliminary Observation, p. 1161)—are for each player i to know the N-tuple of strategies x and his own payoff function u^i. But that is precisely the information a player uses in an uncoupled dynamic—which we have shown *not* to yield Nash equilibrium. This points out the difference between the static and the dynamic frameworks: converging to equilibrium is a more stringent requirement than being in equilibrium.

(i) By their nature, differential equations allow strategies to be conditioned only on the limited information of the past captured by the state variable. It may thus be of interest to investigate the topic of this paper in more general settings.

---

[20] A special family of games may be thought of as giving information on the other players' payoff function (e.g., in two-person zero-sum games and potential games, u^i of one player determines u^j of the other player).

Appendix

We show here how to modify the argument of Section 2 in order to prove Theorem 1. Consider a family of games $\mathcal{U}$ that is a neighborhood of Γ_0, and thus is certain to contain only those games in $\mathcal{U}_0$ that are close to Γ_0. The proof of Lemma 2 uses payoff functions of the form $u^j(x^i, x^j) = -\|x^j - y^j\|^2$ that do not depend on the other player's strategy x^i (i.e., ξ^j is the constant function $\xi^j(z) \equiv y^j$). Since the proof in Section 2 needs the result of Lemma 2 only for y^j in a neighborhood of 0, we will replace the above constant function ξ^j with a function that is constant in a neighborhood of the origin and is close to ϕ.

We will thus construct for each $a \in D$ with $\|a\| < \varepsilon$ a function $\psi_a : D \to D$ such that: (1) $\|\psi_a - \phi\| \leq C\varepsilon$ for some constant $C > 0$; (2) $\psi_a(z) = a$ for all $\|z\| \leq 2\varepsilon$; (3) $\phi(\psi_a(z)) = z$ if and only if $z = \phi(a) = 2a$; and (4) $\psi_b(\psi_a(z)) = z$ if and only if $z = \psi_b(a) = b$. The games corresponding to (ϕ, ψ_{y^j}) and to (ψ_{x^i}, ψ_{y^j}), for $\|y^j\| < \varepsilon/2$ and x^i close to $2y^j$, are therefore in $\mathcal{U}_0$ (by (3) and (4)), are close to Γ_0 (by (1)), and we can use them to obtain the result of Lemma 2 (by (2)).

The ψ functions may be constructed as follows: (i) $\psi_a(z) := a$ for $\|z\| \leq 2\varepsilon$; (ii) $\psi_a(z) := 0$ for $\|z\| = 3\varepsilon$; (iii) $\psi_a(z)$ is a rotation of $\phi(z)$ by the angle ε for $\|z\| \geq 4\varepsilon$; and (iv) interpolate linearly on rays in each one of the two regions $2\varepsilon < \|z\| < 3\varepsilon$ and $3\varepsilon < \|z\| < 4\varepsilon$.

It can be checked that conditions (1)–(4) above are indeed satisfied.[21]

References

Aumann, R. J. and A. Brandenburger (1995), "Epistemic Conditions for Nash Equilibrium," *Econometrica*, 63, 1161–1180.

Crawford, V. P. (1985), "Learning Behavior and Mixed-Strategy Nash Equilibrium," *Journal of Economic Behavior and Organization*, 6, 69–78.

Foster, D. and R. V. Vohra (1997), "Calibrated Learning and Correlated Equilibrium," *Games and Economic Behavior*, 21, 40–55.

---

[21] (1) and (2) are immediate. For (3), let $z = \phi(w)$ and $w = \psi_a(z)$. If $\|z\| \leq 3\varepsilon$, then $w = \alpha a$ for some $\alpha \in [0, 1]$, therefore $\|w\| < \varepsilon$ and $z = \phi(w) = 2w$, so in fact $\|z\| < 2\varepsilon$ and $w = a$. If $\|z\| > 3\varepsilon$, then, denoting by $\theta(x)$ the angle of x and recalling that ϕ rotates by an angle between 0 and $\pi/4$, we have $\theta(z) - \theta(w) = \theta(\phi(w)) - \theta(w) \in [0, \pi/4]$, whereas $\theta(w) - \theta(z) = \theta(\psi_a(z)) - \theta(z) = \theta(\phi(z)) + \varepsilon - \theta(z) \in [\varepsilon, \pi/4 + \varepsilon]$, a contradiction. (4) is proven in a similar way.

Foster, D. and H. P. Young (1998), "On the Non-Convergence of Fictitious Play in Coordination Games," *Games and Economic Behavior*, 25, 79–96.

_____ (2001), "On the Impossibility of Predicting the Behavior of Rational Agents," *Proceedings of the National Academy of Sciences U.S.A.*, 98, 12848–12853.

_____ (2003), "Learning, Hypothesis Testing, and Nash Equilibrium," *Games and Economic Behavior*, 45, 73–96.

Fudenberg, D. and D. K. Levine (1998), *Theory of Learning in Games*. Cambridge, MA: MIT Press.

_____ (1999), "Conditional Universal Consistency," *Games and Economic Behavior*, 29, 104–130.

Gaunesdorfer, A. and J. Hofbauer (1995), "Fictitious Play, Shapley Polygons, and the Replicator Equation," *Games and Economic Behavior*, 11, 279–303.

Hart, S. and A. Mas-Colell (2000), "A Simple Adaptive Procedure Leading to Correlated Equilibrium," *Econometrica*, 68, 1127–1150. [Chapter 2]

_____ (2003), "Regret-Based Continuous-Time Dynamics," *Games and Economic Behavior*, 45, 375–394. [Chapter 5]

Hofbauer, J. and W. H. Sandholm (2002), "On the Global Convergence of Stochastic Fictitious Play," *Econometrica*, 70, 2265–2294.

Hofbauer, J. and K. Sigmund (1998), *Evolutionary Games and Population Dynamics*. Cambridge: Cambridge University Press.

Hurwicz, L. (1986), "Incentive Aspects of Decentralization," in *Handbook of Mathematical Economics*, Vol. III, ed. by K. J. Arrow and M. D. Intriligator. Amsterdam: Elsevier/North-Holland, pp. 1441–1482.

Jordan, J. (1993), "Three Problems in Learning Mixed-Strategy Nash Equilibria," *Games and Economic Behavior*, 5, 368–386.

Saari, D. G. and C. P. Simon (1978), "Effective Price Mechanisms," *Econometrica*, 46, 1097–1125.

Samuelson, L. (1997), *Evolutionary Games and Equilibrium Selection*. Cambridge, MA: MIT Press.

Shapley, L. S. (1964), "Some Topics in Two-Person Games," in *Advances in Game Theory*, Annals of Mathematics Studies 52, ed. by M. Dresher, L. S. Shapley, and A. W. Tucker. Princeton: Princeton University Press, pp. 1–28.

Vasin, A. (1999), "On Stability of Mixed Equilibria," *Nonlinear Analysis*, 38, 793–802.

Vega-Redondo, F. (1996), *Evolution, Games, and Economic Behavior*. Oxford, UK: Oxford University Press.

Weibull, J. W. (1995), *Evolutionary Game Theory*. Cambridge, MA: MIT Press.

Young, H. P. (1998), *Individual Strategy and Social Structure*. Princeton: Princeton University Press.

Chapter 8

STOCHASTIC UNCOUPLED DYNAMICS AND NASH EQUILIBRIUM*

Sergiu Hart and Andreu Mas-Colell

In this paper we consider dynamic processes, in repeated games, that are subject to the natural informational restriction of uncoupledness. We study the almost sure convergence of play (the period-by-period behavior as well as the long-run frequency) to Nash equilibria of the one-shot stage game, and present a number of possibility and impossibility results. Basically, we show that if in addition to random experimentation some recall, or memory, is introduced, then successful search procedures that are uncoupled can be devised. In particular, to get almost sure convergence to pure Nash equilibria when these exist, it suffices to recall the last two periods of play.

1. Introduction

A dynamic process in a multi-player setup is *uncoupled* if the moves of every player do not depend on the payoff (or utility) functions of the other players. This is a natural informational requirement, which holds in most models.

Originally published in *Games and Economic Behavior*, 57 (2006), 286–303.

*Previous versions: October 2004, May 2005.
JEL classification: C7; D83.
Keywords: uncoupled; Nash equilibrium; stochastic dynamics; finite recall; finite memory; finite automaton; exhaustive experimentation.
Research partially supported by a grant of the Israel Academy of Sciences and Humanities and a grant of the Fundacion BBVA.
We thank Bob Aumann, Dean Foster, Fabrizio Germano, Sham Kakade, Gábor Lugosi, Yishay Mansour, Abraham Neyman, Yosi Rinott, Jeff Shamma, Benjy Weiss, Peyton Young, the associate editor, and the referees for their comments and suggestions. Part of this research was conducted during the Conference on Mathematical Foundations of Learning Theory at the Centre de Recerca Matemàtica in Barcelona, June 2004.

In Hart and Mas-Colell (2003b) we introduce this concept and show that uncoupled stationary dynamics cannot always converge to Nash equilibria, even if these exist and are unique. The setup was that of deterministic, stationary, continuous-time dynamics.

It is fairly clear that the situation may be different when *stochastic* moves are allowed, since one may then try to carry out some version of exhaustive search: keep randomizing until by pure chance a Nash equilibrium is hit, and then stop there. However, this is not so simple: play has a decentralized character, and no player can, alone, recognize a Nash equilibrium. The purpose of this paper is, precisely, to investigate to what extent Nash equilibria can be reached when considering dynamics that satisfy the restrictions of our previous paper: uncoupledness and stationarity. As we shall see, one can obtain positive results, but these will require that, in addition to the ability to perform stochastic moves of an experimental nature, the players retain some memory from the past plays.

Because we allow random moves, it is easier to place ourselves in a discrete-time framework. Thus, we consider the repeated play of a given stage game, under the standard assumption that each player observes the play of all players; as for payoffs, each player knows only his own payoff function. We start by studying a natural analog of the approach of our earlier paper; that is, we assume that in determining the random play at time $t+1$ the players recall only the information contained in the current play of all players at time t; i.e., past history does not matter. We call this the case of 1-*recall*. We shall then see that the result of our earlier paper is recovered: convergence of play to Nash equilibrium cannot be ensured under the hypotheses of uncoupledness, stationarity, and 1-recall (there is an exception for the case of generic two-player games with at least one pure Nash equilibrium).

Yet, the exhaustive search intuition can be substantiated if we allow for (uncoupled and stationary) strategies with longer recall. Perhaps surprisingly, to guarantee almost sure convergence of play to pure Nash equilibria when these exist, it suffices to have 2-*recall*: to determine the play at $t+1$ the players use the information contained in the plays of all players at periods t and $t-1$. In general, when Nash equilibria may be mixed, we show that convergence of the long-run empirical distribution of play to (approximate) equilibria can be guaranteed using longer, but finite, recall. Interestingly, however, this does not suffice to obtain the almost sure convergence of the period-by-period behavior probabilities. As it turns out, we can get this too within the broader context of finite memory (i.e., finite-state automata).

In conclusion, one can view this paper as contributing to the demarcation of the border between those classes of dynamics for which convergence to Nash equilibrium can be obtained and those for which it cannot.

The paper is organized as follows. Section 2 presents the model and defines the relevant concepts. Convergence to pure Nash equilibria is studied in Section 3, and to mixed equilibria, in Section 4 (with a proof relegated to Appendix A). We conclude in Section 5 with some comments and a discussion of the related literature, especially the work of Foster and Young (2003, 2006).

2. The Setting

A basic static (one-shot) *game* is given in strategic (or normal) form, as follows. There are $N \geq 2$ *players*, denoted $i = 1, 2, \ldots, N$. Each player i has a finite set of *actions* A^i; let $A := A^1 \times A^2 \times \cdots \times A^N$ be the set of action combinations. The *payoff function* (or *utility function*) of player i is a real-valued function $u^i : A \to \mathbb{R}$. The set of *randomized* or *mixed* actions of player i is the probability simplex over A^i, i.e., $\Delta(A^i) = \{x^i = (x^i(a^i))_{a^i \in A^i} : \sum_{a^i \in A^i} x^i(a^i) = 1$ and $x^i(a^i) \geq 0$ for all $a^i \in A^i\}$; as usual, the payoff functions u^i are multilinearly extended, so $u^i : \Delta(A^1) \times \Delta(A^2) \times \cdots \times \Delta(A^N) \to \mathbb{R}$.

We fix the set of players N and the action sets A^i, and identify a game by its payoff functions $U = (u^1, u^2, \ldots, u^N)$.

For $\varepsilon \geq 0$, a *Nash ε-equilibrium* x is an N-tuple of mixed actions $x = (x^1, x^2, \ldots, x^N) \in \Delta(A^1) \times \Delta(A^2) \times \cdots \times \Delta(A^N)$, such that x^i is an ε-*best reply* to x^{-i} for all i; i.e., $u^i(x) \geq u^i(y^i, x^{-i}) - \varepsilon$ for every $y^i \in \Delta(A^i)$ (we write $x^{-i} = (x^1, \ldots, x^{i-1}, x^{i+l}, \ldots, x^N)$ for the combination of mixed actions of all players except i). When $\varepsilon = 0$ this is a *Nash equilibrium*, and when $\varepsilon > 0$, a Nash *approximate* equilibrium.

The dynamic setup consists of a repeated play, at discrete-time periods $t = 1, 2, \ldots$, of the static game U. Let $a^i(t) \in A^i$ denote the action of player i at time[1] t, and put $a(t) = (a^1(t), a^2(t), \ldots, a^N(t)) \in A$ for the

---

[1] More precisely, the actual *realized* action (when randomizations are used).

combination of actions at t. We assume that there is standard monitoring: at the end of period t each player i observes everyone's realized action, i.e., $a(t)$.

A *strategy*[2] f^i of player i is a sequence of functions $(f_1^i, f_2^i, \ldots, f_t^i, \ldots)$, where, for each time t, the function f_t^i assigns a mixed action in $\Delta(A^i)$ to each history $h_{t-1} = (a(1), a(2), \ldots, a(t-1))$. A strategy profile is $f = (f^1, f^2, \ldots, f^N)$.

A strategy f^i of player i has *finite recall* if there exists a positive integer R such that only the history of the last R periods matters: for each $t > R$, the function f_t^i is of the form $f_t^i(a(t-R), a(t-R+1), \ldots, a(t-1))$; we call this *$R$-recall*.[3] Such a strategy is moreover *stationary* if the ("calendar") time t does not matter: $f_t^i \equiv f^i(a(t-R), a(t-R+1), \ldots, a(t-1))$ for all $t > R$.

Strategies have to fit the game being played. We thus consider a *strategy mapping*, which, to every game (with payoff functions) U, associates a strategy profile $f(U) = (f^1(U), f^2(U), \ldots, f^N(U))$ for the repeated game induced by $U = (u^1, u^2, \ldots, u^N)$. Our basic requirement for a strategy mapping is *uncoupledness*, which says that the strategy of each player i may depend only on the ith component u^i of U, i.e., $f^i(U) \equiv f^i(u^i)$. Thus, for any player i and time t, the strategy f_t^i has the form $f_t^i(a(1), a(2), \ldots, a(t-1); u^i)$. Finally, we shall say that a strategy mapping has *R-recall* and is *stationary* if, for any U, the strategies $f^i(U)$ of all players i have R-recall and are stationary.

3. Pure Equilibria

We start by considering games that possess *pure* Nash equilibria (i.e., Nash equilibria $x = (x^1, x^2, \ldots, x^N)$ where each x^i is a pure action in A^i).[4] Our first result generalizes the conclusion of Hart and Mas-Colell (2003b). We show that with 1-recall—that is, if actions depend only on the current play and not on past history—we cannot hope in all generality to converge, in an uncoupled and stationary manner, to pure Nash equilibria when these exist.

---

[2] We use the term "strategy" for the repeated game and "action" for the one-shot game.
[3] The recall of a player consists of past realized pure actions only, not of mixed actions played (not even his own).
[4] From now on, "game" and "equilibrium" will always refer to the one-shot stage game.

	α	β	γ
α	1,0	0,1	1,0
β	0,1	1,0	1,0
γ	0,1	0,1	1,1

Fig. 1.   A non-generic two-player game.

Theorem 1. *There are no uncoupled, 1-recall, stationary strategy mappings that guarantee almost sure convergence of play to pure Nash equilibria of the stage game in all games where such equilibria exist.*[5]

Proof. The following two examples, the first with $N = 2$ and the second with $N = 3$, establish our result. We point out that the second example is *generic*—in the sense that the best reply is always unique—while the first is not; this will matter in the sequel.

The first example is the two-player game of Fig. 1. The only pure Nash equilibrium is (γ, γ). Assume by way of contradiction that we are given an uncoupled, 1-recall, stationary strategy mapping that guarantees convergence to pure Nash equilibria when these exist. Note that at each of the nine action pairs, at least one of the two players is best-replying. Suppose the current state $a(t)$ is such that player 1 is best-replying (the argument is symmetric for player 2). We claim that player 1 will play at $t + 1$ the same action as in t (i.e., player 1 will not move). To see this consider a new game where the utility function of player 1 remains unaltered and the utility function of player 2 is changed in such a manner that the current state $a(t)$ is the only pure Nash equilibrium of the new game. It is easy to check that in our game this can always be accomplished (for example, to make (α, γ) the unique Nash equilibrium, change the payoff of player 2 in the (α, γ) and (γ, α) cells to, say, 2). The strategy mapping has 1-recall, so it must prescribe to the first player not to move in the new game (otherwise convergence to the unique pure equilibrium would be violated there). By uncoupledness, therefore, player 1 will not move in the original game either.

It follows that (γ, γ) can never be reached when starting from any other state: if neither player plays γ currently then only one player (the one who

---

[5] "Almost sure convergence of play to pure Nash equilibria" means that almost every play path consists of a pure Nash equilibrium being played from some point on.

	α	β	γ	α	β	γ	α	β	γ
α	0,0,0	0,4,4	2,1,2	4,0,4	4,4,0	3,1,3	2,2,1	3,3,1	0,0,0
β	4,4,0	4,0,4	3,1,3	0,4,4	0,0,0	2,1,2	3,3,1	2,2,1	0,0,0
γ	1,2,2	1,3,3	0,0,0	1,3,3	1,2,2	0,0,0	0,0,0	0,0,0	6,6,6
		α			β			γ	

Fig. 2.   A generic three-player game.

is not best-replying) may move; if only one plays γ then the other player cannot move (since in all cases it is seen that he is best-replying). This contradicts our assumption.

The second example is the three-player game of Fig. 2. There are three players $i = 1, 2, 3$, and each player has three actions α, β, γ. Restricted to α and β we essentially have the game of Jordan (1993) (see Hart and Mas-Colell 2003b, Section III), where every player i tries to mismatch the player $i-1$ (the predecessor of player 1 is player 3): he gets 0 if he matches and 4 if he mismatches. If all three players play γ then each one gets 6. If one player plays γ and the other two do not, the player that plays γ gets 1 and the other two get 3 each if they mismatch and 2 each if they match. If two players play γ and the third one does not then each one gets 0.

The only pure Nash equilibrium of this game is (γ, γ, γ). Suppose that we start with all players playing α or β, but not all the same; for instance, (α, β, α). Then players 2 and 3 are best-replying, so only player 1 can move in the next period (this follows from uncoupledness as in the previous example). If he plays α or β then we are in exactly the same position as before (with, possibly, the role of mover taken by player 2). If he moves to γ then the action configuration is (γ, β, α), at which both players 2 and 3 are best-replying and so, again, only player 1 can move. Whatever he plays next, we are back to situations already contemplated. In summary, every configuration that can be visited will only have at most one γ, and therefore the unique pure Nash equilibrium (γ, γ, γ) will never be reached. $\square$

Remark. In the three-player example of Fig. 2, starting with (α, β, α), the empirical joint distribution of play cannot approach the distribution of a mixed Nash equilibrium, because neither (α, α, α) nor (β, β, β) will ever be visited—but these action combinations have positive probability in every mixed Nash equilibrium (there are two such equilibria: in the first each player plays $(1/2, 1/2, 0)$, and in the second each plays $(1/4, 1/4, 1/2)$).

As we noted above, the two-player example of Fig. 1 is not generic. It turns out that in the case of only two players, *genericity*—in the sense that every player's best reply to pure actions is always unique—does help.

Proposition 2. *There exist uncoupled, 1-recall, stationary strategy mappings that guarantee almost sure convergence of play to pure Nash equilibria of the stage game in every two-player generic game where such equilibria exist.*

Proof. We define the strategy mapping of player 1 (and similarly for player 2) as follows. Let the state (i.e., the previous period's play) be $a = (a^1, a^2) \in A = A^1 \times A^2$. If a^1 is a best reply to a^2 according to u^1, then player 1 plays a^1; otherwise, player 1 randomizes uniformly[6] over all his actions in A^1.

With these strategies, it is clear that each pure Nash equilibrium becomes an absorbing state of the resulting Markov chain.[7] Moreover, as we shall show next, from any other state $a = (a^1, a^2) \in A$ there is a positive probability of reaching a pure Nash equilibrium in at most two steps. Indeed, since a is not a Nash equilibrium, then at least one of the players, say player 2, is not best-replying. Therefore there is a positive probability that in the next period player 2 plays $\bar{a}^2$, where $(\bar{a}^1, \bar{a}^2)$ is a pure Nash equilibrium (which is assumed to exist), whereas player 1 plays the same a^1 as last period (this has probability one if player 1 was best-replying, and positive probability otherwise). The new state is thus $(a^1, \bar{a}^2)$. If $a^1 = \bar{a}^1$, we have reached the pure Nash equilibrium $\bar{a}$. If not, then player 1 is not best-replying—it is here that our genericity assumption is used: the *unique* best reply to $\bar{a}^2$ is a^1—so now there is a positive probability that in the next period player 1 will play $\bar{a}^1$ and player 2 will play $\bar{a}^2$, and thus, again, the pure Nash equilibrium $\bar{a}$ is reached.

Therefore an absorbing state—i.e., a pure Nash equilibrium—must eventually be reached with probability one.   □

Interestingly, if we allow for longer recall the situation changes and we can present positive results for general games. In fact, for the case where pure Nash equilibria exist the contrast is quite dramatic, since allowing just one more period of recall suffices.

---

[6]I.e., the probability of each $a^1 \in A^1$ is $1/|A^1|$. Of course, the uniform distribution may be replaced—here as well as in all other constructions in this paper—by any probability distribution with full support (i.e., such that every action has positive probability).

[7]A standard reference for Markov chains is Feller (1968, Chapter XV).

Theorem 3. *There exist uncoupled, 2-recall, stationary strategy mappings that guarantee almost sure convergence of play to pure Nash equilibria of the stage game in every game where such equilibria exist.*

Proof. Let the state—i.e., the play of the previous two periods—be $(a', a) \in A \times A$. We define the strategy mapping of each player i as follows:

- If $a' = a$ (i.e., if *all* players have played exactly the same actions in the past two periods) and a^i is a best reply of player i to a^{-i} according to u^i, then player i plays a^i (i.e., he plays the same action yet again);
- In all other cases, player i randomizes uniformly over A^i.

To prove our result, we partition the state space $S = A \times A$ of the resulting Markov chain into four regions:

$S_1 := \{(a, a) \in A \times A : a \text{ is a Nash equilibrium}\};$

$S_2 := \{(a', a) \in A \times A : a' \neq a \text{ and } a \text{ is a Nash equilibrium}\};$

$S_3 := \{(a', a) \in A \times A : a' \neq a \text{ and } a \text{ is not a Nash equilibrium}\};$

$S_4 := \{(a, a) \in A \times A : a \text{ is not a Nash equilibrium}\}.$

Clearly, each state in S_1 is absorbing. Next, we claim that all other states are transient: there is a positive probability of reaching a state in S_1 in finitely many periods. Indeed:

- At each state (a', a) in S_2 all players randomize; hence there is a positive probability that next period they will play a—and so the next state will be (a, a), which belongs to S_1.
- At each state (a', a) in S_3 all players randomize; hence there is a positive probability that next period they will play a pure Nash equilibrium $\bar{a}$ (which exists by assumption)—and so the next state will be $(a, \bar{a})$, which belongs to S_2.
- At each state (a, a) in S_4 at least one player is not best-replying and thus is randomizing; hence there is a positive probability that the next period play will be some $a' \neq a$—and so the next state will be (a, a'), which belongs to $S_2 \cup S_3$.

In all cases there is thus a positive probability of reaching an absorbing state in S_1 in at most three steps. Once such a state (a, a), where a is a pure Nash equilibrium, is reached (this happens eventually with probability one), the players will continue to play a every period.[8]   □

Thus extremely simple strategies may nevertheless guarantee convergence to pure Nash equilibria. The strategies defined above may be viewed as a combination of search and testing. The search is a standard random search; the testing is done individually, but in a coordinated manner: the players wait until a certain "pattern" (a repetition) is observed, at which point each one applies a "rational" test (he checks whether or not he is best-replying). Finally, the pattern is self-replicating once the desired goal (a Nash equilibrium) is reached. (This structure will appear again, in a slightly more complex form, in the case of mixed equilibria; see the proofs of Proposition 4 and Theorem 5 below.)

4. Mixed Equilibria

We come next to the general case (where only the existence of mixed Nash equilibria is guaranteed). Here we consider first the long-run frequencies of play, and then the period-by-period behavior probabilities. The convergence will be to approximate equilibria. To this effect, assume that there is a bound M on payoffs; i.e., the payoff functions all satisfy $|u^i(a)| \leq M$ for all action combinations $a \in A$ and all players i.

Given a history of play, we shall denote by Φ_t the empirical frequency distribution in the first t periods: $\Phi_t[a] := |\{1 \leq \tau \leq t : a(\tau) = a\}|/t$ for each $a \in A$, and similarly $\Phi_t[a^i] := |\{1 \leq \tau \leq t : a^i(\tau) = a^i\}|/t$ for each i and $a^i \in A$. We shall refer to $(\Phi_t[a])_{a \in A} \in \Delta(A)$ as the *empirical joint distribution of play*,[9] and to $(\Phi_t[a^i])_{a^i \in A^i} \in \Delta(A^i)$ as the *empirical marginal distribution of play of player i* (up to time t).

Proposition 4. *For every M and $\varepsilon > 0$ there exists an integer R and an uncoupled, R-recall, stationary strategy mapping that guarantees, in every*

---

game with payoffs bounded by M, *almost sure convergence of the empirical marginal distributions of play to Nash ε-equilibria; i.e., for almost every history of play there exists a Nash ε-equilibrium of the stage game* $x = (x^1, x^2, \ldots, x^N)$ *such that, for every player* i *and every action* $a^i \in A^i$,

$$\lim_{t \to \infty} \Phi_t[a^i] = x^i(a^i). \tag{1}$$

Of course, different histories may lead to different ε-equilibria (i.e., x may depend on the play path). The length of the recall R depends on the precision ε and the bound on payoffs M (as well as on the number of players N and the number of actions $|A^i|$).

Proof. Given $\varepsilon > 0$, let K be such that[10]

$$\left[\|x^i - y^i\| \leq \frac{1}{K} \text{ for all } i \right] \implies \left[|u^i(x) - u^i(y)| \leq \varepsilon \text{ for all } i \right] \tag{2}$$

for $x^i, y^i \in \Delta(A^i)$ and $|u^i(a)| \leq M$ for all $a \in A$. Let $\bar{y} = (\bar{y}^1, \bar{y}^2, \ldots, \bar{y}^N)$ be a Nash 2ε-equilibrium, such that all probabilities are multiples of $1/K$ (i.e., $K\bar{y}^i(a^i)$ is an integer for all a^i, and all i). Such a $\bar{y}$ always exists: take a $1/K$-approximation of a Nash equilibrium and use (2). Given such a Nash 2ε-equilibrium $\bar{y}$, let $(\bar{a}_1, \bar{a}_2, \ldots, \bar{a}_K) \in A \times A \times \cdots \times A$ be a fixed sequence of action combinations of length K whose marginals are precisely $\bar{y}^i$ (i.e., each action a^i of each player i appears $K\bar{y}^i(a^i)$ times in the sequence $(\bar{a}_1^i, \bar{a}_2^i, \ldots, \bar{a}_K^i)$).

Take $R = 2K$. The construction parallels the one in the proof of Theorem 3. A state is a history of play of length $2K$, i.e., $s = (a_1, a_2, \ldots, a_{2k})$ with $a_k \in A$ for all k. The state s is K-periodic if $a_{K+k} = a_k$ for all $k = 1, 2, \ldots, K$. Given s, for each player i we denote by $z^i \in \Delta(A^i)$ the frequency distribution of the last K actions of i, i.e., $z^i(a^i) := |\{K+1 \leq k \leq 2K : a_k^i = a^i\}|/K$ for each $a^i \in A^i$; put $z = (z^1, z^2, \ldots, z^N)$.

We define the strategy mapping of each player i as follows:

- If the current state s is K-periodic and z^i is a 2ε-best reply to z^{-i}, then player i plays $a_1^i = a_{K+1}^i$ (i.e., continues his K-periodic play).
- In all other cases player i randomizes uniformly over A^i.

---

[10]We use the maximum (ℓ^∞) norm on $\Delta(A^i)$, i.e., $\|x^i - y^i\| := \max_{a^i \in A^i} |x^i(a^i) - y^i(a^i)|$; it is easy to check that $K \geq M \sum_i |A^i|/\varepsilon$ suffices for (2).

Partition the state space S consisting of all sequences over A of length $2K$ into four regions:

$S_1 := \{s$ is K-periodic and z is a Nash 2ε-equilibrium$\}$;

$S_2 := \{s$ is not K-periodic and z is a Nash 2ε-equilibrium$\}$;

$S_3 := \{s$ is not K-periodic and z is not a Nash 2ε-equilibrium$\}$;

$S_4 := \{s$ is K-periodic and z is not a Nash 2ε-equilibrium$\}$.

We claim that the states in S_1 are persistent and K-periodic, and all other states are transient. Indeed, once a state s in S_1 is reached, the play moves in a deterministic way through the K cyclic permutations of s, all of which have the same z—and so, for each player i, his empirical marginal distribution of play will converge to z^i. At a state s in S_2 every player randomizes, so there is a positive probability that everyone will play K-periodically, leading in $r = \max\{1 \leq k \leq K : a_{K+k} \neq a_k\}$ steps to S_1. At a state s in S_3, there is a positive probability of reaching S_2, in $K + 1$ steps: in the first step the play is some $a \neq a_{K+1}$, and, in the next K steps, a sequence $(\bar{a}_1, \bar{a}_2, \ldots, \bar{a}_K)$ corresponding to a Nash 2ε-equilibrium. Finally, from a state in S_4 there is a positive probability of moving to a state in $S_2 \cup S_3$ in one step.   □

Proposition 4 is not entirely satisfactory, because it does not imply that the empirical *joint* distributions of play converge to joint distributions induced by Nash approximate equilibria. For this to happen, the joint distribution needs to be (in the limit) the product of the marginal distributions (i.e., independence among the players' play is required). But this is not the case in the construction in the proof of Proposition 4 above, where the players' actions become "synchronized"—rather than independent—once an absorbing cycle is reached. A more refined proof is thus needed to obtain the stronger conclusion of the following theorem on the convergence of the joint distributions.

Theorem 5. *For every M and $\varepsilon > 0$ there exists an integer R and an uncoupled, R-recall, stationary strategy mapping that guarantees, in every game with payoffs bounded by M, the almost sure convergence of the empirical joint distributions of play to Nash ε-equilibria; i.e., for almost every history of play there exists a Nash ε-equilibrium of the stage game $x = (x^1, x^2, \ldots, x^N)$ such that, for every action combination $a = (a^1, a^2, \ldots, a^N) \in A$,*

$$\lim_{t \to \infty} \Phi_t[a] = \prod_{i=1}^{N} x^i(a^i). \tag{3}$$

Moreover, there exists an almost surely finite stopping time[11] T after which the occurrence probabilities $\Pr[a(t) = a|h_T]$ *also converge to Nash ε-equilibria; i.e., for almost every history of play and every action combination* $a = (a^1, a^2, \ldots, a^N) \in A$,

$$\lim_{t \to \infty} \Pr[a(t) = a|h_T] = \prod_{i=1}^{N} x^i(a^1), \tag{4}$$

where x is the same Nash ε-equilibrium of (3).

As before, x and T may depend on the history; T is the time when some ergodic set is reached. Since the proof of Theorem 5 is relatively intricate, we relegate it to Appendix A. Of course, (1) follows from (3). Note that neither (4) nor its marginal implications,

$$\lim_{t \to \infty} \Pr[a^i(t) = a^i|h_T] = x^i(a^i) \tag{5}$$

for all i, hold for the construction of Proposition 4 (again, due to periodicity).

Now (5) says that, after time T, the overall probabilities of play converge almost surely to Nash ε-equilibria. It does not say the same, however, about the actual play or *behavior probabilities* $\Pr[a^i(t) = a^i|h_{t-1}] = f^i(h_{t-1})(a^i)$ (where $h_{t-1} = (a(1), a(2), \ldots, a(t-1))$). We next show that this cannot be guaranteed in general when the recall is finite.

Theorem 6. *For every small enough[12] ε > 0, there are no uncoupled, finite recall, stationary strategy mappings that guarantee, in every game, the almost sure convergence of the behavior probabilities to Nash ε-equilibria of the stage game.*

Proof. Choose a stage game U with a unique, completely mixed Nash equilibrium, and assume that a certain pure action combination, call it $\bar{a} \in A$, is such that $\bar{a}^1$ is the unique best reply of player 1 to $\bar{a}^{-1} = (\bar{a}^2, \ldots, \bar{a}^N)$. Let U' be another game where the payoff function of player 1 is the same as in U, and the payoff function of every other player $i \neq 1$ depends only on a^i and has a unique global maximum at $\bar{a}^i$. Then $\bar{a}$ is the unique Nash

---

[11]I.e., T is determined by the past only: if $T = t$ for a certain play path $h = (a(1), \ldots, a(t), a(t+1), \ldots)$, then $T = t$ for any other play path $h' = (a(1), \ldots, a(t), a'(t+1), \ldots)$ that is identical to h up to and including time t. This initial segment of history $(a(1), a(2), \ldots, a(T))$ is denoted h_T.

[12]I.e., for all $\varepsilon < \varepsilon_0$ (where ε_0 may depend on N and $(|A^i|)_{i=1}^N$).

equilibrium of U'. Take $\varepsilon > 0$ small enough so that all Nash ε-equilibria of U are completely mixed, and moreover there exists $\rho > 0$ such that, for any two Nash ε-equilibria x and y of U and U', respectively, we have $0 < x^i(\bar{a}^i) < \rho < y^i(\bar{a}^i)$ for all players i (recall that $x^i(\bar{a}^i) < 1 = y^i(\bar{a}^i)$ when x and y are the unique Nash equilibria of U and U', respectively).

We argue by contradiction and assume that for some R there is an uncoupled, R-recall, stationary strategy mapping f for which the stated convergence does in fact obtain.

Consider now the history $\bar{\mathbf{a}} = (\bar{a}, \bar{a}, \ldots, \bar{a})$ of length R that consists of R repetitions of $\bar{a}$. The behavior probabilities have been assumed to converge (a.s.) to Nash ε-equilibria, which, in both games, always give positive probability to the actions $\bar{a}^i$. Hence the state $\bar{\mathbf{a}}$ has a positive probability of occurring after any large enough time T. Therefore, in particular at this state $\bar{\mathbf{a}}$, the behavior probabilities must be close to Nash ε-equilibria. Now all Nash ε-equilibria x of U satisfy $x^1(\bar{a}^1) < \rho$, so the behavior probability of player 1 at state $\bar{\mathbf{a}}$ must also satisfy this inequality, i.e., $f^1(\bar{\mathbf{a}}; u^1)(\bar{a}^1) < \rho$. But the same argument applied to U' (where player 1 has the same payoff function u^1 as in U) implies $f^1(\bar{\mathbf{a}}; u^1)(\bar{a}^1) > \rho$ (since this inequality is satisfied by all Nash ε-equilibria of U'. This contradiction proves our claim. $\qquad\square$

The impossibility result of Theorem 6 hinges on the finite recall assumption. Finite recall signifies that the distant past is irrelevant to present behavior (two histories that differ only in periods beyond the last R periods will generate the same mixed actions).[13] Hence, finite recall is a special, though natural, way to get the past influencing the present through a finite set of parameters. But it is not the only framework with this implication. What would happen if, while retaining the desideratum of a limited influence from the past, we were to broaden our setting by moving from a finite recall to a "finite memory" assumption? It turns out that we then obtain a positive result: the period-by-period behavior probabilities can also be made to converge almost surely.

Specifically, a strategy of player i has *finite memory* if it can be implemented by an automaton with finitely many states, such that, at each period t, its input is $a(t) \in A$, the N-tuple of actions actually played, and its output is $x^i(t+1) \in \Delta(A^i)$, the mixed action to be played next period. To facilitate comparison with finite recall, we shall measure the

---

[13] See Aumann and Sorin (1989) for a discussion of bounded recall.

size of the memory by the number of elements of A it can contain; thus R-*memory* means that the memory can contain any $(a(1), a(2), \ldots, a(R))$ with $a(k) \in A$ for $k = 1, 2, \ldots, R$ (i.e., the automaton has $|A|^R$ states).

Theorem 7. *For every M and $\varepsilon > 0$ there exists an integer R and an uncoupled, R-memory, stationary strategy mapping that guarantees, in every game with payoffs bounded by M, the almost sure convergence of the behavior probabilities to Nash ε-equilibria; i.e., for almost every history of play there exists a Nash ε-equilibrium of the stage game $x = (x^1, x^2, \ldots, x^N)$ such that, for every action $a^i \in A^i$ of every player $i \in N$,*

$$\lim_{t \to \infty} \Pr[a^i(t) = a^i | h_{t-1}] = x^i(a^i). \tag{6}$$

Since the players randomize independently at each stage, (6) implies

$$\lim_{t \to \infty} \Pr[a(t) = a | h_{t-1}] = \prod_{i=1}^{N} x^i(a^i) \tag{7}$$

for every $a \in A$, from which it follows, by the law of large numbers, that the empirical joint distributions of play also converge almost surely, i.e., (3).

Proof. We modify the construction in Proposition 4 as follows. Let $R = 2K + 1$; a state is now $\tilde{s} = (a_0, a_1, a_2, \ldots, a_{2K}$ with $a_k \in A$ for $k = 0, 1, \ldots, 2K$. Let $s = (a_1, a_2, \ldots, a_{2K})$ be the last $2K$ coordinates of $\tilde{s}$ (so $\tilde{s} = (a_0, s)$); the frequencies z^i are still determined by the last K coordinates $a_{K+1}, \ldots, a_{2K}$.

There will be two "modes" of behavior. In the first mode the strategy mappings are as in the Proof of Proposition 4, except that now the recall has length $2K + 1$, and that whenever s (the play of the last $2K$ periods) is K-periodic and z^i is not a 2ε-best reply to z^{-i}, player i plays an action that is *different* from a_1^i (rather than a randomly chosen action); i.e., i "breaks" the K-periodic play. This guarantees that a K-periodic state $\tilde{s}$ (the play of the last $2K + 1$ periods) is reached *only when* z is a Nash 2ε-equilibrium. When this occurs the strategies move to the second mode, where in every period player i plays the mixed action z^i, and the state remains fixed (i.e., it is no longer updated).

Formally, we define the strategy mapping and the state-updating rule for each player i as follows. Let the state be $\tilde{s} = (a_0, a_1, \ldots, a_{2k}) = (a_0, s)$;

then:

- **Mode I:** $\tilde{s}$ *is not K-periodic.*

    — If s is K-periodic and z^i is a 2ε-best reply to z^{-l}, then player i plays a_1^i (which equals a_{K+1}^i; i.e., he continues his K-periodic play).

    — If s is K-periodic and z^i is not a 2ε-best reply to z^{-l}, then player i randomizes uniformly over $A^i \backslash \{a_1^i\}$ (i.e., he "breaks" his K-periodic play).

    — If s is not K-periodic, then player i randomizes uniformly over A^i.
    In all three cases, let a be the N-tuple of actions actually played; then the new state is $\tilde{s}' = (a_1, \ldots, a_{2K}, a)$.

- **Mode II:** $\tilde{s}$ is K-*periodic.*
    Player i plays the mixed action z^i, and the new state is $\tilde{s}' = \tilde{s}$ (i.e., unchanged).

- *The starting state* is any $\tilde{s}$ that is not K-periodic (Mode I).

It is easy to check that once a block of size K is repeated twice, either the frequencies z constitute a Nash 2ε-equilibrium—in which case next period the cyclical play continues and we get to Mode II—or they do not—in which case the cycle is broken by at least one player (and the random search continues). Once Mode II is reached, which happens eventually a.s., the states of all players stay constant, and each player plays the corresponding frequencies forever after.   □

5. Discussion and Comments

This section includes some further comments, particularly on the relevant literature.

(a) *Foster and Young*: The present paper is not the first one where, within the span of what we call uncoupled dynamics, stochastic moves and the possibility of recalling the past have been brought to bear on the formulation of dynamics leading to Nash equilibria. The pioneers were Foster and Young (2003), followed by Foster and Young (2006), Kakade and Foster (2008), and Germano and Lugosi (2007).

The motivation of Foster and Young and our motivation are not entirely the same. They want to push to its limits the "learning with experimentation" paradigm (which does not allow direct exhaustive search procedures that, in our terminology, are not of an uncoupled

nature). We start from the uncoupledness property and try to demarcate the border between what can and what cannot be done with such
dynamics.

(b) *Convergence*: Throughout this paper we have sought a strong form
of convergence, namely, almost sure convergence to a point.[14] One could
consider seeking weaker forms of convergence (as has been done in the
related literature): almost sure convergence to the convex hull of the set
of Nash ε-equilibria, or convergence in probability, or "$1 - \varepsilon$ of the time
being an ε-equilibrium," and so on. Conceivably, the use of weaker
forms of convergence may have a theoretical payoff in other aspects of
the analysis.

(c) *Stationarity*: With stationary finite recall (or finite memory) strategies,
no more than convergence to approximate equilibria can be expected.
Convergence to exact equilibria requires non-stationary strategies with
unbounded recall; see Germano and Lugosi (2007) for such a result.

Another issue is that non-stationarity may allow the transmission
of arbitrarily large amounts of information through the time dimension (for instance, a player may signal his payoff function through his
actions in the first T periods), thus effectively voiding the uncoupledness assumption.

(d) *State space*: Theorems 1 and 3 show how doubling the size of the recall
(from 1 to 2) allows for a positive result. More generally, the results
of this paper may be viewed as a study of convergence to equilibrium
when the common state space is larger than just the action space,
thus allowing, in a sense, more information to be transmitted. Shamma
and Arslan (2005) have introduced procedures in the continuous-time
setup (extended to discrete time in Arslan and Shamma 2004) that
double the state space, and yield convergence to Nash equilibria for
some specific classes of games. Interestingly, convergence to correlated equilibria in the continuous-time setup was also obtained with
a doubled state space, consisting of the current as well as the cumulative average play; see Hart and Mas-Colell (2003, Theorem 5.1 and
Corollary 5.2).

(e) *Unknown game*: Suppose that the players observe, not the history of
play, but only their own realized payoffs; i.e., for each player i and
time t the strategy is $f_t^i(u^i(a(1)), u^i(a(2)), \ldots, u^i(a(t-1)))$ (in fact,

_____

[14]The negative results of Theorems 1 and 6 also hold for certain weaker forms of
convergence.

the player may know nothing about the game being played but his set of actions). What results can be obtained in this case? It appears that, for any positive result, experimentation even at (apparent) Nash equilibria will be indispensable. This suggests, in particular, that the best sort of convergence to hope for, in a stationary setting, is some kind of convergence in probability as mentioned in Remark (b). On this point see Foster and Young (2006).

(f) *Which equilibrium?* Among multiple (approximate) equilibria, the more "mixed" an equilibrium is, the higher the probability that the strategies of Section 4 will converge to it. Indeed, the probability that uniform randomizations yield in a K-block frequencies $(k_1, k_2, \ldots, k_r)$ is proportional to $K!/(k_1!k_2 \ldots k_r!)$, which is lowest for pure equilibria (where $k_1 = K$) and highest for[15] $p_1 = p_2 = \cdots = p_r = K/r$.

(g) *Correlated equilibria*: We know that there are uncoupled strategy mappings with the property that the empirical joint distributions of play converge almost surely to the set of correlated equilibria (see Foster and Vohra 1997; Hart and Mas-Colell 2000; Hart 2005; and the book of Young 2004). Strictly speaking, those strategies do not have finite recall, but enjoy a closely related property: they depend (in a stationary way) on a finite number of summary, and easily updatable, statistics from the past. The results of these papers differ from those of the present paper in several respects. First, the convergence there is to a set, whereas here it is to a point. Second, the convergence there is to correlated equilibria, whereas here it is to Nash equilibria. And third, the strategies there are natural, adaptive, heuristic strategies, while in this paper we are dealing with forms of exhaustive search (see (h) below). An issue for further study is to what extent the contrast can be captured by an analysis of the speeds of convergence (which appears to be faster for correlated equilibria).

(h) *Adaptation vs. experimentation*: Suppose we were to require in addition that the strategies of the players be "adaptive" in one way or another. For example, at time t player i could randomize only over actions that improve i's payoff given some sort of "expected" behavior of the other players at t, or over actions that would have yielded a better payoff if played at $t-1$, or if played every time in the past that the action at $t-1$ was played, or if played every time in the past (these last two are in the

---

[15]For large K, an approximate comparison can be made in terms of entropies: equilibria with higher entropy are more likely to be reached than those with lower entropy.

style of "regret-based" strategies; see Hart 2005 for a survey). What kind of results would then be plausible? Note that such adaptive or monotonicity-like conditions severely restrict the possibilities of "free experimentation" that drive the positive results obtained here. Indeed, even the weak requirement of never playing the currently worst action rules out convergence to Nash equilibria: for example, in the Jordan (1993) example where each player has two actions, this requirement leads to the best-reply dynamic, which does not converge to the unique Nash equilibrium.

Thus, returning to the issue, raised at the end of the Introduction, of distinguishing those classes of dynamics for which convergence to Nash equilibria can be obtained from those for which it cannot, "exhaustive experimentation" appears as a key ingredient in the former.

Appendix A. Proof of Theorem 5

As pointed out in Section 4, the problem with our construction in the proof of Proposition 4 is that it leads to periodic and synchronized behavior. To avoid this we introduce small random perturbations, independently for each player: once in a while, there is a positive probability of repeating the previous period's action rather than continuing the periodic play.[16] To guarantee that these perturbations do not eventually change the frequencies of play (our players cannot use any additional "notes" or "instructions"[17]), we use three repetitions rather than two and make sure that the basic periodic play can always be recognized from the R-history.

Proof of Theorem 5. We take $R = 3K$, where $K > 2$ is chosen so as to satisfy (2). Consider sequences $\mathbf{b} = (b1, b_2, \ldots, b_{3K})$ of length $3K$ over an arbitrary finite set B (i.e., $b_k \in B$ for all k). We distinguish two types of such sequences:

- **Type E** ("Exact"): The sequence is K-periodic, i.e., $b_{K+k} = b_k$ for all $1 \leq k \leq 2K$. Thus $\mathbf{b}$ consists of three repetitions of the *basic* K-sequence $\mathbf{c} := (b1, b_2, \ldots, b_K)$.

- **Type D** ("Delay"): The sequence is not of type E, and there is $2 \leq d \leq 3K$ such that $b_d = b_{d-1}$ and if we drop the element b_d from the sequence

---

[16]This kind of perturbation was suggested by Benjy Weiss. The randomness is needed to obtain (4) (one can get (3) using deterministic mixing in appropriately long blocks).

[17]As would be the case were the strategies of finite memory, as in Theorem 7.

b then the remaining sequence $\mathbf{b}_{-d} = (b_1, \ldots, b_{d-1}, b_{d+1}, \ldots, b_{3K})$ of length $3K-1$ is K-periodic. Again, let **c** denote the *basic K-sequence*,[18] so $\mathbf{b}_{-d}$ consists of three repetitions of **c**, except for the missing last element. Think of b_d as a "delay" element.

We claim that the basic sequence of a sequence **b** of type D is uniquely defined. Indeed, assume that $\mathbf{b}_{-d}$ and $\mathbf{b}_{d'}$ are both K-periodic, with corresponding basic sequences $\mathbf{c} = (c_1, c_2, \ldots, c_k)$ and $\mathbf{c}' = (c'_1, c'_2, \ldots, c'_K)$, and $d < d'$. If $d \geq K + 1$, then the first K coordinates of **b** determine the basic sequence: $(b_1, b_2, \ldots, b_K) = \mathbf{c} = \mathbf{c}'$. If $d' \leq 2K$, then the last K coordinates determine this: $(b_{2K+1}, b_{2K+2}, \ldots, b_{3K}) = (c_K, c_1, \ldots, c_{K-1}) = (c'_K, c'_1, \ldots, c'_{K-1})$, so again $\mathbf{c} = \mathbf{c}'$. If neither of these two hold, then $d \leq K$ and $d' \geq 2K + 1$. Without loss of generality assume that we took d' to be maximal such that $\mathbf{b}_{-d'}$ is K-periodic, and let $d' = 2K + r$ (where $1 \leq r \leq K$). Now $b_{d'-1} = c'_{r-1}$ and $b_{d'} = c_{r-1}$, so $c_{r-1} = c'_{r-1}$ (since $b_{d'} = b_{d'-1}$). But $c_{r-1} = b_{K+r} = c'_r$ (since $d < K + r < d'$; if $r = 1$ put $r - 1 \equiv K$), so $c'_{r-1} = c'_r$. If $d' < 3K$, this last equality implies that $b_{d'} = b_{d'+1}$, so $\mathbf{b}_{-(d'+1)}$ is also K-periodic (with the same basic sequence $\mathbf{c}'$), contradicting the maximality of d'. If $d' = 3K$, the equality becomes $c'_{K-l} = c'_K$, which implies that the sequence **b** is in fact of type E (it consists of three repetitions of $\mathbf{c}'$), again a contradiction.

Given a sequence **b** of type E or D, the frequency distribution of its basic K-sequence **c**, i.e., $w \in \Delta(B)$ where $w(b) := |\{1 \leq k \leq K : c_k = b\}|/K$ for each $b \in B$, will be called the *basic frequency distribution* of **b**.

To define the strategies, let $\mathbf{a} = (a_1, a_2, \ldots, a_{3k}) \in A \times A \times \cdots \times A$ be the state—a history of action combinations of length $3K$—and put $\mathbf{a}^i := (a_1^i, a_2^i, \ldots, a_{3K}^i)$ for the corresponding sequence of actions of player i. When $\mathbf{a}^i$ is of type E or D, we denote by $y^i \in \Delta(A^i)$ its basic frequency distribution. If for each player i the sequence $\mathbf{a}^i$ is of type E or D,[19] we shall say that the state **a** is *regular*; otherwise we shall call **a** *irregular*.

The strategy of player i is defined as follows.

---

[18] As we shall see immediately below, **c** is well defined (even though d need not be unique).

[19] Some players' sequences may be E, and others', D (moreover, they may have different d's); therefore **a** itself need not be E or D.

(∗) If the state **a** is regular, the basic frequency y^i is a 4ε-best reply
   to the basic frequencies of the other players $y^{-i} = (y^j)_{j \neq i}$, and the
   sequence $\mathbf{a}^i$ is of type E, then with probability $1/2$ play a_1^i (i.e.,
   continue the K-periodic play), and with probability $1/2$ play a_{3K}^i
   (i.e., introduce a "delay" period by repeating the previous period's
   action).

(∗∗) If the state **a** is regular, the basic frequency y^i is a 4ε-best reply
   to the basic frequencies of the other players $y^{-i} = (y^j)_{j \neq i}$, and the
   sequence $\mathbf{a}^i$ is of type D, then play the last element of the basic
   sequence **c**, which is a_K^i if $d > K$ and a_{K+1}^i if $d \leq K$ (i.e., continue
   the K-periodic play).

(∗∗∗) In all other cases randomize uniformly over A^i.

As in the proof of Proposition 4, given a state **a**, for each player i let $z^i \in \Delta(A^i)$ denote the frequency distribution of $(a_{2K+1}^i, a_{2K+2}^i, \ldots, a_{3K}^i)$, the last K actions of i, and put $z = (z^1, z^2, \ldots, z^N)$; also, $y = (y^1, y^2, \ldots, y^N)$ is the N-tuple of the basic frequency distributions. We partition the state space S, which consists of all sequences **a** of length $3K$ over A, into four regions:

$S_1 := \{\mathbf{a}$ is regular and y is a Nash 4ε-equilibrium$\}$;
$S_2 := \{\mathbf{a}$ is irregular and z is a Nash 2ε-equilibrium$\}$;
$S_3 := \{\mathbf{a}$ is regular and y is not a Nash 4ε-equilibrium$\}$;
$S_4 := \{\mathbf{a}$ is irregular and z is not a Nash 2ε-equilibrium$\}$.

We analyze each region in turn.

Claim 1. *All states in S_1 are ergodic.*[20]

Proof. Let $\mathbf{a} \in S_1$. For each player i, let $\mathbf{c}^i = (c_1^i, c_2^i, \ldots, c_K^i)$ be the basic sequence of $\mathbf{a}^i$, with y^i the corresponding basic frequency distribution. The strategies are such that the sequence of i in the next period is also of type E or D (by (∗) and (∗∗)), with basic sequence that is the cyclical permutation of $\mathbf{c}^i$ by one step $(c_2, \ldots, c_K, c_1)$, except when $\mathbf{a}^i$ is of type D with $d = 2$, in which case it remains unchanged. Therefore the basic frequency distribution y^i does not change, and the new state is also in S_1. Hence S_1 is a closed set, and once it is reached the conditions of regularity and 4ε-best-replying

---

[20]I.e., aperiodic and persistent; see Feller (1968, Sections XV.4–6) for these and the other Markov chain concepts that we use below.

for each player will always continue to be automatically satisfied; thus each player's play in S_1 depends only on whether his *own* sequence is of type E or D (again, see $(*)$ and $(**)$). Therefore in the region S_1 the play becomes independent among the players, and the Markov chain restricted to S_1 is the product of N independent Markov chains, one for each player. Specifically, the state space S_1^i of the Markov chain of player i consists of all sequences of length $3K$ over A^i that are of type E or D, and the transition probabilities are defined as in $(*)$ and $(**)$, according to whether the sequence is of type E or D, respectively.

We thus analyze each i separately. Let $\mathbf{a}^i$ be a $3K$-sequence in S_1^i with basic sequence $\mathbf{c}^i$. The closure of $\mathbf{a}^i$ (i.e., the minimal closed set containing $\mathbf{a}^i$) consists of all $\tilde{\mathbf{a}}^i$ in S_1^i whose basic sequence is one of the K cyclical permutations of $\mathbf{c}^i$: any such state can be reached from any other in finitely many steps (for instance, it takes at most $3K - 1$ steps to get to a sequence of type E, then at most $K - 1$ steps to the appropriate cyclical permutation, and then another $3K$ steps to introduce a delay and wait until it reaches the desired place in the sequence).

Next, the states in S_1^i are aperiodic. Indeed, if the basic sequence $\mathbf{c}^i$ is constant (i.e., $\mathbf{c}^i = (\hat{a}^i, \hat{a}^i, \ldots, \hat{a}^i)$ for some $\hat{a}^i \in A^i$), then the constant sequence $(\hat{a}^i, \hat{a}^i, \ldots, \hat{a}^i)$ of length $3K$ is an absorbing state (since the next play of i will always be $\hat{a}^i$ by $(*)$), and thus aperiodic. If $\mathbf{c}^i$ is not constant, then assume without loss of generality that $c_1^i \neq c_K^i$ (if not, take an appropriate cyclical permutation of $\mathbf{c}^i$, which keeps us in the same minimal closed set). Let $\mathbf{a}^i$ be the sequence of type E that consists of three repetitions of $\mathbf{c}^i$. Starting at $\mathbf{a}^i$, there is a positive probability that $\mathbf{a}^i$ is reached again in K steps, by always making the first choice in $(*)$ (i.e., playing K-periodically, with no delays). However, there is also a positive probability of returning to $\mathbf{a}^i$ in $3K + 1$ steps, by always making the first choice in $(*)$, *except* for the initial choice which introduces a delay (after $3K$ additional steps the delay coordinate is no longer part of the state and we return to the original sequence[21] $\mathbf{a}^i$). But K and $3K + 1$ are relatively prime, so the state $\mathbf{a}^i$ is aperiodic. Therefore, every minimal closed set contains an aperiodic state, and all states are aperiodic.

Returning to the original Markov chain (over N-tuples of actions), the product of what we have shown is that, to each combination of basic sequences $(\mathbf{c}^1, \mathbf{c}^2, \ldots, \mathbf{c}^N)$ whose frequency distributions constitute a Nash

---

[21] The condition $c_1^i \neq c_K^i$ is needed in order for the delay action, c_K^i, to be different from the K-periodic action, c_1^i.

4ε-equilibrium, there corresponds an ergodic set[22] consisting of all states with basic sequences that are, for each i, some cyclic permutation of $\mathbf{c}^i$. The set S_1 is precisely the union of all these ergodic sets. □

The next three claims show that all states outside S_1 are transient: from any such state there is a positive probability of reaching S_1 in finitely many steps.

Claim 2. *Starting from any state in S_2, there is a positive probability that a state in S_1 is reached in at most $2K$ steps.*

Proof. Let $\mathbf{a} \in S_2$. Since at state $\mathbf{a}$ case $(***)$ applies to every player i, there is a positive probability that i will play a^i_{2K+1} (i.e., play K-periodically). If the new state $\mathbf{a}'$ is regular, then for each i the frequency distribution y^i of the resulting basic sequence either equals the frequency distribution z^i of the last K periods, or differs from it by $1/K$ (in the maximum norm). But z^i is a 2ε-best reply to z^{-i}, which implies that y^i is a 4ε-best reply to y^{-i} by (2)—and so $\mathbf{a}' \in S_1$. If the new state $\mathbf{a}'$ is irregular, then $\mathbf{a}' \in S_2$. Again, at $\mathbf{a}'$ there is a positive probability that every player will play K-periodically. Continuing in this way, we must at some point reach a regular state—since after $2K$ such steps the sequence of each player is surely of type E—a state that is therefore in S_1. □

Claim 3. *Starting from any state in S_3, there is a positive probability that a state in S_2 is reached in at most $5K + 1$ steps.*

Proof. Let $\mathbf{a} \in S_3$. There is a positive probability that every player will continue to play his basic sequence K-periodically, with no delays (this has probability $1/2$ in $(*)$, 1 in $(**)$, and $1/|A^i|$ in $(***)$). After at most $3K+1$ steps, we get a sequence of type E for every player (since all the original delay actions are no longer part of the state). During these steps the basic frequencies y^i did not change, so there is still one player, say player 1, such that y^1 is not a 4ε-best reply to y^{-1}. So case $(***)$ applies to player 1, and thus there is a positive probability that he will next play an action $\hat{a}^1$ that satisfies $1 - y^1(\hat{a}^1) > 1/K$ (for instance, let $\hat{a}^1 \in A^1$ have minimal frequency in y^1, then $y^1(\hat{a}^1) \le 1/|A^1| \le 1/2$ and so[23] $1 - y^1(\hat{a}^1) \ge 1/2 > 1/K$). The

---

[22]I.e., a minimal closed and aperiodic set.

[23]This is where $K > 2$ is used. Note that $|A^1| \ge 2$, since otherwise player 1 would always be best-replying.

sequence of every other player $i \neq 1$ is of type E, so with positive probability i plays a^i_{2K+1} (this has probability $1/2$ if ($*$) and $1/|A^i|$ if ($***$)), and thus the sequence of i remains of type E. With positive probability this continues for K periods for all players $i \neq 1$. As for player 1, note that y^{-1} does not change (since all other players $i \neq 1$ play K-periodically and so their y^i does not change). If at any point during these K steps the sequence of player 1 turns out to be of type E or D, then it contains at least two repetitions of the original basic sequence (recall that we started with three repetitions, and we have made at most K steps), so the basic frequency is still y^1; but y^1 is *not* a 4ε-best reply to y^{-1}, so we are in case ($***$). If the sequence is not of type E or D then of course we are in case ($***$)—so case ($***$) always applies during these K steps, and there is a positive probability that player 1 will always play $\hat{a}^1$.

But after K periods the sequence of player 1 is for sure neither of type E nor D: the frequency of $\hat{a}^1$ in the last K periods equals 1, and in the middle K periods it equals $y^1(\hat{a}^1)$, and these differ by more than $1/K$ (whereas in a sequence of type E or D, any two blocks of length K may differ in frequencies by at most $1/K$). Now these two K-blocks remain part of the state for K more periods, during which the sequence of player 1 can thus be neither E nor D, and so the state is irregular. Hence case ($***$) applies to all players during these K periods, and there is a positive probability that each player i plays a K-sequence whose frequency is $\bar{y}^i$, where $(\bar{y}^1, \bar{y}^2, \ldots, \bar{y}^N)$ is a Nash 2ε-equilibrium (see the beginning of the proof of Proposition 4). So, finally, after at most $(3K + 1) + K + K$ steps, a state in S_2 is reached. $\qquad\square$

Claim 4. *Starting from any state in S_4, there is a positive probability that a state in $S_1 \cup S_2 \cup S_3$ is reached in at most K steps.*

Proof. At a $\in S_4$, case ($***$) applies to every player. There is therefore a positive probability that each player i plays according to a K-sequence with frequency $\bar{y}^i$, where again $(\bar{y}^1, \bar{y}^2, \ldots, \bar{y}^N)$ is a fixed Nash 2ε-equilibrium. Continue in this way until either a regular state (in S_1 or S_3) is reached, or, if not, then after at most K steps the state is in S_2. $\qquad\square$

Combining the four claims implies that almost surely one of the ergodic sets, all of which are subsets of S_1, will eventually be reached. Let T denote the period when this happens—so T is an almost surely finite stopping time—and let $q_T \in S_1$ be the reached ergodic state. It remains to prove (3) and (4).

Let $Q \subset S_1$ be an ergodic set. As we saw in the proof of Claim 1, all states in Q have the same basic frequency distributions $(y^1, y^2, \ldots, y^N)$. The independence among players in S_1 implies that $Q = Q^1 \times Q^2 \times \cdots \times Q^N$, where $Q^i \subset S_1^i$ is an ergodic set for the Markov chain of player i. For each $\mathbf{a}^i \in Q^i$, let $x_{\mathbf{a}^i}^i \in \Delta(A^i)$ be the frequency distribution of all $3K$ coordinates of $\mathbf{a}^i$, then $\|x_{\mathbf{a}^i}^i - y^i\| \leq 1/(3K)$ (they may differ when the sequence contains a delay). Let μ^i be the unique invariant probability measure on Q^i; then the average frequency distribution $x^i := \sum_{\mathbf{a}^i \in Q^i} \mu^i(\mathbf{a}^i) x_{\mathbf{a}^i}^i \in \Delta(A^i)$ also satisfies $\|x^i - y^i\| \leq 1/(3K)$. But $(y^1, y^2, \ldots, y^N)$ is a Nash 4ε-equilibrium, and so $(x^1, x^2, \ldots, x^N)$ is a Nash 6ε-equilibrium (by (2)).

Once the ergodic set Q^i has been reached, i.e., $q_T^i \in Q^i$, the probability of occurrence of each state $\mathbf{a}^i = (a_1^i, a_2^i, \ldots, a_{3K}^i)$ in Q^i converges to its invariant probability (see Feller 1968, Section XV.7):

$$\lim_{t \to \infty} \Pr[a^i(t+k) = a_k^i \quad \text{for } k = 1, 2, \ldots, 3K | q_T^i \in Q^i] = \mu^i(\mathbf{a}^i).$$

Projecting on the kth coordinate yields, for every $a^i \in A^i$,

$$\lim_{t \to \infty} \Pr[a^i(t) = a^i | q_T^i \in Q^i] = \sum_{\mathbf{a}^i \in Q^i: a_k^i = a^i} \mu^i(\mathbf{a}^i),$$

so, in particular, the limit on the left-hand side exists. Averaging over $k = 1, 2, \ldots, 3K$ yields on the right-hand side $\sum_{\mathbf{a}^i \in Q^i} \mu^i(\mathbf{a}^i) x_{\mathbf{a}^i}^i(a^i)$, which equals $x^i(a^i)$; this proves (5), from which (4) follows by independence.

A similar argument applies to the limit of the long-run frequencies Φ_t. This completes the proof of Theorem 5. $\square$

References

Arslan, G. and J. S. Shamma (2004), "Distributed Convergence to Nash Equilibria with Local Utility Measurements," in *43rd IEEE Conference on Decision and Control*, December 2004, Vol. 2, pp. 1538–1543.

Aumann, R. J. and S. Sorin (1989), "Cooperation and Bounded Recall," *Games and Economic Behavior*, 1, 5–39.

Feller, W. (1968), *An Introduction to Probability and Its Applications*, Vol. 1, 3rd edn. New York: Wiley.

Foster, D. P. and R. V. Vohra (1997), "Calibrated Learning and Correlated Equilibrium," *Games and Economic Behavior*, 21, 40–55.

Foster, D. P. and H. P. Young (2003), "Learning, Hypothesis Testing, and Nash Equilibrium," *Games and Economic Behavior*, 45, 73–96.

——— (2006), "Regret Testing: Learning to Play Nash Equilibrium without Knowing You Have an Opponent," *Theoretical Economics*, 1, 341–367.

Germano, F. and G. Lugosi (2007), "Global Nash Convergence of Foster and Young's Regret Testing," *Games and Economic Behavior*, 60, 135–154.

Hart, S. (2005), "Adaptive Heuristics," *Econometrica*, 73, 1401–1430. [Chapter 11]

Hart, S. and A. Mas-Colell (2000), "A Simple Adaptive Procedure Leading to Correlated Equilibrium," *Econometrica*, 68, 1127–1150. [Chapter 2]

―――― (2003a), "Regret-Based Continuous-Time Dynamics," *Games and Economic Behavior*, 45, 375–394. [Chapter 5]

―――― (2003b), "Uncoupled Dynamics Do not Lead to Nash Equilibrium," *American Economic Review*, 93, 1830–1836. [Chapter 7]

Jordan, J. (1993), "Three Problems in Learning Mixed-Strategy Nash Equilibria," *Games and Economic Behavior*, 5, 368–386.

Kakade, S. M. and D. P. Foster (2008), "Deterministic Calibration and Nash Equilibrium," *Journal of Computer and System Sciences*, 74, 115–130.

Shamma, J. S. and G. Arslan (2005), "Dynamic Fictitious Play, Dynamic Gradient Play, and Distributed Convergence to Nash Equilibria," *IEEE Transactions on Automatic Control*, 50, 312–327.

Young, H. P. (2004), *Strategic Learning and Its Limits*. Oxford, UK: Oxford University Press.

Chapter 9

UNCOUPLED AUTOMATA AND PURE NASH EQUILIBRIA*

Yakov Babichenko

We study the problem of reaching a pure Nash equilibrium in multi-person games that are repeatedly played under the assumption of uncoupledness: *every player knows only his own payoff function*. We consider strategies that can be implemented by finite-state automata, and characterize the minimal number of states needed in order to guarantee that a pure Nash equilibrium is reached in every game where such an equilibrium exists.

1. Introduction

We study the problem of reaching Nash equilibria in multi-person games, where the players play the same game repeatedly. The main assumption, called *uncoupledness* (see Hart and Mas-Colell 2003), is that every player knows only his own utility function. The resulting play of the game yields an *uncoupled dynamic*.

Hart and Mas-Colell (2003) show that if the game is played in continuous time, and the moves of every player are deterministic, then uncoupled dynamics cannot always lead to Nash equilibria. In Hart and Mas-Colell (2006) they show that the situation is different when stochastic moves are allowed and the game is played in discrete time: if the players know the history of play,[1] then there are uncoupled strategies that lead to a Nash equilibrium. The question is whether it is necessary to know the whole

Originally published in *International Journal of Game Theory*, 39 (2010), 483–502.

*Based on the author's M.Sc. thesis (2007), written under the supervision of Sergiu Hart.

Keywords: automaton; Nash equilibrium; uncoupledness.

I wish to thank Sergiu Hart for his support and guidance, and Noam Nisan for helpful discussions.

[1]I.e., the past actions of all the players.

191

history in order to reach a Nash equilibrium. The answer is no. Hart and Mas-Colell (2006, Theorems 4 and 5) showed that under the assumption of uncoupledness, convergence of the long-run empirical distribution of play to a (pure or mixed) Nash equilibrium can be guaranteed by using only the history of the last R periods of play, for some finite R. This is called a *finite-recall strategy*. Although finite-recall uncoupled strategies can guarantee convergence of the distribution of play to a Nash equilibrium, Hart and Mas-Colell (2006, Theorem 6) show that this *cannot* hold for the period-by-period behavior probabilities. If, however, instead of finite recall one uses *finite memory* (e.g., finitely many periods of history but not necessarily the last ones), then the convergence of the behavior can be guaranteed as well (Hart and Mas-Colell 2006, Theorem 7).

This leads us to the study of uncoupled strategies with finite memory, i.e., *finite-state automata*. In this paper, we deal with convergence to pure Nash equilibria in games that have such equilibria. Hart and Mas-Colell (2006, Theorem 3) show that in order to guarantee convergence to pure Nash equilibria one needs recall of size $R = 2$. Since finite recall is a special case of finite automata, the question we address here concerns the *minimum number of states required* for uncoupled finite automata to reach a pure Nash equilibrium. There are four classes of finite-state automata: the actions in every state can be deterministic (pure) or stochastic (mixed), and the transitions between states can be deterministic or stochastic. We will analyze each of the four classes in turn.

Section 2 presents the model, defines the relevant concepts, and states the total results of the paper. Since the results are different for two-player games than for games with more than two players, we consider two-player games in Section 3 and n-player games for $n \geq 3$ in Section 4. In Sections 3 and 4, moreover, we discuss each of the four automata classes separately. Appendix A and Appendix B contain the proofs of Theorems 6 and 7, respectively.

2. The Model

2.1. *The game*

A basic static (one-shot) game Γ is given in strategic (or normal) form as follows. There are $n \geq 2$ players, denoted $i = 1, 2, \ldots, n$. Each player i has a finite set of pure actions $A^i = \{a_1^i, \ldots, a_{m^i}^i\}$; let $A := A^1 \times A^2 \times \cdots \times A^n$ be the set of action combinations. The payoff function (or utility function) of player i is a real-valued function $u^i : A \to \mathbb{R}$. The set of mixed (or randomized) actions of player i is the probability simplex over A^i, i.e., $\Delta(A^i) = \{x^i = (x^i(a^i_j))_{j=1,\ldots,m^i} : \Sigma^{m^i}_{j=1} x^i(a^i_j) = 1$ and $x^i(a^i_j) \geq 0$ for $j = 1, \ldots, m^i\}$; payoff functions u^i are multilinearly extended, and so $u^i : \Delta(A^1) \times \Delta(A^2) \times \cdots \times \Delta(A^n) \to \mathbb{R}$.

We fix the set of players n and the action sets A^i, and identify a game by its n-tuple of payoff functions $U = (u^1, u^2, \ldots, u^n)$. Let $\mathcal{U}^i$ be the set of payoff functions of player i, and $\mathcal{U} := \mathcal{U}^1 \times \cdots \times \mathcal{U}^n$.

Denote the actions of all the players except player i by a^{-i}, i.e., $a^{-i} = (a^1_{j_1}, \ldots, a^{i-1}_{j_{i-1}}, a^{i+1}_{j_{i+1}}, \ldots, a^n_{j_n})$, and denote the set of actions of all the players except player i by $A^{-i} = A^1 \times \cdots \times A^{i-1} \times A^{i+1} \times \cdots \times A^n$. An action $a^i_j \in A^i$ will be called a best reply to a^{-i} if $u^i(a^i_j, a^{-i}) \geq u^i(a^i_k, a^{-i})$ for every $a^i_k \in A^i$. A pure Nash equilibrium is an action combination $a = (a^1_{j_1}, a^2_{j_2}, \ldots, a^n_{j_n}) \in A$, such that $a^i_{j_i}$ is a best reply to a^{-i} for all i.

For every game U, let $\widetilde{U} = (\widetilde{u}^1, \widetilde{u}^2, \ldots, \widetilde{u}^n)$ denote the resulting *best-reply game*, which is defined by

$$\widetilde{u}^i(a) = \begin{cases} 1, & \text{if } a^i \text{ is a best reply to } a^{-i} \\ 0, & \text{otherwise} \end{cases}$$

Note that a is a pure Nash equilibrium of U if and only if it is a pure Nash equilibrium of $\widetilde{U}$.

2.2. The dynamic setup

The dynamic setup consists of the repeated play, at discrete-time periods $t = 1, 2, \ldots$, of the static game U. Let $a^i(t) \in A^i$ denote the action of player i at time t, and put $a(t) = (a^1(t), a^2(t), \ldots, a^n(t)) \in A$ for the combination of actions at t. We assume that there is standard monitoring: at the end of period t each player i observes everyone's action, i.e., $a(t)$; when the choices are random, the players observe only the realized actions $a(t)$.

2.3. Automata

An *automaton*[2] *for player* i is a 4-tuple $\Lambda^i := \langle \Psi^i, \mathfrak{s}^i_0, f^i, g^i \rangle$. Ψ^i is the set of *states*; $\mathfrak{s}^i_0 \in \Psi^i$ is the *starting state*; $f^i : \Psi^i \to \Delta(A^i)$ is the *action function*; and $g^i : A \times \Psi^i \to \Delta(\Psi^i)$ is the *transition function*. Let $\mathcal{A}^i$

---

[2]This is short for "a strategy implemented by an automaton."

denote the set of all automata of player i. An automaton $\Lambda^i \in \mathcal{A}^i$ will be called *a pure-action automaton* if the actions in all states are pure, i.e.,[3] $\mathrm{Im}(f^i) \subset A^i$. Otherwise it will be called a *mixed-action automaton*. An automaton $\Lambda^i \in \mathcal{A}^i$ will be called a *deterministic-transition automaton* if all the transitions are deterministic, i.e., $\mathrm{Im}(g^i) \subset \Psi^i$. Otherwise it will be called a *stochastic-transition automaton*. An automaton $\Lambda^i \in \mathcal{A}^i$ will be called a k^i-*automaton* if it has k^i states, i.e., $|\Psi^i| = k^i$.

Let $(\Lambda^1, \Lambda^2, \ldots, \Lambda^n)$ be n automata, where Λ^i is a k^i-automaton for player i. The play proceeds as follows. At time $t = 1$ every player i is at his starting state $\mathfrak{s}_0^i$, and plays an action $a^i(1)$ according to the probability distribution $f^i(\mathfrak{s}_0^i)$. Let the realized actions of all the players be $a(1) := (a^1(1), \ldots, a^n(1))$. Then every player i moves to a new state according to the transition probabilities $g^i(a(1), \mathfrak{s}_0^i)$. Now assume that at time t player i is in state $s^i \in \Psi^i$, and hence at time $t + 1$ player i plays an action $a^i(t)$ according to the probability distribution $f^i(s^i)$. The actions of all the players are $a(t+1)$, and every player i then moves to a new state according to the transition probabilities $g^i(a(t+1), s^i)$.

2.4. *Strategy mappings*

Let $\varphi : \mathcal{U} \to \mathcal{A}^1 \times \cdots \times \mathcal{A}^n$ be a mapping that associates to every game $U = (u^1, \ldots, u^n) \in \mathcal{U}$ an n-tuple of automaton strategies $\varphi(U) = (\varphi^1(U), \ldots, \varphi^n(U))$ (with $\varphi^i(U)$ the automaton of player i). We will call the mapping φ *uncoupled* if, for each player i, the ith coordinate φ^i of φ depends only on u^i, i.e., $\varphi^i : \mathcal{U}^i \to \mathcal{A}^i$ (rather than $\varphi^i : \mathcal{U} \to \mathcal{A}^i$). That is, φ^i associates to each payoff function $u^i \in \mathcal{U}^i$ of player i an automaton $\varphi^i(u^i) \in \mathcal{A}^i$, and

$$\varphi(U) = \varphi(u^1, u^2, \ldots, u^n) = (\varphi^1(u^1), \varphi^2(u^2), \ldots, \varphi^n(u^n)).$$

We will refer to $\varphi^i : \mathcal{U}^i \to \mathcal{A}^i$ as an *uncoupled strategy mapping to automata for player* i; thus φ^i "constructs" an automaton for player i by considering u^i only.[4] If $\varphi^i(u^i) \in \mathcal{A}^i$ is an automaton of size (at most) k^i for every payoff function $u^i \in \mathcal{U}^i$, we will say that φ^i is an *uncoupled strategy mapping to* k^i-*automata*.

Finally, we will say that the mapping φ is a *Pure Nash mapping*, or *PN-mapping* for short, if the strategies $\varphi(U)$ yield almost sure convergence

---

[3]We identify A^i with the unit vectors in $\Delta(A^i)$.

[4]We assume that every player i knows his index i.

of play to a pure Nash equilibrium in every game $U \in \mathcal{U}$ where such an equilibrium exists.

2.5. The results

Clearly, every finite-recall strategy is in particular a finite-automaton strategy. Indeed, a strategy with recall of size R can be implemented by an automaton of size $|A|^R = (\prod_{i=1}^{n} m^i)^R$ (i.e., one state for each possible recall). Therefore, by Theorem 3 in Hart and Mas-Colell (2006), there is uncoupled PN-mapping to automata of size $(\prod_{i=1}^{n} m^i)^2$. The question we address here is whether there is uncoupled PN-mapping to automata with fewer states.

Our purpose is thus to characterize minimal numbers $k^1, \ldots, k^n$ such that there exists uncoupled PN-mapping where, for each i, the range is k^i-automata. We will analyze each of the four cases (pure or mixed-action automata, and deterministic or stochastic-transition automata) separately.

The results are the following:

For two-player games $(n = 2)$:

- There exists uncoupled PN-mapping to automata of sizes:

	mixed actions	pure actions
stochastic transitions	$\begin{cases} m^1 \\ m^2 + 1 \end{cases}$ or $\begin{cases} m^1 + 1 \\ m^2 \end{cases}$	$\begin{cases} m^1 \\ m^2 + 1 \end{cases}$ or $\begin{cases} m^1 + 1 \\ m^2 \end{cases}$
deterministic transitions	$\begin{cases} m^1 + 2 \\ m^2 + 2 \end{cases}$	$\begin{cases} 4m^1 + O(1) \\ 4m^2 + O(1) \end{cases}$

- There is no uncoupled PN-mapping to automata of sizes m^1, m^2.

For n-player games $(n \geq 3)$:

- There exists uncoupled PN-mapping to automata of sizes:

	mixed actions	pure actions
stochastic transitions	$2m^i$	$2m^i$
deterministic transitions	$2m^i + 3$	$O(m^i + n \log n)$

- Let $k^1, k^2, \ldots, k^n$ such that $\forall i = 1, \ldots, n : k^i < 2m^i$. Then there is no uncoupled PN-mapping to automata of sizes $k^1, k^2, \ldots, k^n$.

3. Two-Player Games

3.1. *Stochastic transitions and mixed actions*

We will show that there exists an uncoupled PN-mapping where the range for player 1 is (m^1+1)-automata and the range for player 2 is m^2-automata or, symmetrically, the range for player 1 is m^1-automata and the range for player 2 is $(m^2 + 1)$-automata. On the other hand, we will show that there is no PN-mapping where the ranges of the players are smaller.

Theorem 1. *Let $k^1 \geq m^1$ and $k^2 \geq m^2 + 1$. Then, for each player $i = 1, 2$, there exists an uncoupled strategy mapping to k^i-automata with stochastic transitions and mixed actions that guarantees almost sure convergence of play to a pure Nash equilibrium of the stage game in every game where such an equilibrium exists.*

Proof. We define the mapping φ as follows:

Given a game $U = (u^1, u^2)$, the automaton $\varphi^1(u^1) = \Lambda^1 \in \mathcal{A}^1$ is constructed as follows:

Denote $\varphi^1(u^1) = \Lambda^1 = \langle \Psi^1, \mathfrak{s}_0^1, f^1, g^1 \rangle$ when Λ^1 is an m^1-automaton. We denote the states of Λ^1 by $\Psi^1 = \{s_1^1, \ldots, s_{m^1}^1\}$.

$$\mathfrak{s}_0^1 := s_1^1.$$

$$f^1(s_i^1) := a_i^1 \equiv (0, \ldots, 0, \overset{i}{1}, 0, \ldots, 0).$$

$$g^1(a, s_i^1) = g^1((a_i^1, a^2), s_i^1)$$

$$:= \begin{cases} s_i^1 \equiv (0, \ldots, 0, \overset{i}{1}, 0, \ldots, 0) & \text{if } a_i^1 \text{ is a best reply to } a^2 \\ \left(\dfrac{1}{m^1}, \ldots, \dfrac{1}{m^1} \right) & \text{otherwise} \end{cases}$$

In state s_i^1 player 1 plays action a_i^1. He stays in this state if a_i^1 is a best reply to the action of player 2; otherwise he moves randomly to any one of the m^1 states with equal probability $\frac{1}{m^1}$. Note that whether an action of player 1 is a best reply or not depends only on his payoff function; therefore, Λ^1 depends on u^1 only.

Now we construct the automaton $\varphi^2(u^2) = \Lambda^2 \in \mathcal{A}^2$ as follows:

Denote $\Lambda^2 = \langle \Psi^2, \mathbf{s}_0^2, f^2, g^2 \rangle$ when Λ^2 is an $(m^2 + 1)$-automaton. We denote the states of Λ^2 by $\Psi^2 = \{s_0^2, s_1^2, \ldots, s_{m^1}^2\}$.

$$\mathbf{s}_0^2 := s_0^2.$$

$$f^2(s_j^2) := \begin{cases} \left(\dfrac{1}{m^2}, \ldots, \dfrac{1}{m^2} \right) & j = 0 \\[2mm] a_j^2 & j \geq 1 \end{cases}$$

$$g^2(a, s_j^2) = g^2((a^1, a_j^2), s_j^2)$$

$$:= \begin{cases} \left(\dfrac{1}{m^2 + 1}, \ldots, \dfrac{1}{m^2 + 1} \right) & j = 0 \\[2mm] s_j^2 & j \geq 1 \text{ and } a_j^2 \text{ is a best reply to } a^1 \\[2mm] \left(\dfrac{1}{m^2 + 1}, \ldots, \dfrac{1}{m^2 + 1} \right) & j \geq 1 \text{ and } a_j^2 \text{ is not a best reply to } a^1 \end{cases}$$

In the state s_0^2, player 2 plays the mixed action $(\frac{1}{m^2}, \ldots, \frac{1}{m^2})$, and moves to any of the $m^2 + 1$ states with probability $\frac{1}{m^2+1}$.

In the states s_i^2, $i \geq 1$, player 2 plays action a_i^2. He stays in this state if a_i^2 is a best reply to the action of player 1; otherwise he moves to the state s_0^2.

Now we will prove that (φ^1, φ^2) is a PN-mapping.

We partition the space $\Psi^1 \times \Psi^2$ of the automata states into four regions:

$P_1 := \{(s_i^1, s_j^2), 1 \leq i \leq m^1, 1 \leq j \leq m^2 : \widetilde{u}^1(a_i^1, a_j^2) = 1, \widetilde{u}^2(a_i^1, a_j^2) = 1\}$; i.e., in this case (a_i^1, a_j^2) is a pure Nash equilibrium.

$P_2 := \{(s_i^1, s_0^2), 1 \leq i \leq m^1\}$.

$P_3 := \{(s_i^1, s_j^2), 1 \leq i \leq m^1, 1 \leq j \leq m^2 : \widetilde{u}^2(a_i^1, a_j^2) = 0\}$.

$P_4 := \{(s_i^1, s_j^2), 1 \leq i \leq m^1, 1 \leq j \leq m^2 : \widetilde{u}^1(a_i^1, a_j^2) = 0, \widetilde{u}^2(a_i^1, a_j^2) = 1\}$.

These four regions clearly cover the space $\Psi^1 \times \Psi^2$. In fact, player 2 can be in the state s_0^2 (P_2) or in any other state ($P_1 \cup P_3 \cup P_4$). If player 2 is not in the state s_0^2, then the action of player 2 can be a best reply ($P_1 \cup P_4$) or not (P_3). If it is a best reply, then the action of player 1 can be a best reply (P_1) or not (P_4).

Players 1 and 2 stay at the same state if their action is a best reply; i.e., each state in P_1 is absorbing. We will prove that there is a positive probability of reaching a state from P_1, in finitely many periods, from any other state $s \in \Psi^1 \times \Psi^2$.

Definition 2. An action $a_j^i \in A^i$ of player i will be called *dominant* if for every $a^{-i} \in A^{-i}$ a_j^i is a best reply to a^{-i}.

$s = (s_i^1, s_0^2) \in P_2$: The actions are $(a_i^1, (\frac{1}{m^2}, \ldots, \frac{1}{m^2})) = (f^1(s_i^1), f^2(s_0^2))$. If a_i^1 is a dominant action, then denote by a_l^2 an action that is a best reply to a_i^1. Player 2 moves to s_l^2 with probability $\frac{1}{m^2+1}$. Then $(s_i^1, s_l^2) \in P_1$. If a_i^1 is not a dominant action, then denote by a_k^2 an action such that a_i^1 is not a best reply to it, and then with probability $\frac{1}{m^2}$ player 2 plays action a_k^2. Now both players move randomly over all their states and with positive probability they will get to P_1.

$s = (s_i^1, s_j^2) \in P_3$: The actions are (a_i^1, a_j^2); a_j^2 is not a best reply. Therefore, player 2 moves to s_0^2. Denote the state to which player 1 moves by s_k^1. Then $(s_k^1, s_0^2) \in P_2$.

$s = (s_i^1, s_j^2) \in P_4$: The actions are (a_i^1, a_j^2); a_j^2 is a best reply, and a_i^1 is not. Therefore, player 2 stays in s_j^2, whereas player 1 moves to s_k^1 with probability $\frac{1}{m^1}$, where a_k^1 is a best reply of player 1 to a_j^2. Now either $(s_k^1, s_j^2) \in P_1$ or $(s_k^1, s_j^2) \in P_3$, depending on whether a_j^2 is a best reply to a_k^1.

In each of the above cases there is positive probability of reaching an absorbing state in P_1 in at most 3 steps.

Definition 3. A game U will be called a *full game* if, for every action $a_j^i \in A^i$ of every player i, there exists $a^{-i} \in A^{-i}$ such that a_j^i is a best reply to a^{-i}.

We prove a general result about n-player full games that will be useful in the sequel.

Lemma 4. *Let $\varphi = (\varphi^1, \ldots, \varphi^n)$ be an uncoupled strategy mapping that guarantees almost sure convergence of play to a pure Nash equilibrium of the stage game in every game where such an equilibrium exists. Then for every full game $U = (u_1, \ldots, u_n)$ and for every player i, there exist m^i nonempty sets of states $B_1^i, \ldots, B_{m^i}^i$ in Ψ^i (the set of states of the automaton $\varphi^i(u^i) = \Lambda^i$) such that in every state $s_k^i \in B_j^i$ player i plays a_j^i (with probability 1), and stays in B_j^i (with probability 1) if his action is a best reply to the actions of the other players.*

Proof. (By contradiction). Assume that there exists a full game U s.t. for player 1 the set B_j^1 does not exist (or is empty). U is a full game, and so there exists $a^{-1} \in A^{-1}$ such that a_j^1 is a best reply to it. Consider the game

$$\overline{U} = (\overline{u^1}, \overline{u^2}, \ldots, \overline{u^n}) \text{ when } \overline{u^1} := u^1, \text{ and } \overline{u^i}(a) := \begin{cases} 1 \text{ if } a = (a_j^1, a^{-1}) \\ 0 \text{ otherwise} \end{cases}.$$

The only Nash equilibrium of $\overline{U}$ is (a_j^1, a^{-1}). By uncoupledness we get $\Lambda^1 = \varphi^1(u^1) = \varphi^1(\overline{u^1}) = \overline{\Lambda^1}$. If (a_j^1, a^{-1}) has been played, the next period player 1 will not play a_k^1 with probability 1 (otherwise the set B_j^1 could not be empty), and the equilibrium in the game $\overline{U}$ will never be reached with probability 1 (in contradiction to the assumption). □

Theorem 5. *Let* $k^1 = m^1$ *and* $k^2 = m^2$. *Then there are no uncoupled strategy mappings to* k^i-*automata with stochastic transitions and mixed actions, that guarantees almost sure convergence of play to a pure Nash equilibrium of the stage game in every game where such an equilibrium exists.*

Proof. (By contradiction). Let U be a full game. Consider the sets $B_1^i, \ldots, B_{m^i}^i$ in Λ^i (see Lemma 3). By assumption $|\Lambda^i| \leq m^i$. On the one hand, $|B_j^i| \geq 1$, and, on the other hand, $\sum_j |B_j^i| \leq |\Lambda^i| = m^i$, and so $|B_j^i| = 1$ and $\cup_j B_j^i = \Psi^i$. In other words, every B_j^i includes exactly one state in which player i plays a_j^i and stays there if a_j^i is a best reply, and there are no other states. Therefore, the strategy of player i is such that if his action is a best reply to the action of the other player, then in the next step he plays the same action. Hart and Mas-Colell (2006, Proof of Theorem 1) show that such a strategy cannot always lead to a pure Nash equilibrium, contradicting our assumption. □

3.2. *Stochastic transitions and pure actions*

We will show the result of Theorem 1 continues to hold when the automata are restricted to pure-action automata. As was shown in Theorem 5, however, there is no PN-mapping where the ranges of the players are smaller.

Theorem 6. *Let* $k^1 \geq m^1$ *and* $k^2 \geq m^2 + 1$. *Then, for each player* $i = 1, 2$, *there exists an uncoupled strategy mapping to* k^i-*automata with stochastic transitions and pure actions, that guarantees almost sure convergence of* *play to a pure Nash equilibrium of the stage game in every game where such an equilibrium exists.*

The proof is relegated to Appendix A.

3.3. *Deterministic transitions and mixed actions*

We will show that there exists an uncoupled PN-mapping where the range for player i is $(m^i + 2)$-automata. Clearly, every deterministic-transition automaton is a particular case of a stochastic-transition automaton, and so Theorem 5 holds here as well.

Theorem 7. *Let $k^i \geq m^i + 2$. Then for each player i there exists an uncoupled strategy mapping to k^i-automata with deterministic transitions and mixed actions, that guarantees almost sure convergence of play to a pure Nash equilibrium of the stage game in every game where such an equilibrium exists.*

The proof is relegated to Appendix B.

3.4. *Deterministic transitions and pure actions*

In Theorem 13, we will show for general n-player games that there exists an uncoupled PN-mapping where the range for player i is $(O(m^i + n \log n))$-automata, where m^i is the number of actions of player i and n is the number of players. In the case of 2 players, the construction of the automata in the proof of Theorem 13 proves the existence of an uncoupled PN-mapping where the range for player i is $(4m^i + O(1))$-automata.

Another uncoupled PN-mapping, specifically for 2 players, has a range of $(5m^1 + m^2 - 5)$-automata for player 1, and $(5m^2 + 2m^1 - 9)$-automata for player 2. We will not show this construction here but the idea is to go through all the possible actions (a_i^1, a_j^2) in some "economical" way.

4. *n*-**Player Games** $(n \geq 3)$

4.1. *Stochastic transitions, and pure or mixed actions*

We will show that there exists an uncoupled PN-mapping where the range for player i is $2m^i$-automata. On the other hand, we will show that there is no uncoupled PN-mapping whose range is smaller for all players.

Theorem 8. *Let $k^i \geq 2m^i$. Then for each player i there exists an uncoupled strategy mapping to k^i-automata with stochastic transitions and pure actions, that guarantees almost sure convergence of play to a pure Nash equilibrium of the stage game in every game where such an equilibrium exists.*

Proof. Let us introduce the mappings $\varphi^i(u^i) = \Lambda^i$ given a payoff function $U = (u^1, \ldots, u^n)$.

Denote the states of Λ^i by $\Psi^i = \{s^i_{1,0}, s^i_{1,1}, s^i_{2,0}, s^i_{2,1}, \ldots, s^i_{m^i,0}, s^i_{m^i,1}\}$. The states $s^i_{j,0}$ will be called 0-*states*; the states $s^i_{j,1}$, 1-*states*.

Definition 9. Given a state $s = (s^1, \ldots, s^n) \in \Psi^1 \times \cdots \times \Psi^n$ we will say that player i is *fit* at s if

- player i is at a 0-state and player $i + 1 (\mathrm{mod}\, n)$ is at a state $s^i_{j,k}$ for $k \in \{0, 1\}$ and $j \neq 1$, or
- player i is at a 1-state and player $i + 1 (\mathrm{mod}\, n)$ is at a state $s^i_{1,k}$ for $k \in \{0, 1\}$.

In every state $s^i_{j,l}$ player i plays action a^i_j. If a^i_j is a best reply to what the other players played, and player i fits player $i + 1 (\mathrm{mod}\, n)$, player i stays in $s^i_{j,l}$. Otherwise he moves to any one of the $2m^i$ states with equal probability $\frac{1}{2m^i}$.

Let the starting states be $\mathfrak{s}^i_0 := s^i_{1,0}$.

To prove that these automata reach a pure Nash equilibrium we partition the space $\Psi^1 \times \cdots \times \Psi^n$ of the automata states into $n + 2$ regions:

$P_1 := \{(s^1_{k_1,l_1}, \ldots, s^n_{k_n,l_n}), 1 \leq k_i \leq m^i, l_i = 0, 1 : (a^1_{k_1}, \ldots, a^n_{k_n})$ is a pure Nash equilibrium and all the players are fit$\}$.

Note that for every pure Nash equilibrium $(a^1_{k_1}, \ldots, a^n_{k_n})$ there is a state $s \in P_1$ where the players play $(a^1_{k_1}, \ldots, a^n_{k_n})$. Take $s^i_{k_i,l_i}$ with $l_i = 1$ when $k_{i+1} = 1$ and $l_i = 0$ otherwise.

For $0 \leq r \leq n - 1$, $P_{2,r} := \{(s^1_{k_1,l_1}, \ldots, s^n_{k_n,l_n})$: there exist exactly r players that are fit$\}$.

$P_3 := \{(s^1_{k_1,l_1}, \ldots, s^n_{k_n,l_n})$: all the players are fit, but $(a^1_{k_1}, \ldots, a^n_{k_n})$ is not a pure Nash equilibrium$\}$.

Clearly each state in P_1 is absorbing. Next we claim that a state in P_1 is reached with positive probability, in finitely many periods, from any other state $s \in \Psi^1 \times \cdots \times \Psi^n$.

$s \in P_{2,0}$: all the players are not fit, and so all the players move randomly over all their states, and there is a positive probability of reaching P_1.

For $1 \leq r \leq n-1$: $s = (s^1_{k_1,l_1}, \ldots, s^n_{k_n,l_n}) \in P_{2,r}$. Assume player i is fit, but player[5] $i+1$ is not. Such i exist, because we have a circle of players of which some are fit, and some are not. There is a positive probability that all the players except $i+1$ will stay at their states, and player $i+1$ (who moves randomly because he is not fit) will move in the following way: if $k_{i+1} = 1$ then he moves to $s^{i+1}_{2,l_{i+1}}$, and if $k_{i+1} \geq 2$, then he moves to $s^{i+1}_{1,l_{i+1}}$. Now all the players except i and $i+1$ remain fit/not fit, as they were before, because neither they nor the next player change their state. Player $i+1$ does not change his l_{i+1}, player $i+2$ does not change his state, and so player $i+1$ stays not fit, as he was before. Player i was fit but after the move of player $i+1$, he is not fit. The only player that changes his fitness is player i, and they get to $P_{2,r-1}$. By induction, with positive probability they get to $P_{2,0}$ in r steps.

$s = (s^1_{k_1,l_1}, \ldots, s^n_{k_n,l_n}) \in P_3$: The action is $a = (a^1_{k_1}, \ldots, a^n_{k_n})$ and it is not a pure Nash equilibrium; therefore, there exists player i s.t. $a^i_{k_i}$ is not a best reply to a^{-i}, and player i moves randomly over all the states. Hence, there is a positive probability that all the players except i will stay at their states and player i will move to $s^i_{k_i,1-l_i}$. Now all the players except player i stay fit, but player i is not fit. And so $(s^1_{k_1,l_1}, \ldots, s^i_{k_i,1-l_i}, \ldots, s^n_{k_n,l_n}) \in P_{2,n-1}$.

From any state s there is a positive probability of reaching an absorbing state in P_1 in at most $n+2$ steps. $\qquad \square$

Theorem 10. *Let $n \geq 4$, and let $k^1, \ldots, k^n$ satisfy $k^i < 2m^i$ for all $i = 1, \ldots, n$ (except, perhaps, for one of them). Then there is no uncoupled strategy mapping to k^i-automata with stochastic transitions and mixed actions, that guarantees almost sure convergence of play to a pure Nash equilibrium of the stage game in every game where such an equilibrium exists.*

Proof. Assume on the contrary that such a strategy mapping exists, and that $k^i < 2m^i$ for all $i = 2, \ldots, n$. By Lemma (2) in Λ^i there exist m^i nonempty sets of states $B^i_1, \ldots, B^i_{m^1}$ s.t. in every state $s^i_j \in B^i_k$ player i plays a^i_k. $|\Lambda^i| < 2m^i$, and by the pigeon hole principle there exists $k(i)$ s.t. $|B^i_{k(i)}| = 1$. Therefore, every player i has a state $s^i_{k(i)}$ where he plays $a^i_{k(i)}$, and he stays there if it is a best reply. Consider a four-player game where every player has 2 actions.

---

[5]From here till the end of the proof, we will write $i+1$ instead of $i+1(\mathrm{mod}\ n)$.

Consider the following utility function of players 2, 3, 4:

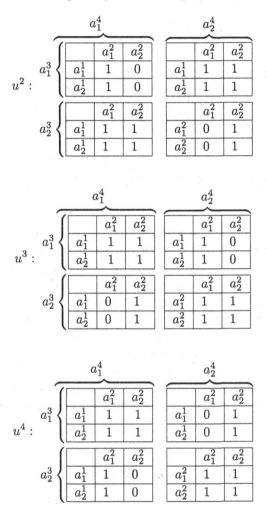

Player $i = 2, 3, 4$ gets 1 if he plays the same action as one of the players 2, 3, 4 (except himself). Otherwise he gets 0.

The strategy mapping $\varphi^i : u^i \to \mathcal{A}^i$ constructs an automaton. As mentioned, there exists a state $s^i_{k(i)}$ where player i plays $a^i_{k(i)}$, and he stays there if it is a best reply. There are 2 actions for every player, and 3 actions $a^i_{k(i)}$ $i = 2, 3, 4$. So there exist 2 players i, j who have the same action $a^i_{k(i)}, a^j_{k(j)}$, where $k(i) = k(j)$. Because of the symmetry of the functions

u^2, u^3, u^4, assume $k(i) = k(j) = 1$, and assume that the two players are players 3 and 4.

Let us now consider the following game:

$\Gamma_1 :$

	a_1^4				a_2^4		
			a_1^2	a_2^2		a_1^2	a_2^2
a_1^3		a_1^1	1,0,1,1	0,1,1,1	a_1^1	0,0,1,0	0,0,0,1
		a_2^1	0,1,1,1	1,0,1,1	a_2^1	0,0,1,0	0,0,0,1
			a_1^2	a_2^2		a_1^2	a_2^2
a_2^3		a_1^1	0,0,0,1	0,0,1,0	a_1^2	1,1,1,1	1,1,1,1
		a_2^1	0,0,0,1	0,0,1,0	a_2^2	1,1,1,1	1,1,1,1

Player 3 and player 4 have the utility functions u^3 and u^4, respectively. Therefore, the automaton that their strategy mapping constructs includes states s_1^3, s_1^4, where they play action a_1^3, a_1^4, respectively, and stay there if it is a best reply. If players 3 and 4 get to the states $s_{k(3)}^3$ and $s_{k(4)}^4$, then the pure Nash equilibrium will never be reached.

For a larger number of actions the same proof works, if we take all the actions $a_2^i, \ldots, a_{m^i}^i$ to be identical to the action a_2^i in this proof.

For a larger number of players we take the utility functions of players 2, 3, and 4 to be the same as in the case of 4 players and independent of the actions of the other players $(1, 5, \ldots, n)$. And in the game Γ_1 the utility functions of players $5, \ldots, n$ will be 1 if players 3 and 4 played the same action, and 0 otherwise. □

4.2. *Deterministic transitions and mixed actions*

We will show that there exists an uncoupled PN-mapping such that the range for player i is $(2m^i + 3)$-automata. Clearly, every deterministic-transition automaton is a particular case of a stochastic-transition automaton, and so Theorem 10 holds here as well.

Theorem 11. *Let $k^i \geq 2m^i + 3$. Then for each player i there exists an uncoupled strategy mapping to k^i-automata with deterministic transitions and mixed actions, that guarantees almost sure convergence of play to a pure Nash equilibrium of the stage game in every game where such an equilibrium exists.*

Proof. We introduce the mappings $\varphi^i(u^i) = \Lambda^i$ given a payoff function $U = (u^1, \ldots, u^n)$:

As in Theorem 8, we will use the states $\{s_{1,0}^i, s_{1,1}^i, s_{2,0}^i, s_{2,1}^i, \ldots, s_{m^i,0}^i, s_{m^i,1}^i\}$. These are the same states, except that their transitions are deterministic. Denote the states of Λ^i by $\Psi^i = \{s_1^i, s_2^i, s_3^i\} \cup \{s_{1,0}^i, s_{1,1}^i, s_{2,0}^i, s_{2,1}^i, \ldots, s_{m^i,0}^i, s_{m^i,1}^i\}$.

The states $s_{j,l}^i$ are similar to the states $s_{j,l}^i$ in Theorem 8. In every state $s_{j,l}^i$ player i plays action a_j^i. If a_j^i is a best reply to what the other players played, and player i fits, then player i stays there (exactly as before). Otherwise he moves to s_1^i.

In the state s_1^i player i plays $\left(\frac{1}{m^i}, \ldots, \frac{1}{m^i}\right)$. If he played a_1^i he stays in s_1^i. If he played a_2^i he moves to s_2^i. If he played a_j^i $j \neq 1, 2$ he moves to $s_{j,0}^i$.

In the state s_2^i player i plays $\left(\frac{1}{m^i}, \ldots, \frac{1}{m^i}\right)$. If he played a_3^i he stays in s_1^i. If he played a_4^i he moves to s_3^i. If he played a_j^i $j \neq 3, 4$ he moves to $s_{j,1}^i$.

In the state s_3^i player i plays $\left(\frac{1}{m^i}, \ldots, \frac{1}{m^i}\right)$. If he played a_1^i he moves to $s_{1,0}^i$. If he played a_2^i he moves to $s_{2,0}^i$. If he played a_3^i he moves to $s_{3,1}^i$. If he played a_4^i he moves to $s_{4,1}^i$. If he played a_5^i he stays in s_3^i. If he played a_j^i $j \geq 6$, he moves to s_1^i.

Let the starting states be $\mathfrak{s}_0^i := s_1^i$.

The proof of the claim that this mapping is a PN-mapping is proven similarly to Theorem 7 with 2 players. $\square$

4.3. *Deterministic transitions and pure actions*

We will show that there exists an uncoupled PN-mapping such that the range for player i is $O(m^i + n \log n)$-automata. Clearly, every deterministic-transition automaton is a particular case of a stochastic-transition automaton, and so Theorem 10 holds here as well.

Lemma 12. *For every $m, n \in \mathbb{N}$ there exist n different prime numbers $p_1, \ldots, p_n$, $p_i \geq m$ for every i, such that $p_i = O(m + n \log n)$ for every i.*

Proof. We know that in $\{1, 2, \ldots, m\}$ there exist at most $\alpha \frac{m}{\log m}$ prime numbers (α is a constant).

We also know that in $\{1, 2, \ldots, \beta(n + \alpha \frac{m}{\log m}) \log(n + \alpha \frac{m}{\log m})\}$ there exist at least $n + \alpha \frac{m}{\log m}$ prime numbers (β is a constant).

Therefore, in $\{m + 1, m + 2, \ldots, \beta(n + \alpha \frac{m}{\log m}) \log(n + \alpha \frac{m}{\log m})\}$ there exist at least n prime numbers, and we can take different prime numbers $p_1, \ldots, p_n$ s.t. $\forall i : m < p_i < \beta(n + \alpha \frac{m}{\log m}) \log(n + \alpha \frac{m}{\log m})$.

To complete the proof we will show that $\beta(n+\alpha\frac{m}{\log m})\log(n+\alpha\frac{m}{\log m}) = O(m + n\log n)$.

– If $m \le n$ then

$$\beta\left(n + \alpha\frac{m}{\log m}\right)\log\left(n + \alpha\frac{m}{\log m}\right) = O((1 + \alpha)n\log((1 + \alpha)n))$$

$$= O(n\log n) = O(m + n\log n).$$

– If $n < m \le n\log n$, then

$$\frac{m}{\log m} \le 2n\frac{\log n}{2\log m} = 2n\frac{\log n}{\log m^2} \le 2n$$

$$\Longrightarrow \beta\left(n + \alpha\frac{m}{\log m}\right)\log\left(n + \alpha\frac{m}{\log m}\right)$$

$$= O((1 + 2\alpha)n\log((1 + 2\alpha)n)) = O(n\log n) = O(m + n\log n).$$

– If $n\log n < m \le n^2$, then

$$n \le \frac{m}{\log m}\frac{\log m}{\log n} \le \frac{m}{\log m}\frac{\log n^2}{\log n}$$

$$\le 2\frac{m}{\log m} \Longrightarrow \beta\left(n + \alpha\frac{m}{\log m}\right)\log\left(n + \alpha\frac{m}{\log m}\right)$$

$$= O\left((2 + \alpha)\frac{m}{\log m}\log\left((2 + \alpha)\frac{m}{\log m}\right)\right)$$

$$= O\left(\frac{m}{\log m}(\log m - \log\log m)\right) = O(m) = O(m + n\log n).$$

– If $n^2 < m$, then

$$\beta\left(n + \alpha\frac{m}{\log m}\right)\log\left(n + \alpha\frac{m}{\log m}\right)$$

$$= O\left((1 + \alpha)\frac{m}{\log m}\log\left((1 + \alpha)\frac{m}{\log m}\right)\right)$$

$$= O\left(\frac{m}{\log m}(\log m - \log\log m)\right) = O(m) = O(m + n\log n).$$

In any case $\beta\left(n + \alpha\frac{m}{\log m}\right)\log\left(n + \alpha\frac{m}{\log m}\right) = O(m + n\log n)$.   □

Theorem 13. *Let $k^i \geq O(m + n \log n)$, where $m = \max\{m^i\}$. Then for each player i there exists an uncoupled strategy mapping to k^i-automata with deterministic transitions and pure actions, that guarantees almost sure convergence of play to a pure Nash equilibrium of the stage game in every game where such an equilibrium exists.*

Proof. Let $p_1, \ldots, p_n$ be different prime numbers s.t. $\forall i \; p_i > m^i$.[6] By Lemma 12 we can take $p_1, \ldots, p_n$ s.t. $p_i = O(m + n \log n)$.

We will show that there exists a PN-mapping $(\varphi^1, \ldots, \varphi^n)$ such that the range of φ^i is $(4p_i + 3)$-automata, and in doing so we will have concluded our proof.

We introduce the mappings $\varphi^i(u^i) = \Lambda^i$, given a payoff function $U = (u^1, \ldots, u^n)$:

The automaton Λ^i consists of a starting state, a state after that, and p_i regions. The first region has five states. The other regions have the same structure of four states. Denote the p_i regions of Λ^i by $Q_1^i = \{s_{j,1}^i, s_{j,2}^i, \ldots, s_{j,5}^i\}$, and $Q_j^i = \{s_{j,1}^i, \ldots, s_{j,4}^i\}$ for $j = 2, \ldots, p_i$. Denote the starting state by $\mathfrak{s}_0^i := s_1^i$ and the state after it by s_2^i.

s_1^i : player i plays a_1^i. If $(a_1^1, \ldots, a_1^n)$ was played he moves to $s_{1,1}^i$ (the starting state of Q_1^i). Otherwise, if a_1^i is a best reply, then he stays at s_1^i; if a_1^i is not a best reply, then he moves to s_2^i.

s_2^i : player i plays a_2^i and in any case moves to $s_{1,1}^i$ (the starting state of Q_1^i).

These two starting states guarantee us two things: First, if the equilibrium is $(a_1^1, \ldots, a_1^n)$ then the players will stay there. Second: if it is not, then all the players get to $s_{1,1}^i$ simultaneously. We check the actions $(a_1^1, \ldots, a_1^n)$ separately, because in the continuation of the construction of the automata we will use the fact that $(a_1^1, \ldots, a_1^n)$ is not an equilibrium.

The regions $Q_1^i, Q_2^i, \ldots, Q_{p_i}^i$ are arranged in a circle when the players move to the region $Q_{j+1 (\mathrm{mod}\, p_i)}^i$ from the previous region Q_j^i.

---

[6]Every player i has to choose his number p_i to construct his automaton. But he doesn't know the numbers of the other players: $p_1, \ldots, p_{i-1}, p_{i+1}, \ldots, p_n$. Yet we have to ensure that every player will choose a different number. Therefore, we have to define a choice function: $\chi : \mathbb{N}^n \to (PRIME)^n$, known to the players. The choice function chooses for every $(m^1, \ldots, m^n)$ n prime numbers: $\chi(m^1, \ldots, m^n)$, and then every player i will choose the number $p_i := (\chi(m^1, \ldots, m^n))_i$.

As an example of choice function, let $\{w_k\}_{k=1}^{\infty}$ be the sequence of all the prime numbers in increasing order. For every i let $k(i)$ be the minimal number such that $\begin{cases} w_{k(i)} \geq m^i \\ k(i) = i (\mathrm{mod}\, n) \end{cases}$, and define $p_i := w_{k(i)}$.

For $j \neq 1$ the construction of Q_j^i is the following:

The starting state $s_{j,1}^i$: player i plays a_j^i. If player $i+1 \pmod n$ played[7] a_1^{i+1}, player i moves to $s_{j,2}^i$; otherwise he moves to $s_{j,3}^i$. If $(a_1^1, \ldots, a_1^n)$ was played, he moves to[8] $s_{j+1,1}^i$.

$s_{j,2}^i$: player i plays a_j^i. If it is a best reply, and also player $i+1$ played a_1^{i+1} (the same action as before) then player i stays there. Otherwise he moves to $s_{j,4}^i$. If $(a_1^1, \ldots, a_1^n)$ was played, he moves to $s_{j+1,1}^i$.

$s_{j,3}^i$: player i plays a_j^i. If it is a best reply, and also player $i+1$ played a_k^{i+1} $k \geq 2$ (the same action as before), player i stays there. Otherwise he moves to $s_{j,4}^i$. If $(a_1^1, \ldots, a_1^n)$ was played, he moves to $s_{j+1,1}^i$.

$s_{j,4}^i$: player i plays a_1^i. If $(a_1^1, \ldots, a_1^n)$ was played, he moves to $s_{j+1,1}^i$; otherwise he stays at $s_{j,4}^i$.

For $j = 1$ the construction of Q_1^i is quite similar. The only difference is that the state $s_{j,4}^i$, changed by two $s_{1,4}^i, s_{1,5}^i$:

$s_{1,4}^i$: player i plays a_2^i. In any case, he moves to $s_{1,5}^i$.

$s_{1,5}^i$: player i plays a_1^i. If $(a_1^1, \ldots, a_1^n)$ was played, he moves to $s_{j+1,1}^i$. Otherwise he stays in $s_{1,5}^i$.

For $m^i < j \leq p_i$ let $Q_j^i := Q_{m^i}^i$ denoted that the act in every $s_{j,k}^i m^i < j \leq p_i, k = 1, \ldots, 4$, is identical to the act in $s_{m^i,k}^i$.

In the state $s_{j,1}^i$ player i informs player $i-1$ what he played.

If the players located at $Q_{k_1}^1, \ldots, Q_{k_n}^n$, and $(a_{k_1}^1, \ldots, a_{k_n}^n)$ is an equilibrium, then all the players will stay at $s_{j,2}^i, s_{j,3}^i$.

In the state $s_{j,4}^i$ player i plays the "opposite" action than he played before and informs player $i-1$ that it is not an equilibrium.

In the state $s_{j,4}^i$ for $j \neq 1$, and $s_{1,5}^i$ for $j = 1$ player i waits until all the players are informed that it is not an equilibrium. Then all the players can move to their next region simultaneously.

To summarize, if the players are located at $Q_{k_1}^1, \ldots, Q_{k_n}^n$, and $(a_{k_1}^1, \ldots, a_{k_n}^n)$ is an equilibrium, then all the players stay at the equilibrium all the time. Otherwise they all will move to $Q_{k_1+1}^1, \ldots, Q_{k_n+1}^n$ simultaneously.

Since the players move from Q_j^i to Q_{j+1}^i simultaneously, by the Chinese Remainder theorem it follows that they will visit every $(Q_{k_1}^1, \ldots, Q_{k_n}^n)$: $1 \leq k_i \leq m^i$, until they become stuck in some $(Q_{k_1}^1, \ldots, Q_{k_n}^n)$ for which $(a_{k_1}^1, \ldots, a_{k_n}^n)$ is an equilibrium. Therefore, if a pure equilibrium exists, the automata will eventually reach it. □

---

[7]From here till the end of the proof, we will write $i \pm 1$ instead of $i \pm 1 \pmod n$.

[8]From here till the end of the proof, we will write $j + 1$ instead of $j + 1 \pmod{p_i}$.

Appendix A. Proof of Theorem 6

Proof. We define the mapping φ as follows:

Given a game $U = (u^1, u^2)$, the automaton $\varphi^1(u^1) = \Lambda^1 \in \mathcal{A}^1$ is constructed as follows:

Denote $\varphi^1(u^1) = \Lambda^1 = \langle \Psi^1, \mathfrak{s}_0^1, f^1, g^1 \rangle$ when Λ^1 is an m^1-automaton. We denote the states of Λ^1 by $\Psi^1 = \{s_1^1, \ldots, s_{m^1}^1\}$.

$\mathfrak{s}_0^1 := s_1^1$.

In state s_i^1 player 1 plays action a_i^1. He stays in this state if a_i^1 is a best reply to the action of player 2; otherwise he moves randomly to any one of the m^1 states with equal probability $\frac{1}{m^1}$.

In order to define the mapping φ^2, we start by considering the following action b^2 of player 2: for every action a_j^2 of player 2, let $\#BR(a_j^2)$ be the number of 1s in the column a_j^2 in the table $\tilde{u}^2$; i.e., $\#BR(a_j^2) := |\{a_i^1|\tilde{u}^2(a_i^1, a_j^2) = 1\}| = |\{a_i^1|a_j^2$ is a best reply to $a_i^1\}|$. Consider an action a_k^2 with a maximal number of 1s in its column: $\#BR(a_k^2) = \max\{\#BR(a_j^2)|a_j^2 \in A^2\}$. Without loss of generality assume $k = 1$; i.e., the first column of $\tilde{u}^2$ has no fewer 1s than any other column. Denote this action by $b^2 := a_1^2$.

Now we construct the automaton $\varphi^2(u^2) = \Lambda^2 \in \mathcal{A}^2$ as follows:

Denote $\Lambda^2 = \langle \Psi^2, \mathfrak{s}_0^2, f^2, g^2 \rangle$ when Λ^2 is an $(m^2 + 1)$-automaton. We denote the states of Λ^2 by $\Psi^2 = \{s_0^2, s_1^2, \ldots, s_{m^1}^2\}$.

$\mathfrak{s}_0^2 := s_0^2$.

In the state s_0^2, player 2 plays the action b^2, and moves to any of the $m^2 + 1$ states with probability $\frac{1}{m^2+1}$.

In the states s_i^2, $i \geq 1$, player 2 plays action a_i^2. He stays in this state if a_i^2 is a best reply to the action of player 1; otherwise he moves randomly to any one of the $m^2 + 1$ states with equal probability $\frac{1}{m^2+1}$.

Now we will prove that (φ^1, φ^2) is PN-mapping. Let us consider two cases:

Case 1. For every $i = 1, \ldots, m^1 : \tilde{u}^2(a_i^1, b^2) = 1$; i.e., b^2 is a dominant action.

We partition the space $\Psi^1 \times \Psi^2$ of the automata states into four regions:

$P_1 := \{(s_i^1, s_j^2), 1 \leq i \leq m^1, 1 \leq j \leq m^2 : \tilde{u}^1(a_i^1, a_j^2) = 1, \tilde{u}^2(a_i^1, a_j^2) = 1\}$;
  i.e., in this case (a_i^1, a_j^2) is a pure Nash equilibrium.
$P_2 := \{(s_i^1, s_0^2), 1 \leq i \leq m^1\}$
$P_3 := \{(s_i^1, s_j^2), 1 \leq i \leq m^1, 1 \leq j \leq m^2 : \tilde{u}^2(a_i^1, a_j^2) = 0\}$
$P_4 := \{(s_i^1, s_j^2), 1 \leq i \leq m^1, 1 \leq j \leq m^2 : \tilde{u}^1(a_i^1, a_j^2) = 0, \tilde{u}^2(a_i^1, a_j^2) = 1\}$

These four regions clearly cover the space $\Psi^1 \times \Psi^2$ (because player 2 can be in the state s_0^2 (P_2) or in any other state ($P_1 \cup P_3 \cup P_4$). The action of player 2 can be a best reply ($P_1 \cup P_4$) or not (P_3). If it is a best reply, then the action of player 1 can be a best reply (P_1) or not (P_4).

Players 1 and 2 stay at the same state if their action is a best reply; i.e., each state in P_1 is absorbing. We will prove that there is a positive probability of reaching a state from P_1, in finitely many periods, from any other state $s \in \Psi^1 \times \Psi^2$.

$s = (s_i^1, s_0^2) \in P_2$: The actions are $(a_i^1, b^2) = (f^1(s_i^1), f^2(s_0^2))$. Whether a_i^1 is a best reply or not, player 1 has a positive probability (1 or $\frac{1}{m^1}$ respectively) to move to s_k^1, where a_k^1 is a best reply of player 1 to b^2. Player 2 will move to the state s_1^2 with probability $\frac{1}{m^2+1}$, where b^2 is a best reply to every action of player 1, in particular to a_k^1. Now $(s_k^1, s_1^2) \in P_1$.

$s = (s_i^1, s_j^2) \in P_3$: The actions are (a_i^1, a_j^2); a_j^2 is not a best reply. Therefore, player 2 moves to s_0^2 with probability $\frac{1}{m^2+1}$. Denote the state to which player 1 moves by s_k^1. Then $(s_k^1, s_0^2) \in P_2$.

$s = (s_i^1, s_j^2) \in P_4$: The actions are (a_i^1, a_j^2); a_j^2 is a best reply, a_i^1 is not. Therefore, player 2 stays in s_j^2, and player 1 moves to s_k^1 with probability $\frac{1}{m^1}$, where a_k^1 is a best reply of player 1 to a_j^2. Now either $(s_k^1, s_j^2) \in P_1$ or $(s_k^1, s_j^2) \in P_3$, depending on whether a_j^2 is a best reply to a_k^1.

In all the cases there is a positive probability of reaching an absorbing state in P_1 in at most 3 steps.

Case 2. There exists an action of player 1, say a_l^1, such that b^2 is not a best reply to it; i.e., there exists $l = 1, \ldots, m^1$ such that $\widetilde{u}^2(a_l^1, b^2) = 0$.

Before proving the theorem, we will prove the following simple claim about the configuration of 0s and 1s in $\widetilde{u}^2$. This claim will be useful later.

Claim 1. If there exists $l = 1, \ldots, m^1$, such that $\widetilde{u}^2(a_l^1, b^2) = 0$, then:

(a) Let $a_j^2 \in A^2$. Then there exists $a_i^1 \in A^1$ such that $\widetilde{u}^2(a_i^1, a_j^2) = 0$.
(b) Let $a_i^1 \in A^1$. Then there exists $a_j^2 \in A^2$ such that $\widetilde{u}^2(a_i^1, a_j^2) = 1$.

(c) Let $a_i^1 \in A^1, a_j^2 \in A^2$, such that $\begin{cases} \widetilde{u}^2(a_i^1, a_j^2) = 1 \\ \widetilde{u}^2(a_i^1, b^2) = 0 \end{cases}$. Then there exists

$a_k^1 \in A^1$ such that $\begin{cases} \widetilde{u}^2(a_k^1, a_j^2) = 0 \\ \widetilde{u}^2(a_k^1, b^2) = 1 \end{cases}$.

Proof.

(a) Otherwise there would be a column in $\tilde{u}^2$ that includes 1s only. By the assumption in this case there exist $i = 1, \ldots, m^1$, such that $\tilde{u}^2(a_i^1, b^2) = 0$, which contradicts the fact that b^2 is the column with the maximal number of 1s.

(b) There is some action $a_j^2 \in A^2$ that is a best reply to the action $a_i^1 \in A^1$. This action satisfies $\tilde{u}^2(a_i^1, a_j^2) = 1$.

(c) Otherwise the number of 1s in the j-th column would be bigger than in the first column. $\qquad\square$

We partition the space $\Psi^1 \times \Psi^2$ of the automata states into regions: For every $x, y, z \in \{0, 1\}$ put:

$$P_{xyz} := \{(s_i^1, s_j^2), 1 \leq i \leq m^1, 1 \leq j \leq m^2 : \tilde{u}^1(a_i^1, a_j^2)$$
$$= x, \tilde{u}^2(a_i^1, a_j^2) = y, \tilde{u}^2(a_i^1, b^2) = z\}$$

and

$$Q_{xy} := \{(s_i^1, s_0^2), 1 \leq i \leq m^1 : \tilde{u}^1(a_i^1, b^2) = x, \tilde{u}^2(a_i^1, b^2) = y\}.$$

In words Q for a region we mean that player 2 is in the state s_0^2, and by P we mean that he is not. There are three indices for region P, and two Boolean indices for region Q. The first index in P, Q corresponds to player 1 and indicates whether his action is a best reply (1) or not (0). The second index in P, Q corresponds similarly to player 2. The third index in P indicates whether action b^2 is a best reply to the action that player 1 played (1) or not (0).

Clearly $(\cup_{x,y,z \in \{0,1\}} P_{xyz}) = \Psi^1 \times (\Psi^2 \backslash \{s_0^2\})$ and $(\cup_{x,y \in \{0,1\}} Q_{xy}) = \Psi^1 \times \{s_0^2\}$, since we are considering all the possibilities. Therefore, $(\cup_{x,y,z \in \{0,1\}} P_{xyz}) \cup (\cup_{x,y \in \{0,1\}} Q_{xy}) = \Psi^1 \times \Psi^2$.

Consider the region $P_{11\bullet}$. By the definition of P_{xyz} we can see that $P_{11\bullet} = \{(s_i^1, s_j^2), 1 \leq i \leq m^1, 1 \leq j \leq m^2 : \tilde{u}^1(a_i^1, a_j^2) = \tilde{u}^2(a_i^1, a_j^2) = 1\} = \{(s_i^1, s_j^2), 1 \leq i \leq m^1, 1 \leq j \leq m^2 : (a_i^1, a_j^2) \text{ is a pure Nash equilibrium}\}$.

Players 1 and 2 stay at the same state if their action is a best reply; i.e., each state in $P_{11\bullet}$ is absorbing. If we can show that a state from $P_{11\bullet}$ is reached with positive probability, in finitely many periods, from all the other regions that we have already defined, then we will have concluded our proof.

$P_{00\bullet}$: This is the region of all the states where the actions of both players are not a best reply, and so both players move randomly over all their states; therefore, they reach a pure Nash equilibrium with probability $\frac{1}{m^1} \cdot \frac{1}{m^2+1}$; i.e., they reach $P_{11\bullet}$.

$Q_{0\bullet}$: Player 1's action is not a best reply, and player 2 is in state s_0^2. Therefore, as before, both players move randomly over all their states, and they reach $P_{11\bullet}$ with probability $\frac{1}{m^1} \cdot \frac{1}{m^2+1}$, as before.

Q_{11}: Player 1's action is a best reply and so he stays in the same state. Player 2 is in state s_0^2 and so he moves randomly over all his states. Player 2 will move to s_1^2 with probability $\frac{1}{m^2+1}$, and they get to $P_{11\bullet}$.

P_{101}: Player 1 stays, and player 2 randomizes. Since $\widetilde{u}^2(a_i^1, b^2) = 1$ player 2 will move to s_0^2 with probability $\frac{1}{m^2+1}$, and they get to Q_{01} or Q_{11}.

P_{010}: Player 1 randomizes and player 2 stays. By claim 1(c) there exists $a_k^1 \in A^1$ such that $\begin{cases} \widetilde{u}^2(a_k^1, a_j^2) = 0 \\ \widetilde{u}^2(a_k^1, b^2) = 1 \end{cases}$. Player 1 will move to s_k^1 with probability $\frac{1}{m^1}$, and they get to P_{001} or P_{101}.

P_{100}: Player 1 stays and player 2 randomizes. By claim 1(b) there exists $a_j^2 \in A^2$ such that $\widetilde{u}^2(a_i^1, a_j^2) = 1$. Player 2 will move to s_j^2 with probability $\frac{1}{m^2+1}$. Note that a_i^1 does not change and so still $\widetilde{u}^2(a_i^1, b^2) = 0$ and they get to $P_{0,1,0}$ or P_{110} (i.e., $P_{\bullet 10}$).

P_{011}: Player 1 randomizes and player 2 stays. By claim 1(a) there exists $a_k^1 \in A^1$ such that $\widetilde{u}^2(a_k^1, a_j^2) = 0$. Player 1 will move to s_k^1 with probability $\frac{1}{m^1}$, and they get to $P_{00\bullet} \cup P_{100} \cup P_{101} = P_{\bullet 0 \bullet}$.

Q_{10}: Player 1 stays and player 2 randomizes. Player 2 moves to s_1^2 with probability $\frac{1}{m^2+1}$, and they get to P_{100}.

Thus, we have covered all the regions and shown that in at most 5 periods there is a positive probability of reaching the absorbing state $P_{11\bullet}$. The regions cover all the space $\Psi^1 \times \Psi^2$, and so the automata will reach a pure Nash equilibrium when such an equilibrium exists with probability 1. $\qquad\qquad\square$

Appendix B. Proof of Theorem 7

Proof. We define the mapping φ as follows:

Given a game $U = (u^1, u^2)$, the automaton $\varphi^1(u^1) = \Lambda^1 \in \mathcal{A}^1$ is constructed as follows:

Denote the states of Λ^i by $\Psi^i = \{s_{00}^i, s_{01}^i, s_1^i, \ldots, s_{m^1}^i\}$.

In the state s_j^i player i plays action a_j^i. He stays in this state if a_j^i is the best reply to the action of player 2; otherwise he moves to s_{00}^i.

In the state s_{00}^i player i plays $(\frac{1}{m^i}, \ldots, \frac{1}{m^i})$. If he played a_1^i, he stays in s_{00}^i. If he played a_2^i, he moves to s_{01}^i. If he played a_j^i, $j \geq 3$, he moves to s_j^i.

In the state s_{01}^i player i plays $(\frac{1}{m^i}, \ldots, \frac{1}{m^i})$. If he played a_1^i, he moves to s_1^i. If he played a_2^i, he moves to s_2^i. If he played a_3^i, he stays in s_{01}^i. If he played a_j^i $j \geq 4$, he moves to s_{00}^i.

Let the starting state be $\mathfrak{s}_0^i := s_{00}^i$.

The proof that this mapping is a PN-mapping requires consideration of four separate cases: if a pure Nash equilibrium is (a_k^1, a_l^2) then we will consider the cases $\{k \leq 2, l \leq 2\}$, $\{k \leq 2, l > 2\}$, $\{k > 2, l \leq 2\}$, and $\{k > 2, l > 2\}$. We will prove only one case, say $\{k \leq 2, l > 2\}$, since the proofs for the other cases are similar.

We partition the space $\Psi^1 \times \Psi^2$ of the automata states into regions:

$P_1 := \{(s_i^1, s_j^2), 1 \leq i \leq m^1, 1 \leq j \leq m^2 \colon (a_i^1, a_j^2)$ is a pure Nash equilibrium$\}$

$P_2 := \{(s_{01}^1, s_{00}^2)\}$

$P_3 := \{(s_{00}^1, s_{00}^2) \cup (s_{00}^1, s_{01}^2) \cup (s_{01}^1, s_{01}^2)\}$

For every $x \in \{00, 01\}, y \in \{\leq, >\}$ put:

$Q_{x,y} := \{(s_x^1, s_j^2), 1 \leq j \leq m^2 \colon$ there exist $\begin{cases} i \leq 2 \text{ if } y \text{ is } \leq \\ i > 2 \text{ if } y \text{ is } > \end{cases}$ such that a_j^2 is not a best reply to $a_i^1\}$

For every $x \in \{\leq, >\}, y \in \{00, 01\}$ put:

$Q_{x,y} := \{(s_i^1, s_y^2), 1 \leq i \leq m^1 \colon$ there exist $\begin{cases} j \leq 2 \text{ if } x \text{ is } \leq \\ j > 2 \text{ if } x \text{ is } > \end{cases}$ such that a_i^1 is not a best reply to $a_j^2\}$

$P_4 := \{(s_i^1, s_j^2), 1 \leq i \leq m^1, 1 \leq j \leq m^2 \colon (a_i^1, a_j^2)$ is not a pure Nash equilibrium$\}$

Clearly, $P_2 \cup P_3 = \{s_{00}^1, s_{01}^1\} \times \{s_{00}^2, s_{01}^2\}$, $P_1 \cup P_4 = \{s_1^1, \ldots, s_{m^1}^1\} \times \{s_1^2, \ldots, s_{m^2}^2\}$, $\cup_{x \in \{00, 01\}, y \in \{\leq, >\}} Q_{x,y} = \{s_{00}^1, s_{01}^1\} \times \{s_1^2, \ldots, s_{m^2}^2\}$, $\cup_{x \in \{\leq, >\}, y \in \{00, 01\}} Q_{x,y} = \{s_1^1, \ldots, s_{m^1}^1\} \times \{s_{00}^2, s_{01}^2\}$. Therefore, the union of all the regions is $\Psi^1 \times \Psi^2$.

Players 1 and 2 stay at the same state if their action is a best reply; i.e., each state in P_1 is absorbing. If we can show that a state from P_1 is reached with positive probability, in finitely many periods, from all the other regions that we have already defined, then we will have concluded our proof.

P_2: Both players randomize. They play (a_k^1, a_l^2) with positive probability, and then they get to P_1.

P_3: Both players randomize. From every one of the states there is a positive probability of getting to P_2. For example, if they are in the states (s_{00}^1, s_{00}^2), then if they play (a_2^1, a_1^2) they move to $(s_{01}^1, s_{00}^2) \in P_2$.

$Q_{00,\le}$: Player 1 randomizes. With positive probability the actions are (a_i^1, a_j^2) such that a_j^2 is not a best reply to a_i^1, so player 1 stays at $\{s_{00}^1, s_{01}^1\}$ and the action of player 2 is not a best reply. Therefore, player 2 moves to s_{00}^2, and so $(s_{00}^1, s_{00}^2), (s_{01}^1, s_{00}^2) \in P_2 \cup P_3$.

$Q_{01,\le}$: Player 1 randomizes. With positive probability the actions are (a_4^1, a_j^2). If a_j^2 is a best reply to a_4^1, they move to $(s_{00}^1, s_j^2) \in Q_{00,\le}$. Otherwise they move to (s_{00}^1, s_{00}^2).

$Q_{01,>}, Q_{00,>}$: Similar to $Q_{00,\le}, Q_{01,\le}$

$Q_{\le,00}, Q_{\le,01}, Q_{>,01}, Q_{>,00}$: Symmetric to $Q_{00,\le}, Q_{01,\le}, Q_{01,>}, Q_{00,>}$

P_4: The action of one of the players is not a best reply to the action of the other, and so one of them will move to the state s_{00} (i.e., s_{00}^1 or s_{00}^2). Hence, they get to one of the previous regions.

We have covered all the regions and shown that in at most 5 periods there is positive probability of reaching the absorbing state $P_{11\bullet}$. The regions cover all the space $\Psi^1 \times \Psi^2$; therefore, the mapping is a PN-mapping.

In the other cases where the pure Nash equilibrium (a_k^1, a_l^2) satisfies $\{k \le 2, l \le 2\}$, $\{k > 2, l \le 2\}$, or $\{k > 2, l > 2\}$ the only difference is in how the regions P_2 and P_3 are defined. For example, for the case $\{k \le 2, l \le 2\}$, P_2 and P_3 will be defined by $P_2 := \{(s_{01}^1, s_{01}^2)\}$ and $P_3 := \{(s_{00}^1, s_{00}^2) \cup (s_{00}^1, s_{01}^2) \cup (s_{01}^1, s_{00}^2)\}$. $\qquad\square$

References

Hart, S. and A. Mas-Colell (2003), "Uncoupled Dynamics Do Not Lead to Nash Equilibrium," *American Economic Review*, 93, 1830–1836. [Chapter 7]

Hart, S. and A. Mas-Colell (2006), "Stochastic Uncoupled Dynamics and Nash Equilibrium," *Games and Economic Behavior*, 57, 286–303. [Chapter 8]

Chapter 10

HOW LONG TO EQUILIBRIUM? THE COMMUNICATION COMPLEXITY OF UNCOUPLED EQUILIBRIUM PROCEDURES*

Sergiu Hart and Yishay Mansour

We study the question of how long it takes players to reach a Nash equilibrium in *uncoupled* setups, where each player initially knows only his own payoff function. We derive lower bounds on the *communication complexity* of reaching a Nash equilibrium, i.e., on the number of bits that need to be transmitted, and thus also on the required number of steps. Specifically, we show lower bounds that are exponential in the number of players in each one of the following cases: (1) reaching a pure Nash equilibrium; (2) reaching a pure Nash equilibrium in a Bayesian setting; and (3) reaching a mixed Nash equilibrium. We then show that, in contrast, the communication complexity of reaching a correlated equilibrium is polynomial in the number of players.

1. Introduction

Equilibrium is a central concept in interactions between decision-makers. The definition of equilibrium is *static*: it is characterized by the property

Originally published in *Games and Economic Behavior*, 69 (2010), 107–126.

*Yishay Mansour: Tel Aviv University, Israel.

Previous versions: April 2006, December 2006; *Proceedings of the 39th Annual ACM Symposium on Theory of Computing (STOC)* 2007, pp. 345–353 (extended abstract: "The Communication Complexity of Uncoupled Nash Equilibrium Procedures").

JEL classification: C7; D83.

Keywords: uncoupled dynamics; Nash equilibrium; communication complexity; correlated equilibrium; speed of convergence.

The research was partially supported by grants of the Israel Science Foundation (Hart, Mansour) and by an IBM faculty award (Mansour).

We thank Fabrizio Germano, Adam Kalai, Eyal Kushilevitz, Andreu Mas-Colell, Noam Nisan, and the anonymous referees for useful discussions and suggestions, and the Institute for Advanced Studies at the Hebrew University of Jerusalem where some of this work was done.

that the participants ("players") have no incentive to depart from it. No less fundamental, however, are the *dynamic* issues of how such an equilibrium arises (see, e.g., the books of Fudenberg and Levine 1998; Young 2004). Since decisions are assumed to be taken independently by the participants, it is only natural to study dynamics in *decentralized* environments, where each decision-maker has only partial information—for instance, he knows only his own preferences and not those of the other players. As a result, no player can find an equilibrium on his own, and the resulting dynamics become complex and need not converge to a rest-point (i.e., an equilibrium).

Significant progress has been made in understanding the dynamic aspects of one equilibrium concept, that of *correlated equilibrium* (Aumann 1974). A correlated equilibrium obtains when players receive signals before the game is played; these signals, which may be correlated, do not affect the payoffs in the game. Of course, the players may well use these signals when making their strategic choices. To date, there are several efficient algorithms (Blum and Mansour 2007; Cahn 2004; Cesa-Bianchi and Lugosi 2003, 2006; Foster and Vohra 1997; Hart 2005; Hart and Mas-Colell 2000, 2001; Stoltz and Lugosi 2005, 2007; Young 2004) that, in all games, converge fast to (approximate) correlated equilibria.

In contrast, convergence to *Nash equilibrium* is a much more complex and less clear-cut issue.[1] As we have stated above, a natural assumption that most dynamics satisfy is that of *uncoupledness* (Hart and Mas-Colell 2003): each player is assumed to know initially *only his own payoff function*, and not those of the other players. Parallel notions are "informationally decentralized" or "privacy-preserving" in economics, and "distributed" in computer science; in particular, the longstanding study of mechanism design in economics, starting with the work of Hurwicz (1960), has led to significant concepts and insights concerning the aggregation of information (see Jordan 2008 for a short survey).

The existing results on uncoupled dynamics and Nash equilibria are as follows. On the one hand, it has been shown that it is impossible for uncoupled dynamics that are deterministic and continuous[2] always to converge to a Nash equilibrium, even when it is unique (Hart and Mas-Colell

---

[1] A Nash equilibrium is a fixed point of a nonlinear function, whereas a correlated equilibrium is a solution of finitely many linear inequalities. This may be one reason—though not the only one—that it appears to be more difficult to converge to the former than to the latter.

[2] Continuous with respect to both actions and time.

2003). On the other hand, there are a number of uncoupled dynamics that converge to Nash equilibria in general games; these dynamics use various techniques such as hypothesis-testing, regret-testing, and other variants of exhaustive or stochastic search (Foster and Kakade 2004; Foster and Young 2003, 2006; Germano and Lugosi 2007; Hart and Mas-Colell 2006; Young 2004).[3] Since all these dynamics perform some form of search over all action combinations, it follows that the number of steps until a Nash equilibrium is reached is exponential in the number of players (when the number of actions of each player is kept fixed). In this paper we will show that this is a general phenomenon rather than a deficiency of the existing literature: *the lower bound on the speed of convergence to Nash equilibria is exponential in the number of players.*

To make this precise, define a *Nash equilibrium procedure* as a dynamic process whereby the players reach a Nash equilibrium, whether *pure* or *mixed.* We study the number of steps needed before the procedure terminates at the appropriate equilibrium. Again, we are considering uncoupled procedures: each player's payoff function is private, initially known only to him. We use the theory of *communication complexity* (see Kushilevitz and Nisan 1997) to derive lower bounds on the amount of communication, measured in terms of the number of transmission bits—and thus also the number of steps—needed by the players in order to reach a Nash equilibrium. This important connection was first observed in Conitzer and Sandholm (2004), where various lower bounds for *two*-person games are derived (as the number of actions increases). Here we analyze general n-person games.

Our results provide lower bounds that are exponential in the number of players (we keep the number of actions of each player bounded, e.g., two) for the communication complexity in each of the following cases:

(1) reaching a pure Nash equilibrium—in general games, and also in the restricted class of games having the "finite improvement property" (Section 3 and Appendix A);
(2) reaching a pure Nash equilibrium in a Bayesian setup (Section 4); and
(3) reaching a mixed Nash equilibrium (Section 5).

---

[3]Dynamics of the "best-reply" variety have been studied in Bayesian setups where players possess certain probabilistic beliefs about the payoff functions of the other players (see, e.g., Jordan 1991; Kalai and Lehrer 1993); however, additional coordination between the players is needed to obtain convergence to Nash equilibria (cf. Section 4 in Jordan 1991 and footnote 20 in Hart 2005).

We also exhibit simple procedures that yield upper bounds that are also exponential (Section 6).

These exponential lower bounds may seem unsurprising, given that the size of the input (i.e., the players' private payoff functions) is also exponential. We thus analyze the communication complexity of reaching *correlated equilibria*, and we show that it is, in contrast, only *polynomial* in the number of players (Section 7 and Appendix B). Therefore, the exponential communication complexity of Nash equilibrium procedures is a result of the equilibrium requirement, and *not* of the size of the input.

In summary, this paper may be viewed as providing further evidence of the intrinsic difficulty of reaching Nash equilibria, in contrast to correlated equilibria.[4]

2. Preliminaries

2.1. *Game-theoretic setting*

The basic setting is as follows. There are $n \geq 2$ players, $i = 1, 2, \ldots, n$. Each player i has a finite set of actions A_i with[5] $|A_i| \geq 2$, and the joint action space is $A = \Pi_{i=1}^{n} A_i$. Let Δ_i denote the set of probability distributions over A_i and put $\Delta = \Pi_{i=1}^{n} \Delta_i$. Most of the games we introduce will be *binary-action games*, where the action space of each player i is $A_i = \{0, 1\}$, and so $A = \{0, 1\}^n$; in this case a mixed action of player i is given by $0 \leq p_i \leq 1$, interpreted as the probability that $a_i = 1$.

Each player i has a payoff (or utility) function u_i which maps A to the real numbers, i.e., $u_i : A \to \mathbb{R}$. We extend u_i to Δ in a multilinear way, by defining $u_i(p_1, \ldots, p_n) = \mathrm{E}[u_i(a_1, \ldots, a_n)]$ for each $(p_1, \ldots, p_n) \in \Delta$, where the expectation E is taken with respect to the product distribution $p_1 \times \cdots \times p_n$ on A. We denote this game by $G = (n, \{A_i\}_i, \{u_i\}_i)$.

For a joint action $a = (a_1, \ldots, a_i, \ldots, a_n) \in A$, let $a^{-i} = (a_1, \ldots, a_{i-1}, a_{i+1}, \ldots, a_n)$ be the joint action of all players *except* player i. For each player i, the (pure) *best-reply correspondence* maps a joint action a^{-i} of the other players to the set $\mathrm{BR}(a^{-i}; u_i) = \arg\max_{a_i \in A_i} u_i(a_i, a^{-i})$. A joint action $a \in A$ is *a pure Nash equilibrium* if $u_i(a) \geq u_i(b_i, a^{-i})$ for every player i and any action $b_i \in A_i$; or equivalently, $a_i \in \mathrm{BR}(a^{-i}; u_i)$ for all i.

---

[4]See Hart and Mas-Colell (2006, Section 5(g)), particularly the last sentence there.
[5]The number of elements of a finite set Z is denoted $|Z|$.

Similarly, a combination of mixed actions $p \in \Delta$ is a *mixed Nash equilibrium* if $u_i(p) \geq u_i(q_i, p^{-i})$ for every player i and any $q_i \in \Delta_i$.

Finally, we define the concepts of "improvement step" and "improvement path." Given a joint action $a \in A$, an *improvement step* of player i is an action $b_i \in A_i$ such that $u_i(b_i, a^{-i}) \geq u_i(a)$; we refer to i as the *improving player*. An *improvement path* is a sequence of improvement steps (where the improvement steps can be performed by different players). A game G has *the finite improvement property* if all the improvement paths are finite[6]; such a game always possesses a pure Nash equilibrium.

2.2. Communication complexity background

In the "classical" setting in communication complexity there are two agents,[7] one holding an input $x \in \{0,1\}^K$ and the other holding an input $y \in \{0,1\}^K$, where K is a finite set. Their task is to compute a joint function of their inputs $f(x,y) \in \{0,1\}$. The agents send messages to one another, and we assume that at the end of the communication they each have the value of $f(x,y)$. The *communication complexity* of a deterministic communication protocol Π for computing $f(x,y)$ is the number of bits sent during the computation of $f(x,y)$ by Π; denote this number of bits by $\mathrm{CC}(\Pi, f, x, y)$. The communication complexity $\mathrm{CC}(\Pi, f)$ of a protocol Π for computing a function f is defined as the worst case over all possible inputs $(x,y) \in \{0,1\}^K \times \{0,1\}^K$, i.e., $\mathrm{CC}(\Pi, f) = \max_{x,y \in \{0,1\}^K} \mathrm{CC}(\Pi, f, x, y)$. Finally, the *communication complexity* $\mathrm{CC}(f)$ *of computing a function* f is the minimum over all protocols Π for computing f, i.e., $\mathrm{CC}(f) = \min_{\Pi} \mathrm{CC}(\Pi, f)$.

A well-studied function in communication complexity is the *disjointness* function. Let S be a finite set; the S-*disjointness function* DISJ_S is defined on the subsets of S (i.e., on $\{0,1\}^S \times \{0,1\}^S$) by $\mathrm{DISJ}_S(S_1, S_2) = 1$ if the two inputs $S_1, S_2 \subset S$ are disjoint sets (i.e., $S_1 \cap S_2 = \emptyset$), and $\mathrm{DISJ}_S(S_1, S_2) = 0$ otherwise. There is a large literature on the communication complexity of the disjointness function (see Kushilevitz and Nisan 1997). We state here one result that will be used to derive bounds in our setting (see Kushilevitz and Nisan 1997, Section 1.3).

---

[6]These are the "generalized ordinal potential games" (Monderer and Shapley 1996).
[7]We call them "agents" to avoid confusion with the players of the game.

Theorem 1. *The communication complexity of the S-disjointness function is $|S|$ bits, i.e., $CC(DISJ_S) = |S|$.*

2.3. Nash equilibrium procedures

A *Nash equilibrium procedure* is a dynamic process by which the players reach a Nash equilibrium of the game, whether pure or mixed (both cases will be considered below). Fix the number of players n and the action spaces A_i; a game G is thus identified with its payoff functions $(u_1, \ldots, u_n)$. Let $\mathcal{G}$ be a family of games to which the procedure should apply. The basic assumption is that of *uncoupledness:* each player knows only his own payoff function u_i (Hart and Mas-Colell 2003, 2006).

We emphasize that we make no assumptions about the players' incentives, since we are interested in *lower bounds*, which give the *minimum* it takes to reach an equilibrium—*no matter what the incentives are.* Indeed, any form of strategic behavior when choosing the messages would be an additional restriction that could only increase the communication complexity (cf. Conitzer and Sandholm 2004).

Formally, the n players who participate in a Nash equilibrium procedure have the following information and capabilities. The "input" of the procedure is a game $G = (u_1, \ldots, u_n)$ in the family $\mathcal{G}$. Initially, each player i has access only to his own "private" payoff function[8] u_i. In each round $t = 1, 2, \ldots$, every player i performs an action[9,10] $a_{i,t} \in A_i$. At the end of round t all the players observe each other's actions; i.e., they all observe the joint action $(a_{1,t}, \ldots, a_{n,t}) \in A$.

In a *mixed Nash equilibrium procedure* Π for $\mathcal{G}$, the "output" of each player i is a distribution $p_i \in \Delta i$, such that $(p_1, \ldots, p_n) \in \Delta$ is a mixed Nash equilibrium of the game $G = (u_1, \ldots, u_n)$ that was given as input.[11] In *a pure Nash equilibrium procedure* Π for $\mathcal{G}$, the "output" of player i is either (1) a pure action $a_i \in A_i$, or (2) a declaration of "no pure Nash

---

[8]The number of players n, the action spaces A_i, and the set of games $\mathcal{G}$ are fixed and commonly known.

[9]It is natural to consider dynamics in the framework of repeated games, and so we assume without loss of generality that the communication proceeds through actions. Using any other set B_i instead of A_i will only affect the communication complexity by a constant factor (cf. Proposition 2). For binary-action games, $a_{i,t} \in A_i$ just means that the communication of each player in each period is 1 bit.

[10]The procedure is thus *deterministic*; see Section 8.1 for *stochastic* procedures.

[11]Finite games always possess mixed Nash equilibria.

equilibrium." In case (1), the joint output $(a_1, \ldots, a_n) \in A$ is a pure Nash equilibrium of G, whereas in case (2) G has no pure Nash equilibrium. Let PNEP and MNEP denote the collection of pure and mixed Nash equilibrium procedures, respectively.

The communication complexity $CC(\Pi, G)$ of a Nash equilibrium procedure Π applied to a game G is the number of bits communicated until Π terminates when the input is G. Given a family of games $\mathcal{G}$, the communication complexity of a Nash equilibrium procedure Π for the family $\mathcal{G}$ is the worst-case communication complexity of Π over all games $G \in \mathcal{G}$, i.e., $CC(\Pi, \mathcal{G}) = \max_{G \in \mathcal{G}} CC(\Pi, G)$. Next, $CC(\text{PURE}, \mathcal{G})$, the *communication complexity of pure Nash equilibrium procedures* for a family of games $\mathcal{G}$, is the minimal communication complexity of any pure Nash equilibrium procedure Π for the family of games $\mathcal{G}$, i.e., $CC(\text{PURE}, \mathcal{G}) = \min_{\Pi \in \text{PNEP}} CC(\Pi, \mathcal{G})$; similarly, $CC(\text{MIXED}, \mathcal{G}) = \min_{\Pi \in \text{MNEP}} CC(\Pi, \mathcal{G})$ is the *communication complexity of mixed Nash equilibrium procedures* for $\mathcal{G}$. Finally, when the games in the family $\mathcal{G}$ are chosen according to a probability distribution Pr, the *expected communication complexity of pure Nash equilibrium procedures* is $min_{\Pi \in \text{PNEP}} E[CC(\Pi, G)]$, where the expectation E is taken with respect to Pr; we denote this by $E[CC(\text{PURE}, \mathcal{G})]$.

One may measure the communication complexity of Nash procedures also in terms of the number of rounds; this may be more natural from the game-theoretic viewpoint. Formally, the *time communication complexity* $tCC(\Pi, G)$ of a Nash equilibrium procedure Π applied to a game G is the number of time periods until Π terminates. The two communication complexity measures, CC and tCC, are closely related: in each time period the players transmit at least 1 bit and at most $\sum_i \log |A_i| = \log |A|$ bits.[12]

Proposition 2. *The time communication complexity* tCC *and the (bit) communication complexity* CC *satisfy:*

$$\frac{1}{\log |A|} CC \leq tCC \leq CC. \tag{1}$$

(A similar connection for two-player games was observed in Conitzer and Sandholm 2004.)

We are interested in the asymptotic behavior of the communication complexity of Nash equilibrium procedures as the number of players n increases, while the size of the action sets is fixed. Let Γ_s^n be the family

---

[12]Throughout this paper log is always $\log_2$.

of all n-person games where each player has at most s actions, i.e., $|A_i| \leq s$ for all i. We want to estimate the communication complexity of Nash equilibrium procedures on the class Γ_s^n as n increases and s is fixed. Our results will deal with the class Γ_2^n of binary-action games (except for Theorem 4, where we need 4 actions). Since the communication complexity is defined as the worst case over all games, any lower bound for Γ_2^n is clearly also a lower bound for Γ_s^n for every $s \geq 2$; see also Section 8.2.

3. Pure Equilibria

In this section we derive exponential lower bounds on the communication complexity of *pure* Nash equilibrium procedures. Our result is[13]

Theorem 3. *Any pure Nash equilibrium procedure has communication complexity* $\Omega(2^n)$, *i.e., for every* $s \geq 2$,

$$\mathrm{CC}(\mathrm{PURE}, \Gamma_s^n) \geq \mathrm{CC}(\mathrm{PURE}, \Gamma_2^n) = \Omega(2^n).$$

Proposition 2 implies that the *time communication complexity* of pure Nash equilibrium procedures is $\mathrm{tCC}(\mathrm{PURE}, \Gamma_2^n) = \Omega(2^n/n) = \Omega(2^{n-\log n})$.

At this point one may conjecture that restricting the class of games to those that have pure Nash equilibria may decrease the communication complexity. However, this is not so. Even if one considers only the specific class $\mathcal{FIP}_s^n$ of n-person s-action games that have the "finite improvement property" (see Section 2.1) and thus always possess pure Nash equilibria, the lower bound remains exponential. Specifically, for games with $s \geq 4$ actions, we have

Theorem 4. *Any pure Nash equilibrium procedure on the class* $\mathcal{FIP}_s^n$ *of* s-action games with the finite improvement property has communication complexity $\Omega(2^{n/2})$, i.e., for every $s \geq 4$,

$$\mathrm{CC}(\mathrm{PURE}, \mathcal{FIP}_s^n) \geq \mathrm{CC}(\mathrm{PURE}, \mathcal{FIP}_4^n) = \Omega(2^{n/2}).$$

Theorem 3 will be proved in Section 3.2 using a simple reduction from the disjointness problem (recall Theorem 1), whereas Theorem 4 will require a much more complex construction, which is relegated to Appendix A.

---

[13]Notation: $f(n) = \Omega(g(n))$ and $f(n) = O(g(n))$ mean that there exists a constant $C \geq 0$ such that $f(n) \geq Cg(n)$, respectively $f(n) \leq Cg(n)$, for all n.

3.1. *Reductions*

We now show how to reduce the disjointness problem to the problem of finding pure Nash equilibria. Divide the player set $\{1, \ldots, n\}$ into two sets T_1 and T_2 of size $n/2$ each (assume for simplicity that n is even), say $T_1 = \{1, \ldots, n/2\}$ and $T_2 = \{n/2 + 1, \ldots, n\}$. It will be convenient to rename the players such that the players in T_l are (l, i) for $i \in \{1, \ldots, n/2\}$ and $l \in \{1, 2\}$. For any two sets $S_1, S_2 \subset S$—an input of the S-disjointness problem—the *reduction* will define a game $G = (n, \{A_i\}i, \{u_i\}i)$, such that two properties are satisfied:

- *Reducibility*: $S_1 \cap S_2 \neq \emptyset$ if and only if G has a pure Nash equilibrium.
- *Constructibility*: The payoff function of each player (l, i) in T_l is constructible from S_l (i.e., for every $a \in A$ the number $u_{l,i}(a)$ is computable, by a finite algorithm, from a, S_l, and i).

The reducibility property enables us to relate the outcome of a pure Nash equilibrium procedure on G with the outcome of the S-disjointness function on S_1 and S_2. Namely, if the players reach a pure Nash equilibrium in G then the sets S_1 and S_2 are not disjoint, and if they do not reach a pure Nash equilibrium then the sets are disjoint. The constructibility property ensures that given a pure Nash equilibrium procedure Π_{NE} we are able to generate a protocol Π_D for the disjointness problem, with the same communication complexity. More specifically, given Π_{NE} we create a protocol Π_D by having agent $l \in \{1, 2\}$ simulate all the players in T_l (he can do so by the constructibility property). We summarize this in the following claim, which is based on Theorem 1.

Claim 5. *Assume that there exists a reduction from the S-disjointness problem to n-person pure Nash equilibrium procedures that satisfies the reducibility and constructibility properties. Then any pure Nash equilibrium procedure has communication complexity of at least $|S|$ bits.*

3.2. *Matching pennies reduction*

We now provide a simple reduction, which we call the *matching pennies reduction*, and establish Theorem 3.

Take $S = \{0, 1\}^n$; for each $S_1, S_2 \subset S$ the reduction will generate a binary-action game G in Γ_2^n as follows. The action spaces are $A_i = \{0, 1\}$ for all i, and a joint action is thus $a \in A = \{0, 1\}^n$. The payoff $u_{l,i(a)}$ of each player (l, i) in T_l will be high (specifically, 2) if the joint action a lies in the set S_l, and low (specifically, 0) if it does not. In the latter case, two distinguished players in T_l, say $(l, 1)$ and $(l, 2)$, will in addition play a matching pennies game between themselves.

Formally, for $l = 1, 2$, the payoff function $u_{l,i}$ of a player (l, i) in T_l is defined as follows. For $i \geq 3$, put[14]

$$u_{l,i}(a) = \begin{cases} 2, & \text{if } a \in S_l, \\ 0, & \text{if } a \notin S_l; \end{cases}$$

as for players $(l, 1)$ and $(l, 2)$ in T_l, their payoff functions are

$$u_{l,1}(a) = \begin{cases} 2, & \text{if } a \in S_l, \\ 1, & \text{if } a \notin S_l \ \text{ and } \ a_{l,1} = a_{l,2}, \\ 0, & \text{if } a \notin S_l \ \text{ and } \ a_{l,1} \neq a_{l,2}; \end{cases}$$

$$u_{l,2}(a) = \begin{cases} 2, & \text{if } a \in S_l, \\ 0, & \text{if } a \notin S_l \ \text{ and } \ a_{l,1} = a_{l,2}, \\ 1, & \text{if } a \notin S_l \ \text{ and } \ a_{l,1} \neq a_{l,2}. \end{cases}$$

Claim 6. *For $n \geq 4$, the reducibility and constructibility properties hold for the matching pennies reduction.*

Proof. The payoff functions of the players in T_l depend on S_l only, and so the constructibility property holds. For the reducibility property, note that a is a pure Nash equilibrium if and only if $a \in S_1 \cap S_2$ (indeed, if $a \in S_1 \cap S_2$, then every player gets the maximal payoff of 2; otherwise, $a \notin S_l$ for some l, and then either $(l, 1)$ or $(l, 2)$ benefits by deviating). □

We can now prove Theorem 3.

Proof of Theorem 3. Follows from Claims 5 and 6 (recall that $S = \{0, 1\}^n$). □

4. Pure Equilibria in a Bayesian Setting

We now consider a Bayesian setting where the game (i.e., the payoff functions) is chosen according to a *probability distribution* that is known to all players. While the communication complexity of pure Nash equilibrium

---

[14]Alternatively: put $u_{l,i}(a) = 0$ for all $a \in A$ and all $i \geq 3$.

procedures has been shown to be exponential in the worst case, it is conceivable that the *expected* communication complexity will be smaller (where the expectation is taken over the randomized selection of the payoff functions). However, that turns out not to be the case. We will exhibit a simple distribution for which the expected communication complexity of pure Nash equilibrium procedures is exponential. Our result is the following.

Theorem 7. *There exists a probability distribution over games such that any pure Nash equilibrium procedure has expected communication complexity* $\Omega(2^n)$; *i.e., there exists a probability distribution* $\Pr$ *over the family of binary-action games* Γ_2^n *such that*

$$E[CC(\text{PURE}, \Gamma_2^n)] = \Omega(2^n).$$

(Note that Theorem 3 is implied by Theorem 7.) Unlike the results in the previous section, here we will not apply a reduction, but rather provide a direct proof, using techniques from "distributional communication complexity" (see Kushilevitz and Nisan 1997, Sections 1.2 and 3.4).

Some further background from communication complexity is needed at this point. A *combinatorial rectangle* is $\chi = X_1 \times \cdots \times X_n$, where each X_i is a subset of inputs of player i. Every sequence of messages in a communication protocol can be described by a combinatorial rectangle, namely, all inputs generating that sequence of messages. Given a function f of n inputs $x_1, \ldots, x_n$, a combinatorial rectangle χ is called *monochromatic* if $f(x)$ has the same value for all $x = (x_1, \ldots, x_n) \in \chi$. A *minimal covering* of a function f using combinatorial rectangles is the minimum number of monochromatic combinatorial rectangles needed to represent f (i.e., the minimum number of monochromatic rectangles whose union covers the space of all possible inputs). Clearly, the logarithm of this number is a lower bound on the communication complexity of f (since, roughly speaking, every bit of communication can only split combinatorial rectangles into two; for more details see Kushilevitz and Nisan 1997, Chapter 1).

In our setting, the combinatorial rectangles are $\mathcal{U} = U_1 \times \cdots \times U_n$, where each U_i is a set of payoff functions of player i. A monochromatic combinatorial rectangle is labeled by either (1) a pure joint action $a \in A$ (when a is a Nash equilibrium for every game $(u_1, \ldots, u_n) \in \mathcal{U}$), or (2) "no pure Nash equilibrium" (when no game $(u_1, \ldots, u_n) \in \mathcal{U}$ has a pure Nash equilibrium).

Informally, the lower bound on the expected communication complexity of pure Nash equilibrium procedures will be a consequence of the fact that it will be "hard" for the players to agree that there is no pure Nash equilibrium. We will construct a probability distribution over payoff functions such that, first, the probability that there is no pure Nash equilibrium is bounded away from 0 as the number of players n increases. And second, we will show that any combinatorial rectangle that is labeled "no pure Nash equilibrium" has a low probability. This will yield a lower bound on the number of monochromatic combinatorial rectangles, and thus on the communication complexity.

Formally, our probability distribution Pr is defined on the family Γ_2^n of binary-action games (i.e., $A_i = \{0,1\}$ for all i). The payoff function u_i of player i is selected randomly as follows. For every $a^{-i} \in \{0,1\}^{n-1}$, with probability $1/2$ put $u_i(0, a^{-i}) = 0$ and $u_i(1, a^{-i}) = 1$, and with probability $1/2$ put $u_i(0, a^{-i}) = 1$ and $u_i(1, a^{-i}) = 0$; these choices are made independently over all a^{-i} and over all i. Note that for every $a \in \{0,1\}^n$ each player i has a unique best reply, and $\Pr[u_i : a_i \in \mathrm{BR}(a^{-i}; u_i)] = \Pr[u_i : a_i \notin \mathrm{BR}(a^{-i}; u_i)] = 1/2$.

We start by showing that the probability that there are no pure Nash equilibria is bounded away from 0.

Lemma 8. *There exists a constant $\alpha \geq 0$ such that*

$$\Pr[(u_1, \ldots, u_n): \text{ the game } (u_1, \ldots, u_n) \text{ has no pure Nash equilibrium}] \geq \alpha$$

for all $n \geq 2$.

Proof. First, we claim that for every $a \in \{0,1\}^n$, the probability that a is a pure Nash equilibrium is 2^{-n}. Indeed, a is a pure Nash equilibrium if and only if for every player i the payoff function u_i satisfies $a_i \in \mathrm{BR}(a^{-i}; u_i)$. This holds with probability $1/2$, independently, for each player i, and so the probability that a is a pure Nash equilibrium is 2^{-n}.

Second, let N be the number of pure Nash equilibria; then $E[N] = 1$, since there are 2^n joint actions $a \in \{0,1\}^n$ and the probability of each one being a pure Nash equilibrium is 2^{-n}. Therefore $\Pr[N=0] \geq \Pr[N \geq 2]$, since $1 = E[N] = \sum_{k \geq 1} k \Pr[N=k] \geq 1\Pr[N=1] + 2\Pr[N \geq 2] = (1 - \Pr[N=0]) + \Pr[N \geq 2]$.

Third, let $Z = \{a \in \{0,1\}^n : |\{i : a_i = 1\}|$ is even$\}$ be the set of joint actions with an even number of ones, and let ζ be the probability that in Z there are exactly 2 pure Nash equilibria. For any $a, b \in Z$ with $a \neq b$, the event that a is a pure Nash equilibrium is independent of the event that b is a pure Nash equilibrium (a and b differ in at least two coordinates by the definition of Z; hence for every player i we have $a^{-i} \neq b^{-i}$). Therefore the probability that exactly two elements of Z (whose size is $|Z| = 2^{n-1}$) are Nash equilibria is

$$\zeta = \binom{2^{n-1}}{2}(2^{-n})^2(1 - 2^{-n})^{2^{n-1}-2},$$

which is positive for all $n \geq 2$. Moreover, $\zeta \to (1/8)e^{-1/2} > 0$ as $n \to \infty$, and so $\Pr[N = 0] \geq \Pr[N \geq 2] \geq \zeta$ is indeed bounded away from 0. $\qquad\square$

Next we show that every combinatorial rectangle labeled "no pure Nash equilibrium" has low probability.

Lemma 9. *Let $\mathcal{U} = U_1 \times \cdots \times U_n$ be a combinatorial rectangle labeled "no pure Nash equilibrium." Then*

$$\Pr[(u_1, \ldots, u_n) \in \mathcal{U}] \leq 2^{-2^{n-1}}.$$

Proof. First, we claim that for every $a \in \{0,1\}^n$ there exists a player i such that $a_i \notin \mathrm{BR}(a^{-i}; u_i')$ for *all* $u_i \in U_i$. Indeed, otherwise, for every player i we would have $a_i \in \mathrm{BR}(a^{-i}; u_i')$ for some $u_i' \in U_i$. But this would imply that a is a pure Nash equilibrium for $(u_1', \ldots, u_n')$, which belongs to $\mathcal{U}$ since $\mathcal{U}$ is a rectangle—in contradiction to the assumption that every game in $\mathcal{U}$ has no pure Nash equilibrium.

Second, as in the proof of Lemma 8, let Z be the set of all $a \in \{0,1\}^n$ with an even number of ones. Define $Z_i = \{a \in Z : a_i \notin \mathrm{BR}(a^{-i}; u_i)$ for all $u_i \in U_i\}$; then $U_i \subset \{u_i : a_i \notin \mathrm{BR}(a^{-i}; u_i)$ for all $a \in Z_i\}$, and, as we saw just above, $Z = \cup_{i=1}^n Z_i$. Since the events $\{u_i : a_i \notin \mathrm{BR}(a^{-i}; u_i)\}$ are independent for different $a \in Z_i$ (again, $a \neq b$ implies that $a^{-i} \neq b^{-i}$ for $a, b \in Z$) and each one has probability $1/2$, we get

$$\Pr[u_i \in U_i] \leq \prod_{a \in Z_i} \Pr[u_i : a_i \notin \mathrm{BR}(a^{-i}; u_i)] = 2^{-|Z_i|}.$$

Therefore,

$$\Pr[(u_1, \ldots, u_n) \in \mathcal{U}] = \prod_{i=1}^n \Pr[u_i \in U_i] \leq \prod_{i=1}^n 2^{-|Z_i|} \leq 2^{-2^{n-1}},$$

since $\sum_i |Z_i| \geq |Z| = 2^{n-1}$. $\qquad\square$

Combining the two lemmata allows us to prove Theorem 7.

Proof of Theorem 7. By Lemma 8, the total probability of the event that there is no pure Nash equilibrium is bounded from below by $\alpha \geq 0$. By Lemma 9, each combinatorial rectangle labeled "no pure Nash equilibrium" has probability at most $2^{-2^{n-1}}$. Therefore R, the number of such rectangles, satisfies $R \geq \alpha 2^{2^{n-1}}$; this gives a lower bound on the expected communication complexity of $\log R = \Omega(2^n)$ (see the discussion following the statement of Theorem 7, or Section 1.2 in Kushilevitz and Nisan 1997). □

5. Mixed Equilibria

Before we introduce our result for mixed Nash equilibrium procedures, a certain preliminary discussion is in order. In the case of mixed Nash equilibria the values of the payoff functions play a crucial role. Consider the following variant of the matching pennies game

$1,0$	$0,1$
$0,1$	$M,0$

where M is a positive integer. There is a unique Nash equilibrium: $(1/2, 1/2)$ for the row player and $(M/(M+1), 1/(M+1))$ for the column player. Since the parameter M appears only in the payoff function of the row player, and in equilibrium the column player needs to know the precise value of M, it follows that $\log M$ bits have to be communicated. This is a somewhat unsatisfactory result, since the number of bits needed to encode one of the values of the payoff function of the row player is also $\log M$. However, had it been commonly known, for instance, that the payoff functions under consideration have either 1 or M in that entry, then only one bit would have sufficed. We therefore distinguish between two concepts, "magnitude" and "encoding."

Let U_i be a family of payoff functions of player i. The *magnitude* of a rational number ρ is $\mathfrak{mag}(\rho) = \log|M| + \log|K|$, where $\rho = M/K$ is a reduced fraction (i.e., M and K have no common divisor higher than 1), and the magnitude of the family U_i is $\mathfrak{mag}(U_i) = \max_{u_i \in U_i, a \in A} \mathfrak{mag}(u_i(a))$. For each $a \in A$, the *encoding* of the payoff of player i at a is $\mathfrak{enc}(U_i, a) = \log|\{u_i(a): u_i \in U_i\}|$; i.e., the number of bits required to encode the possible values of $u_i(a)$ as u_i varies over U_i; the encoding of the family U_i is $\mathfrak{enc}(U_i) = \max_{a \in A} \mathfrak{enc}(U_i, a)$. For example, if every payoff function u_i in U_i has two values 1 and M (i.e., $u_i(a) \in \{l, M\}$ for all $u_i \in U_i$ and all $a \in A$), then the encoding of U_i is $\mathfrak{enc}(U_i) = 1$ bit, whereas its magnitude is $\mathfrak{mag}(U_i) = \log M$ bits. Finally, if $\mathcal{U} = U_1 \times \cdots \times U_n$ is a family of games, then $\mathfrak{enc}(\mathcal{U}) = \max_{1 \leq i \leq n} \mathfrak{enc}(U_i)$ and $\mathfrak{mag}(\mathcal{U}) = \max_{1 \leq i \leq n} \mathfrak{mag}(U_i)$.

When deriving lower bounds on the communication complexity of mixed Nash equilibrium procedures, one would like the encoding as well as the magnitude to be as low as possible (so that a high complexity will *not* be just a trivial consequence, as in the example above). Specifically, we will construct a large family of games $\mathcal{U}$ that has an encoding of 1 bit and a magnitude of $O(n)$ bits, such that each game in $\mathcal{U}$ will have a different unique Nash equilibrium. This will imply that, in order to reach the correct Nash equilibrium, the number of bits to be transmitted must be at least the logarithm of the size of the family $\mathcal{U}$. Formally, our result is

Theorem 10. *For every* $n \geq 2$ *there exists a family of binary-action games* $\mathcal{U}^n \subset \Gamma_2^n$ *whose encoding is 1 bit and whose magnitude is* $O(n)$ *bits (i.e.,* $\mathrm{enc}(\mathcal{U}^n) = 1$ *and* $\mathrm{mag}(\mathcal{U}^n) = O(n)$*), such that any mixed Nash equilibrium procedure over* $\mathcal{U}^n$ *has communication complexity* $\Omega(2^n)$*, i.e.,*

$$\mathrm{CC}(\mathrm{MIXED}, \mathcal{U}^n) = \Omega(2^n).$$

Our construction is based on a generalization of Jordan's game (Jordan 1993) in which we modify the payoff of one of the players. For $n \geq 2$, the *n-person Jordan game* J_n is a binary-action game with payoff functions $u_i(a) = \mathbf{1}_{\{a_i = a_{i-1}\}}(a)$ for all players $i \neq 2$ (where $a_0 \equiv a_n$) and $u_2(a) = \mathbf{1}_{\{a_2 \neq a_1\}}(a)$ for player 2 (we write $\mathbf{1}_X$ for the indicator function of the event X; e.g., $\mathbf{1}_{\{a_1 = a_n\}}(a) = 1$ if $a_1 = a_n$ and $\mathbf{1}_{\{a_1 = a_n\}}(a) = 0$ otherwise). Thus player 2 wants to "mismatch" the action of player 1, whereas every other player $i \neq 2$ wants to "match" the action of the previous player[15] $i - 1$.

Let f be a real function from $\{0, 1\}^{n-2}$ to the half-open interval $[0, 1)$, i.e., $f: \{0, 1\}^{n-2} \to [0, 1)$; we define the *f-modified Jordan game* $J_n(f)$ by

$$
\begin{aligned}
u_i(a) &= \mathbf{1}_{\{a_i = a_{i-1}\}}(a), \quad \text{for } i \neq 2; \quad \text{and} \\
u_2^f(a) &= \mathbf{1}_{\{a_2 \neq a_1\}}(a) + \mathbf{1}_{\{a_1 = a_2 = 1\}}(a) \cdot f(a_3, \dots, a_n)
\end{aligned}
\tag{2}
$$

(only the payoff of player 2 has been modified).

The following lemma shows that a modified Jordan game has a unique Nash equilibrium, and gives an explicit formula for it. For every function f as above, let

$$\mu(f) = \frac{1}{2^{n-2}} \sum_{(a_3, \dots, a_n) \in \{0,1\}^{n-2}} f(a_3, \dots, a_n)$$

be the average of the values of f; equivalently, this is the expected value of f when every player $i \geq 3$ randomizes uniformly, i.e., $p_i = 1/2$.

---

[15]This game has a unique Nash equilibrium $(1/2, \dots, 1/2)$ (this also follows from Lemma 11 below).

Lemma 11. *The modified Jordan game $J_n(f)$ has a unique Nash equilibrium $(p_1, \ldots, p_n)$, where $p_i = 1/2$ for all players $i \neq 1$, and*[16]

$$p_1 = \frac{1}{2 - \mu(f)}. \tag{3}$$

Proof. Let $(p_1, \ldots, p_n)$ be a Nash equilibrium of $J_n(f)$. Assume that $p_i \neq 1/2$ for some player $i \geq 2$; without loss of generality, $p_i \geq 1/2$. Then player $i + 1$ must be playing purely the action 1, i.e., $p_{i+1} = 1$. Repeating this argument implies that $p_{i+2} = \cdots = p_n = p_1 = 1$. Therefore player 2, who wants to mismatch player 1, must be playing the action 0 (here we use the fact that $0 \leq f < 1$), i.e., $p_2 = 0$. Hence player 3 must be playing 0, and so on; i.e., $p_3 = p_4 = \cdots = p_i = 0$, contradicting $p_i > 1/2$.

So we have $p_i = 1/2$ for all $i \neq 1$. As for player 1, the probability p_1 is determined by the condition that player 2 must be indifferent between his two actions ($p_2 = 1/2$ and so player 2 plays both actions with positive probability), i.e.,

$$u_2^f(p_1, 0, 1/2, \ldots, 1/2) = u_2^f(p_1, 1, 1/2, \ldots, 1/2),$$

or, recalling the definitions of u_2^f and $\mu(f)$,

$$p_1 = p_1 \mu(f) + 1 - p_1.$$

Solving for p_1 yields (3) and completes the proof. □

To construct our family of games, we vary the function f over a set $\mathcal{F}$ of functions; thus, for each $i \neq 2$, the family $U_i = \{u_i\}$ is a singleton, whereas the family $U_2 = \{u_2^f : f \in \mathcal{F}\}$ consists of all payoff functions u_2^f of player 2 that are obtained for all $f \in \mathcal{F}$. The property of the family $\mathcal{F}$ will be that, for each function $f \in \mathcal{F}$, when we substitute f in (3) we get a different value for p_1. The lower bound on the communication complexity will follow from the fact that for each $f \in \mathcal{F}$ the communication to player 1 must be different. (Indeed, player 1 needs to reach a different value of p_1 for each f, and always starts with the same information.) This will imply that the number of bits that have to be communicated is at least $\log |\mathcal{F}|$. To formalize this, we will call a set of functions $\mathcal{F}$ *separating* if for any two functions $f_1 \neq f_2$ in $\mathcal{F}$ we have $\mu(f1) \neq \mu(f_2)$ Thus

---

[16]Recall that p_i stands for the probability of action 1, i.e., $p_i = \Pr[a_i = 1]$.

Claim 12. *Let $\mathcal{U}$ be given as above by a separating set of functions $\mathcal{F}$. Then the communication complexity of any mixed Nash equilibrium procedure on $\mathcal{U}$ is at least $\log|\mathcal{F}|$.*

We now construct our family of functions. For every $x = (x_1, \ldots, x_{n-2})$ in $\{0,1\}^{n-2}$, let $[x]_2 = \sum_{i=1}^{n-2} x_i 2^{n-2-i}$ be the integer corresponding to the binary string x. Let $\mathcal{H}$ be the set of Boolean functions $h : \{0,1\}^{n-2} \to \{0,1\}$. For every $h \in \mathcal{H}$, define a function f_h on $\{0,1\}^{n-2}$ by

$$f_h(x) = h(x) \frac{1}{\mathfrak{prime}([x]_2)}$$

for each $x \in \{0,1\}^{n-2}$, where $\mathfrak{prime}(k)$ is the k-th prime, starting for convenience with $\mathfrak{prime}(0) = 2$ (thus $\mathfrak{prime}(1) = 3, \mathfrak{prime}(2) = 5$, and so on; note that indeed $f_h(x) \in [0,1)$). Let $\mathcal{F}_{\mathcal{H}} = \{f_h : h \in \mathcal{H}\}$. The following lemma shows that $\mathcal{F}_{\mathcal{H}}$ is a separating family.

Lemma 13. *The family $\mathcal{F}_{\mathcal{H}}$ is separating; i.e., for any two Boolean functions $h_1 \neq h_2$ in $\mathcal{H}$ we have $\mu(f_{h_1}) \neq \mu(f_{h_2})$.*

Proof. Let $r_1, \ldots, r_m$ be m distinct prime numbers, and let $\rho = \sum_{i=1}^{m} 1/r_i$. Express ρ as a reduced fraction $\rho = M/K$, and let $R = \prod_{i=1}^{m} r_i$. We claim that $K = R$. Indeed, on the one hand the common denominator in the sum ρ is R, so K divides R. On the other hand, multiplying ρ by $K \prod_{i=2}^{m} r_i$ gives

$$M \prod_{i=2}^{m} r_i = \frac{K \prod_{i=2}^{m} r_i}{r_1} + \sum_{i=2}^{m} K \prod_{j \neq 1, i} r_j;$$

therefore $K \prod_{i=2}^{m} r_i / r_1$ is an integer, and since the r_i are distinct primes, it follows that r_1 divides K. The same holds for all i, and so $\prod_{i=1}^{m} r_i = R$ divides K. Altogether $K = R$ (since each one divides the other), as claimed.

Now, for $l = 1, 2$,

$$2^{n-2} \mu(f_{h_l}) = \sum_{x \in \{0,1\}^{n-2}} f_{h_l}(x) = \sum_{x : h_l(x) = 1} \frac{1}{\mathfrak{prime}([x]_2)} = \frac{M_l}{K_l},$$

where M_l / K_l is a reduced fraction. If $\mu(f_{h_1}) = \mu(f_{h_2})$ then $K_1 = K_2$, or

$$\prod_{x : h_1(x) = 1} \mathfrak{prime}([x]_2) = \prod_{x : h_2(x) = 1} \mathfrak{prime}([x]_2),$$

which implies that the binary functions h_1 and h_2 are identical. $\qquad \square$

Next, the magnitude of $\mathcal{F}_{\mathcal{H}}$ is $O(n)$ bits, since $\mathfrak{prime}(k) = O(k \log k)$ by the Prime Number Theorem, and so $\log(\mathfrak{prime}([x]_2)) \leq \log(\mathfrak{prime}(2^{n-2})) = O(n)$; whereas the encoding of $\mathcal{F}_{\mathcal{H}}$ is just 1 bit, since $f_h(x)$ has only two possible values, $1/\mathfrak{prime}([x]_2)$ and 0. The same therefore holds for the resulting family of games $\mathcal{U} = \mathcal{U}^n := \{J_n(f) : f \in \mathcal{F}_{\mathcal{H}}\}$ (see (2)). We have thus established

Claim 14. *The family $\mathcal{U}$ satisfies* $\mathfrak{enc}(\mathcal{U}) = 1$ *and* $\mathfrak{mag}(\mathcal{U}) = O(n)$.

We can now complete the proof of Theorem 10.

Proof of Theorem 10. There are $2^{2^{n-2}}$ Boolean functions h in H, so $|\mathcal{F}_{\mathcal{H}}| = |H| = 2^{2^{n-2}}$. Combining this with Claims 12 and 14 and Lemma 13 proves Theorem 10. $\square$

6. Upper Bounds

In this section we will show that two rather trivial procedures achieve near-optimal communication complexity. This demonstrates the power of our lower bounds, showing that even trivial procedures achieve a near-optimal performance.

We start with a pure Nash equilibrium procedure on binary-action games Γ_2^n.

Enumeration: Let Π_{ENUM} be the following procedure. All 2^n joint actions are examined in turn, with each player communicating whether or not he is best-replying there. Formally, at every period $t = 0, 1, \ldots, 2^n - 1$, let $b \in \{0,1\}^n$ be such that $[b]_2 = t$ (where $[b]_2 = \sum_{i=1}^n b_i 2^{n-i}$ is the integer corresponding to the binary string b); then player i broadcasts $a_{i,t} = 1$ if b_i is a best reply to b^{-i} (i.e., $u_i(b) \geq u_i(1-b_i, b^{-i})$), and broadcasts $a_{i,t} = 0$ otherwise. If all players broadcasted 1 at a certain period t, then the procedure terminates at that point with the corresponding b (i.e., $[b]_2 = t$) as a pure Nash equilibrium of the game. Otherwise, the procedure ends at period 2^n with "no pure Nash equilibrium."

Since every one of the n players communicates 1 bit in each one of the 2^n periods, we have

Proposition 15. *For every $n \geq 2$,*

$$\text{CC}(\text{PURE}, \Gamma_2^n) \leq \text{CC}(\Pi_{\text{ENUM}}, \Gamma_2^n) = n2^n.$$

For the next procedure, consider first the family of binary-action games that also have binary payoffs, i.e., $u_i(a) \in \{0,1\}$ for all $a \in A = \{0,1\}^n$. For every game G let $\bar{x}(G)$ be a Nash equilibrium of G. Think of $\bar{x}$ as a predetermined and agreed-upon rule that selects a Nash equilibrium for each game (for example, among pure Nash equilibria select that one a that minimizes $[a]_2$; among mixed Nash equilibria, select that one $(p_1, \ldots, p_n)$ that minimizes the p_i lexicographically). The procedure is as follows.

Broadcast: Let Π_{BCAST} be the following procedure. The values of the payoff functions at each one of the 2^n joint actions are broadcast in turn, by all players. Formally, at every period $t = 0, 1, \ldots, 2^n - 1$, let $b \in \{0,1\}^n$ be such that $[b]_2 = t$; then player i communicates $a_{i,t} = u_i(b)$. At period $t = 2^n$ every player can reconstruct the payoff function of every other player: $u_i(b) = a_i, [b]_2$ for each $b \in \{0,1\}^n$, and so they all know the game G, and thus the agreed-upon Nash equilibrium $\bar{x}(G)$.

The communication complexity of this procedure is again $n2^n$. Clearly, the same holds for any family of games $\mathcal{U}_1^n \subset \Gamma_2^n$ whose encoding is 1 bit (recall Section 5; the magnitude of $\mathcal{U}_1^n$ does not matter and may be arbitrarily large), whereas if the encoding is r bits, then the number of periods becomes $r2^n$ (every payoff entry takes r periods to transmit). So we have

Proposition 16. *For every $n \geq 2$ let $\mathcal{U}_r^n \subset \Gamma_r^n$ be a family of binary-action games whose encoding is at most r bits, i.e., $\text{enc}(\mathcal{U}_r^n) \leq r$. Then,*

$$\text{CC}(\text{MIXED}, \mathcal{U}_r^n) \leq \text{CC}(\Pi_{\text{BCAST}}, \mathcal{U}_r^n) \leq rn2^n.$$

7. Correlated Equilibria

In this section we study the communication complexity of reaching a correlated equilibrium, and prove that it is polynomial rather than exponential in the number of players. This shows that the exponential bounds for Nash equilibrium procedures are not due just to the complexity of the input, i.e., to the payoff functions being of exponential size, but rather to the intrinsic complexity of reaching Nash equilibria.

Based on the polynomial-time algorithm of Papadimitriou and Roughgarden (2008) for *computing* correlated equilibria of certain "succinct polynomial games," we derive a correlated equilibrium procedure with polynomial communication complexity, for *all games* with integer payoffs.

Specifically, let $\mathcal{U}_u^n \subset \Gamma_2^n$ be the family of n-person binary-action games with integer payoffs of magnitude at most u bits, i.e., $\max_{1 \leq i \leq n} \mathfrak{mag}(u_i) \leq u$; our correlated equilibrium procedure will have a communication complexity that is polynomial in the number of players n and the magnitude of the payoffs u (for simplicity we again consider only binary-action games; otherwise, it would be polynomial in n, u, and $\max_{1 \leq i \leq n} |A_i|$).

We start by recalling the definition of a correlated equilibrium; see Aumann (1974). Given a game $G = (n, \{A_i\}i, \{u_i\}i)$, a distribution Q over the space of joint actions $A = \prod_{i=1}^{n} A_i$ is (the distribution of) a *correlated equilibrium* of G if for each player i and all actions $b_i, b_i' \in A_i$, we have $\mathrm{E}_Q[u_i(b_i, a^{-i})\mathbf{1}_{\{a_i=b_i\}}] \geq \mathrm{E}_Q[u_i(b_i', a^{-i})\mathbf{1}_{\{a_i=b_i\}}]$ (where E_Q denotes expectation with respect to the distribution Q). Equivalently, consider the "extended game" where, before G is played, a joint action $a = (a_1, \ldots, a_n) \in A$ is randomly chosen according to Q and each player i is given a "recommendation" to play a_i, his coordinate of the chosen a; then Q is a correlated equilibrium of G if and only if the combination of strategies where each player always plays according to his recommendation constitutes a Nash equilibrium of the extended game.

A *correlated equilibrium procedure* Π is defined in the same way as a Nash equilibrium procedure, except that now the output of each player is a distribution Q, such that Q is a correlated equilibrium of the game $G = (u_1, \ldots, u_n)$ that was given as input.[17] Let CEP be the collection of correlated equilibrium procedures. Similarly to CC(MIXED, $\mathcal{G}$) and CC(PURE, $\mathcal{G}$), we define the *communication complexity of correlated equilibrium procedures* for a family of games $\mathcal{G}$ as CC(CORRELATED, $\mathcal{G}$) = $\min_{\Pi \in \text{CEP}} \text{CC}(\Pi, \mathcal{G}) = \min_{\Pi \in \text{CEP}} \max_{G \in \mathcal{G}} \text{CC}(\Pi, G)$.

We come now to the construction of Papadimitriou and Roughgarden (2008), which consists of running an ellipsoid algorithm in the Hart–Schmeidler setup (Hart and Schmeidler 1989). In our communication complexity framework, every player can run internally the computations of the algorithm at no cost. However, since the payoff function u_i is known only to player i, only i can compute his own expected payoffs—which he can then broadcast to all players. The communication complexity counts only the number of bits transmitted, and therefore, as we will see, there is no need to restrict ourselves to "succinct games of polynomial type" as in Papadimitriou and Roughgarden (2008).

---

[17]Finite games always possess correlated equilibria.

We define the procedure Π_{CORR} as follows. All players simulate the algorithm of Papadimitriou and Roughgarden (2008). At each step of the ellipsoid algorithm, an n-tuple of mixed strategies $p = (p_1, \ldots, p_n) \in \Delta = \prod_{i=1}^{n} \Delta_i$ is generated (the whole vector p is computed internally by—and thus known to—each player). Every player i then computes his expected payoff $u_i(p)$ and broadcasts it. In terms of communication complexity, again, the local computation of p and $u_i(p)$ has no cost; only the transmission of $u_i(p)$ counts.

Papadimitriou and Roughgarden (2008) prove, first, that a correlated equilibrium is reached in a number of steps that is bounded by a polynomial in n and u; and second, that the n-tuples of mixed strategies $p \in \Delta$ generated at every step have a magnitude $\mathfrak{mag}(p) = O(nu)$ bits. Therefore, when the payoffs $u_i(a)$ for all $a \in A$ are integers of at most u bits, the expected payoff $u_i(p)$ for $p \in \Delta$ requires at most $O(n\mathfrak{mag}(p) + u + n) = O(n^2 u)$ bits (since it is a weighted sum of 2^n entries). Altogether, this implies that the total number of bits transmitted in the procedure Π_{CORR} is bounded by a polynomial in n and u, and we have shown

Theorem 17. *For every $n \geq 2$ and $u \geq 1$, let $\mathcal{U}_u^n \subset \Gamma_2^n$ be the family of n-person binary-action games with integer payoffs of magnitude at most u, i.e., $\max_{i \leq i \leq n} \mathfrak{mag}(u_i) \leq u$. Then there exists a correlated equilibrium procedure Π_{CORR} whose communication complexity over $\mathcal{U}_u^n$ is polynomial in n and u, i.e.,*

$$\text{CC}(\text{CORRELATED}, \mathcal{U}_u^n) \leq \text{CC}(\Pi_{\text{CORR}}, \mathcal{U}_u^n) \leq polynomial(n, u).$$

In Appendix B we will present further results on the communication complexity of reaching correlated equilibria. Specifically, in the classes of games of Sections 4 and 5 where the communication complexity of reaching Nash equilibria was shown to be exponential, that of correlated equilibria turns out to be quite low; moreover, it is attained by simple procedures. We will also analyze the communication complexity of reaching correlated *approximate* equilibria.

8. Extensions

8.1. *Stochastic procedures*

For simplicity we have discussed only *deterministic* procedures; however, our results carry over to "stochastic" (or "randomized") procedures. In an *equilibrium stochastic procedure*, the players may randomize when choosing their actions (i.e., $a_{i,t} \in A_i$ is chosen according to a probability distribution, which may depend on the history and on player i's payoff function), and the procedure terminates in an appropriate equilibrium (pure, mixed, or correlated, as the case may be). The analysis of stochastic procedures is in terms of the *expected* communication complexity (i.e., the expectation of the number of bits transmitted until termination), which we denote CCS (where S stands for "stochastic").

In general, the communication complexity of deterministic and stochastic protocols are polynomially related (see Kushilevitz and Nisan 1997, Section 2.3), and so we expect the analysis of stochastic procedures also to lead to exponential bounds. In particular, the lower bound on the communication complexity of the $\mathcal{S}$-disjointness problem DISJ$_\mathcal{S}$ remains $\Omega(|\mathcal{S}|)$ also for stochastic protocols (see Kushilevitz and Nisan 1997, Section 3.4), so the result of Theorem 3 holds also for pure Nash equilibrium *stochastic* procedures. As for Theorem 10, the counting argument of Section 5 applies to any procedure, whether deterministic or stochastic. See Appendix B for the analysis of some correlated equilibrium stochastic procedures.

8.2. Larger action spaces

We have mainly considered games in Γ_2^n where every player has 2 actions. What happens when the number of actions increases to $s > 2$? This has two effects. First, the space of joint actions is larger: s^n rather than 2^n; and second, the number of communication bits per message is now $\log s$ rather than 1. Of course, as we noted at the end of Section 2.3, any lower bound for Γ_2^n is also a lower bound for Γ_s^n. However, most of our techniques turn out to be easily extendable to Γ_s^n, and to yield better bounds when $s > 2$. For instance, the matching pennies reduction (Section 3.2) gives a lower bound of $\Omega(s^n)$ when applied to Γ_s^n (recall Theorem 1: the disjointness problem for subsets of a set of size s^n has communication complexity s^n). As for the upper bounds, the enumeration and broadcast procedures Π_{ENUM} and Π_{BCAST} (Section 6) lead to a communication complexity of $O(ns^n \log s)$ and $O(rns^n)$, respectively.

8.3. Nash approximate equilibria

An *approximate* equilibrium requires each player's gain from deviating to be small. Formally, given $\varepsilon > 0$, a *Nash ε-equilibrium* is a combination of mixed actions $p = (p_1, \ldots, p_n) \in \Pi_{i=1}^n \Delta_i = \Delta$ such that $u_i(p) \geq u_i(q_i, p^{-i}) - \varepsilon$ for every player i and any mixed action $q_i \in \Delta_i$ of i. It would be interesting to study the communication complexity of reaching Nash approximate equilibria, and determine whether or not it is also exponential in the number of players.

Appendix A. Potential Game Reduction

In Section 3.2 we provided a reduction—the matching pennies reduction—from procedures for the 2^n-disjointness problem to n-person pure Nash equilibrium procedures. We now construct another reduction, which we call the *potential game reduction*, whose additional property[18] is that whenever the two sets in the disjointness problem intersect, the corresponding game has the finite improvement path property. This reduction will establish Theorem 4.

Before we describe the potential game reduction, it is worthwhile to investigate why an alternative naive reduction fails. Let us start with a few notations, which will be useful later on. The *Hamming distance* $d_H(w, v)$ between two vectors $w, v \in \{0, 1\}^k$ equals the number of coordinates in which they differ; for a set $V \subset \{0, 1\}^k$, put $d_H(w, V) = \min_{v \in V} d_H(w, v)$.

Recall from Section 3.1 that n is assumed even and the set of n players is partitioned into two sets of $n/2$ players, T_1 and T_2; the players of T_ℓ are denoted (ℓ, i) for $i \in \{1, \ldots, n/2\}$ and $\ell \in \{1, 2\}$. Take $\mathcal{S} = \{0, 1\}^{n/2}$, and consider the following reduction from the $\mathcal{S}$-disjointness problem to binary-action games Γ_2^n. Let $S_1, S_2 \subset \mathcal{S} = \{0, 1\}^{n/2}$. For each joint action $a \in \{0, 1\}^n$ define $z = z(a) \in \{0, 1\}^{n/2}$ by $x_i = a_{1,i} \oplus a_{2,i}$ for all $i \in \{1, \ldots, n/2\}$, and let the payoff functions be $u_{\ell,i}(a) = -d_H(z(a), S_\ell)$, for all $a \in \{0, 1\}^n$, $i \in \{1, \ldots, n/2\}$, and $\ell \in \{1, 2\}$. One can show that there exists a pure Nash equilibrium in this game iff $S_1 \cap S_2 \neq \emptyset$. However, improvement paths in these games are not necessarily finite.[19] Our potential game reduction will also use the Hamming distance to drive the joint action to a certain region, but will require a much more complex structure in order to guarantee that all improvement paths are finite.

---

[18] Besides reducibility and constructibility; see Section 3.1.

[19] For example, take S_1, S_2 such that $S_1 \cap S_2 \neq \emptyset$ and there are two vectors w and v with $d_H(w, S_1) < d_H(w, S_2)$, $d_H(v, S_2) < d_H(v, S_1)$, and $d_H(w, v) = 1$. Let i be the index where w and v differ; players $(1, i)$ and $(2, i)$ can then alternate indefinitely in performing improvement steps.

We now present the potential game reduction. Let H be a Hamiltonian cycle in the hypercube $\{0,1\}^{n/2}$. For $x, y \in \{0,1\}^{n/2}$, when y immediately follows x in the cycle H we write $y = next(x)$ and $x = prev(y)$; also, let $r(x) \in \{1, \ldots, n/2\}$ denote the index of the unique bit in which x and $next(x)$ differ. Let

$$L = \{xx : x \in \{0,1\}^{n/2}\} \cup \{yx : y, x \in \{0,1\}^{n/2}, \ y = next(x)\},$$

where zw denotes the concatenation of the strings z and w. Clearly $L \subset \{0,1\}^n$ and $|L| = 2 \cdot 2^{n/2}$. In our games every joint action $a \in A$ will be mapped to some $z(a) \in \{0,1\}^n$, and the payoff of every player will increase as $z(a)$ approaches the set L. A pure Nash equilibrium a, if it exists, will always have $z(a) \in L$.

The players in T_1 have binary actions, i.e., $A_{1,i} = \{0,1\}$, whereas those in T_2 have four actions: $A_{2,i} = \{0,1\} \times \{0,1\}$. For an action $a_{2,i} = (c_{2,i}, d_{2,i}) \in A_{2,i}$, we will refer to $c_{2,i}$ as the *action bit* and to $d_{2,i}$ as the *done bit*. A joint action $a \in A$ can be written $a = (a_1, a_2)$, where $a_1 \in \prod_i A_{1,i}$ and $a_2 \in \prod_i A_{2,i}$ are the joint actions of T_1 and T_2, respectively. Given $a = (a_1, a_2)$, define

$$x_1 \equiv x_1(a_1) = a_1 = (a_{1,1}, \ldots, a_{1,n/2}) \in \{0,1\}^{n/2},$$
$$x_2 \equiv x_2(a_2) = (c_{2,1}, \ldots, c_{2,n/2}) \in \{0,1\}^{n/2},$$
$$d_2 \equiv d_2(a_2) = (d_{2,1}, \ldots, d_{2,n/2}) \in \{0,1\}^{n/2}, \quad \text{and}$$
$$z \equiv z(a) = x_1 x_2 \in \{0,1\}^n$$

($x_1 x_2$ is the concatenation of x_1 and x_2).

We will view L as a cycle that moves from each xx to yx, where $y = next(x)$, and then from yx to yy. As the joint action a changes, so does the resulting $z(a)$. To move $z(a)$ in L between xx and yx one player in T_1, namely $(1, r(x))$, must change his action; we call him the *forward active 1-player* at xx, and also the *backward active 1-player* at yx, and denote him by $r_1(xx) = r_1(yx) = (1, r(x))$. Similarly, the move between yx and yy is controlled by the action bit of one player in T_2, namely $r_2(yx) = r_2(yy) = (2, r(x))$, which we call the *forward active 2-player* at yx, and also the *backward active 2-player* at yy.

A high-level description of our reduction is as follows. Given two subsets S_1 and S_2 of $S = \{0,1\}^{n/2}$, we define the payoff functions of the players such that: (1) all players want to reach L (i.e., have $z(a) \in L$) and stay in it; (2) when in L, only the active players have an incentive to change their actions; (3) if the joint action is xx and $x \in S_1 \cap S_2$ then no active player has an incentive to change his action, and we are at a pure Nash equilibrium; and (4) the payoff functions of the players in T_ℓ depend only on S_ℓ, for $\ell \in \{1, 2\}$.

Specifically, for each player $(1, i)$ in T_1 we define his payoff function

$$u_{1,i}(a) = \begin{cases} -d_H(x_1 x_2, L), & \text{if } x_1 x_2 \notin L, \\ 1, & \text{if } x_1 x_2 \in L \text{ and } x_1 \neq x_2, \\ 2, & \text{if } x_1 x_2 \in L, x_1 = x_2, x_1 \in S_1, \text{ and } d_{r_2(x_1 x_2)} = 1, \\ 0, & \text{if } x_1 x_2 \in L, x_1 = x_2, x_1 \notin S_1 \text{ or } d_{r_2(x_1 x_2)} = 0. \end{cases}$$

Thus, if $z = x_1 x_2 \notin L$ then $u_{1,i}(a)$ is the negative of the Hamming distance from z to the set L (this provides the incentive always to move in the direction of L, and once L is reached not to leave it). If $x_1 = x_2$, $x_1 \in S_1$, and $d_{r_2(x_1 x_2)} = 1$, then $u_{1,i}(a)$ has the maximal value of 2 (this is where the pure Nash equilibria will be, if at all); note that players in T_1 can test $d_{r_2(x_1 x_2)} = 1$ since the identity of the active 2-player $r_2(x_1 x_2) = r_2(z(a))$ is just a function of the joint action a. If $x_1 \neq x_2$ then $u_{1,i}(a) = 1$, and otherwise $u_{1,i}(a) = 0$ (this will cause the players in T_1 to prefer to move from $x_1 = x_2$ to $x_1 \neq x_2$, unless both $x_1 \in S_1$ and $d_{r_2(x_1 x_2)} = 1$).

For each player $(2, i)$ in T_2, we first define an auxiliary function $GoodDone_{2,i}$

$$GoodDone_{2,i}(a) = \begin{cases} 0, & \text{if } x_1 = x_2, (2, i) = r_2(x_1 x_2), \text{ and } d_{2,i} \neq 1_{\{x_2 \in S_2\}}, \\ 1, & \text{otherwise,} \end{cases}$$

and then the payoff function

$$u_{2,i}(a) = \begin{cases} -d_H(x_1 x_2, L), & \text{if } x_1 x_2 \notin L, \\ 0, & \text{if } x_1 x_2 \in L \text{ and } x_1 \neq x_2, \\ 2 \cdot GoodDone_{2,i}(a), & \text{if } x_1 x_2 \in L, x_1 = x_2, \text{ and } x_2 \in S_2, \\ GoodDone_{2,i}(a), & \text{if } x_1 x_2 \in L, x_1 = x_2, \text{ and } x_2 \notin S_2. \end{cases}$$

The idea is that when $x_1 = x_2$ the active 2-player $(2, i) = r_2(x_1 x_2)$ should "signal" through his done bit whether or not $x_2 \in S_2$ (this is needed to let the players in T_1 know when a Nash equilibrium has been reached); if he does not signal correctly he is "penalized" by having $GoodDone_{2,i} = 0$ instead of 1, which decreases his payoff.

Claim 18. *The constructibility and reducibility properties hold for the potential game reduction.*

Proof. By definition of the reduction, the payoffs of the players in T_ℓ depend only on S_ℓ, and so the constructibility property holds. It remains to show that the reducibility property holds.

We will distinguish five types of joint actions a in A and analyze each in turn.

(1) a such that $z(a) = x_1x_2 \in L, x_1 = x_2, x_1 \in S_1, d_{r_2(x_1x_2)} = 1$ and $x_2 \in S_2$—thus $x_1 = x_2 \in S_1 \cap S_2$—is a pure Nash equilibrium, since all players get their maximal payoff of 2 (we have $GoodDone_{2,i}(a) = 1$ for all players $(2, i)$ in T_2). Such an a is obtained from $x = x_1 = x_2 \in S_1 \cap S_2$ by putting $a_{1,i} = x_{(i)}$ (= the i's coordinate of x) for each player $(1, i)$ in T_1, and $a_{2,i} = (c_{2,i}, d_{2,i})$ with action bit $c_{2,i} = x_{(i)}$ and arbitrary done bit $d_{2,i}$ for each player $(2, i)$ in T_2, except for the active 2-player $r_2(x_1x_2)$, whose done bit is $d_{r_2(x_1x_2)} = 1$.

(2) a such that $z(a) = x_1x_2 \notin L$ cannot be a Nash equilibrium since at least one player (ℓ, i), by changing his action, can bring the new $z(a')$ closer to L and thus increase his payoff by 1.

(3) a such that $z(a) = x_1x_2 \in L, x_1 \neq x_2$ cannot be a Nash equilibrium, since the (forward) active 2-player, by changing his action bit and also setting his done bit correctly (to $d_{r_2(x_1x_2)} = 1_{\{x_2 \in S_2\}}$), can increase his payoff from 0 to either 1 or 2.

(4) a such that $z(a) = x_1x_2 \in L, x_1 = x_2$, and either $x_1 \notin S_1$ or $d_{r_2(x_1x_2)} = 0$ cannot be a Nash equilibrium since the active 1-player can increase his payoff from 0 to 1 by changing his action.

(5) a such that $z(a) = x_1x_2 \in L, x_1 = x_2, x_1 \in S_1, d_{r_2(x_1x_2)} = 1$ and $x_2 \notin S_2$ cannot be a Nash equilibrium since $GoodDone_{r_2(x_1x_2)}(a) = 0$ and so the active 2-player $r_2(x_1x_2)$ can increase his payoff from 0 to 1 by changing his done bit to $d_{r_2(x_1x_2)} = 0$.

Now (1)–(5) cover all possibilities, and we have shown that if $S_1 \cap S_2 \neq \emptyset$ then there is a pure Nash equilibrium (case (1)), whereas if $S_1 \cap S_2 = \emptyset$ then there is no pure Nash equilibrium. □

Next, we will show that when $S_1 \cap S_2 \neq \emptyset$ all the improvement paths are finite.

Lemma 19. *Consider an improvement step from a to a'. If $z(a) = x_1x_2 \in L$, then $z(a') \in L$ and the improving player is either the active 1-player $r_1(x_1x_2)$ or the active 2-player $r_2(x_1x_2)$.*

Proof. Improvement steps cannot lead outside L: at a all payoffs are non-negative since $z(a) \in L$, whereas if $z(a') \notin L$ then all payoffs become negative. Only an active player can change his action bit such that $z(a') \in L$; as for the done bit of a player in T_2, it affects his own payoff only when he is the active 2-player. $\square$

Lemma 20. *Consider an improvement step from a to a'. If $z(a) = x_1 x_2 \in L$ and the improving player is the backward active 2-player, then he modifies only his done bit.*

Proof. A backward active 2-player exists only in the case where $x_1 = x_2 = x$; changing his action leads from xx to xy where $y = prev(x)$, and thus makes his payoff 0—which cannot be an improvement. $\square$

Lemma 21. *Consider an improvement path that starts with an improvement step by a player $(2, j)$ in T_2 from a with $z(a) \in L$ to a'. Then from a' until the next improvement step by a player from T_2, if any, we have $GoodDone_{2,i} = 1$ for all players $(2, i) \in T_2$.*

Proof. We distinguish two cases according to the first improvement step. If $(2, j)$ modified only his done bit, then $z(a)$ was of the form xx, and $GoodDone_{2,j}$ changed from 0 to 1. So we are now at $z(a') = xx$ with $GoodDone_{2,i}(a') = 1$ for all players $(2, i) \in T_2$ (for player $(2, i)$, where $i \neq j$, this follows from the definition of $GoodDone_{2,i}$), and that will remain so as long as no player in T_2 moves (since an improvement step by a 1-player leads to yx where, again by definition, $GoodDone_{2,i} = 1$).

If the first improvement step by $(2, j)$ involved the action bit, then Lemma 20 implies that $(2, j)$ was the forward active 2-player, and so $z(a) = yx$ and $z(a') = yy$ (where $y = next(x)$). But unless $(2, j)$ also set the done bit correctly (to $1_{\{x \in S_2\}}$) it could not have been an improvement step, and so $GoodDone_{2,j}(a') = 1$. From here on the argument is identical to the first case. $\square$

Corollary 22. *Consider an improvement path that starts at a with $z(a) \in L$. At each step after the first improvement step performed by some player from T_2, all the players $(2, i)$ in T_2 have $GoodDone_{2,i} = 1$.*

Lemma 23. *Consider an improvement path that starts at a with $z(a) \in L$. Then there are no consecutive improvement steps performed by players from the same set T_ℓ, where $\ell \in \{1, 2\}$.*

Proof. By Lemma 19 only the active players can perform improvement steps, and the path never leaves L.

By way of contradiction, assume that two players in T_1 perform consecutive improvement steps. It must therefore be the same player moving back and forth between xx and yx (where $y = next(x)$), but then his payoff after the two consecutive steps does not change—a contradiction.

If two players in T_2 perform consecutive improvement steps, then, again, it must be the same player, and so his payoff must have first increased from 0 to 1, and then from 1 to 2. But then that second step was from an $xx \in L$ with $x \notin S_2$ (payoff of 1) to a $yy \in L$ with $y \in S_2$ (payoff of 2), which is impossible (since it involves two players modifying their actions). $\square$

Proposition 24. *Consider an improvement path. Then after at most $n+2$ initial steps:* (I) *the path reaches L and never leaves it;* (II) *all the players $(2, i)$ in T_2 have $GoodDone_{2,i} = 1$; and either* (IIIa) *all improvement steps are performed by forward active players or* (IIIb) *a pure Nash equilibrium is reached.*

Proof. Every improvement step outside L decreases the distance to L, so after at most n steps we must reach L, and never leave it (Lemma 19). While in L the improvement steps alternate between T_1 and T_2 (Lemmata 19 and 23), so in at most 2 more steps (in total, no more than $n + 2$ steps from the start), an active player from T_2 performs an improvement step. By Corollary 22, from that time on $GoodDone_{2,i} = 1$ for all the players $(2, i)$ in T_2. Therefore, in particular, no improvement step can change only the done bit, and so the players in T_2 can make only forward moves (Lemma 20). As for T_1, if there is a backward move, then it goes from yx to xx and must increase the payoff of the active 1-player from 1 to 2. Therefore $x \in S_1$ and $d_{r_2(xx)} = 1$, which together with $GoodDone_{r_2(xx)} = 1$ implies that $x \in S_2$, and so the payoffs of all players equal 2, the maximal payoff—and a pure Nash equilibrium has been reached. $\square$

Proposition 25. *If $S_1 \cap S_2 \neq \emptyset$ then the game G generated by the potential game reduction has finite improvement paths.*

Proof. Assume by contradiction that an infinite improvement path exists. Proposition 24 implies that from some point on (IIIa) always holds, and so the improvement steps are all performed by active forward players. Therefore all states in L are traversed in turn, in particular xx for

$x \in S_1 \cap S_2$. At that point the payoff of every player $(2, i)$ in T_2 is 2 (since $GoodDone_{2,i} = 1$ by (II) of Proposition 24); also, $d_{r_2(xx)} = 1_{\{x \in S_2\}} = 1$ (since $GoodDone_{r_2(xx)} = 1$), and so the payoff of every player in T_1 is 2 too, and no further improvement is possible. □

We can now complete the proof of our result.

Proof of Theorem 4. Let $\mathcal{G} \subset \Gamma_4^n$ be the family of games that are obtained in our construction for all $S_1, S_2 \subset \mathcal{S} = \{0, 1\}^{n/2}$; Claims 18 and 5 imply that $CC(\text{PURE}, \mathcal{G}) = \Omega(|\mathcal{S}|) = \Omega(2^{n/2})$. By Proposition 25, every game $G \in \mathcal{G}$ either has the finite improvement path property, i.e., $G \in \mathcal{FIP}_4^n$ (when $S_1 \cap S_2 \neq \emptyset$) or G has no pure Nash equilibrium (when $S_1 \cap S_2 = \emptyset$). Now $CC(\text{PURE}, \mathcal{FIP}_4^n) \geq CC(\text{PURE}, \mathcal{G} \cap \mathcal{FIP}_4^n)$ (the communication complexity is defined as the worst case over all games in the family), and we claim that $CC(\text{PURE}, \mathcal{G} \cap \mathcal{FIP}_4^n) = CC(\text{PURE}, \mathcal{G})$. Indeed, take any pure Nash equilibrium procedure Π over $\mathcal{G} \cap \mathcal{FIP}_4^n$, and let its communication complexity be K bits. We can then use Π over the whole family $\mathcal{G}$, stopping it once K bits have been transmitted without a pure Nash equilibrium having been reached. Since all games in $\mathcal{G}$ having pure Nash equilibrium lie in $\mathcal{FIP}_4^n$ by Proposition 25, this is indeed a pure Nash equilibrium procedure for $\mathcal{G}$, and its communication complexity is also K. So

$$CC(\text{PURE}, \mathcal{FIP}_4^n) \geq CC(\text{PURE}, \mathcal{G} \cap \mathcal{FIP}_4^n) = CC(\text{PURE}, \mathcal{G}) = \Omega(2^{n/2}).$$

This establishes Theorem 4. □

Appendix B. Correlated Equilibria

In this appendix we show that, in the settings where we have obtained lower bounds on the communication complexity of pure and mixed Nash equilibrium procedures, there are simple procedures reaching a correlated equilibrium, whose communication complexity is low (polynomial, or even zero). Moreover, the support of the correlated equilibria that are reached is also low (the *support* of a correlated equilibrium Q is the number of joint actions $a \in A$ for which $Q(a) > 0$).

Finally, we study the communication complexity of reaching correlated approximate equilibria, and show that regret-minimization techniques yield simple polynomial upper bounds.

B.1. Modified Jordan games

We will exhibit a simple distribution that turns out to be a correlated equilibrium for *all* modified Jordan games, as defined in Section 5. For each $n \geq 3$, consider the following four joint actions in A:

$$z_{00} = (0, 1, 0, 0, 0, \ldots, 0),$$
$$z_{01} = (0, 1, 1, 0, 0, \ldots, 0),$$
$$z_{10} = (1, 0, 0, 1, 1, \ldots, 1),$$
$$z_{11} = (1, 0, 1, 1, 1, \ldots, 1)$$

(i.e., $z_{bc} = (b, 1 - b, c, b, b, \ldots, b)$ for $b, c \in \{0, 1\}$), and let Q_n be the distribution that assigns equal probability of $1/4$ to each one of these four joint actions.

Lemma 26. *The distribution Q_n is a correlated equilibrium of the modified Jordan game $J_n(f)$ for every $n \geq 3$ and every $f : \{0, 1\}^{n-2} \to [0, 1)$.*

Proof. For each a in the support of Q_n, i.e., $a \in \{z_{00}, z_{01}, z_{10}, z_{11}\}$, we have the following: every player $i \neq 2, 3, 4$ always matches his predecessor, and so he gets his highest payoff of 1 and will not deviate; also, player 2 always mismatches player 1, and so he will not deviate either (we use here $0 \leq f < 1$). As for players $i = 3, 4$, when $a_i = 0$ it is equally likely (according to Q_n) that $a_{i-1} = 0$ and $a_{i-1} = 1$, and so the payoff of i is $1/2$ and a deviation to $b_i = 1$ does not change his payoff; the same holds when $a_i = 1$.   □

Since there is a correlated equilibrium that does not depend on the specific modification f, no communication is needed to reach it. Therefore we have established the following:

Theorem 27. *For every $n \geq 3$, the communication complexity of reaching a correlated equilibrium over the family of binary-action games $\mathcal{U}^n \subset \Gamma_2^n$ of Theorem 10 is zero, i.e.,*

$$CC(\text{CORRELATED}, \mathcal{U}^n) = 0.$$

B.2. Bayesian setting

Recall the Bayesian setting of Section 4. The probability distribution Pr over the family Γ_2^n of binary-action games is obtained by putting $u_i(0, a^{-i}) = 0$ and $u_i(1, a^{-i}) = 1$ with probability $1/2$, and $u_i(0, a^{-i}) = 1$

and $u_i(1, a^{-i}) = 0$ with probability $1/2$, independently over all i and a^{-i}. For each player i and action $b = 0, 1$, let $B_{i,b} := \{a \in A : a_i = b, u_i(a) = 1\}$ be the set of joint actions where player i plays the action b, and that is his best-reply action.

We start with a preparatory lemma. Let $m = 4n^2$. After the payoff functions $\{u_i\}_i$ have been chosen (according to Pr) and every player i has been informed of his own u_i, let player i select a random subset of size m of $B_{i,0}$, denote it $S_{i,0}$, and a random subset of size m of $B_{i,1}$, denote it $S_{i,1}$, with all joint actions in each $B_{i,b}$ equally likely to be selected. Put $S_i = S_{i,0} \cup S_{i,1}$ and $S = \bigcup_i S_i$; then $S \subset A$ is a random set containing at most $2mn$ joint actions (the same joint action may be selected by different players). Let ζ_S denote the uniform distribution on S (i.e., $\zeta_S(s) = 1/|S|$ for every $s \in S$).

Lemma 28. *ζ_S is a correlated equilibrium in the game $(u_1, \ldots, u_n)$ with probability at least $1 - 2ne^{-n}$.*

Note that there are two randomizations here: first, the payoff functions u_i (according to Pr), and second, the selections S_i (conducted after each player i knows his u_i); we will write Pr for the resulting joint probability on both payoffs and selections.

Proof. In order for ζ_S to be a correlated equilibrium, the $2n$ inequalities

$$\sum_{a \in S : a_i = b} [u_i(b, a^{-i}) - u_i(1 - b, a^{-i})] \geq 0, \tag{B.1}$$

for all i and $b = 0, 1$, must be satisfied. Fix a player i and an action $a_i = b$ in $\{0, 1\}$, and let $T = \{a \in S : a_i = b\}$. For each $a \in A$ such that $a_i = b$, let $X_a = u_i(b, a^{-i}) - u_i(1 - b, a^{-i})$; then $X_a \in \{1, -1\}$, and we want to obtain an upper bound on the probability that $\sum_{a \in T} X_a < 0$, i.e., the corresponding inequality (B.1) is violated.

If $a \in T$ was selected by player i, i.e., $a \in S_i$, then necessarily $a \in S_{i,b}$ and so $X_a = 1$ (since $S_{i,b} \subset B_{i,b}$). If $a \notin S_i$, then a was selected by some other player $j \neq i$, and then it is equally likely that $X_a = 1$ and $X_a = -1$ (since player i's payoff and player j's selection are independent). Let K be the number of elements in $T \backslash S_i$; then $0 \leq K \leq 2m(n-1)$ and $\sum_{a \in T} X_a = m + \sum_{a \in T \backslash S_i} X_a = m + \sum_{k=1}^{K} Y_k$, where the Y_k are independent random variables with $\Pr[Y_k = 1] = \Pr[Y_k = -1] = 1/2$. Applying Hoeffding's inequality (Hoeffding 1963) yields $\Pr[\sum_{a \in T} X_a < 0 \mid K] =$

$\Pr[\sum_{k=1}^{K} Y_k < -m] \leq e^{-m^2/(2K)} \leq e^{-n}$ for every $K \leq 2m(n-1)$ (recall that we took $m = 4n^2$), and so $\Pr[\sum_{a \in T} X_a < 0] \leq e^{-n}$.

The same computation applies to all i and $b = 0, 1$, and so the probability that at least one of the $2n$ inequalities (B.1) is violated is at most $2ne^{-n}$.   □

Consider now the following stochastic procedure Π_{BAYES}. In the first stage, each player i selects a random set $S_i = S_{i,0} \cup S_{i,1}$ of joint actions as above, and then broadcasts it. In the second stage, each player i verifies whether the resulting $S = \bigcup_{i=1}^{n} S_i$ satisfies his two inequalities (B.1) (only u_i matters here), and then broadcasts the result, using one bit. If all inequalities are satisfied, then ζ_S is a correlated equilibrium and the procedure terminates. If not, then in the third stage all players broadcast their complete payoff functions, and then they each compute a correlated equilibrium using the same algorithm, so as to get the same result. Since Π_{BAYES} is a stochastic procedure, for each game G we consider the expected communication complexity $\text{CCS}(\Pi_{\text{BAYES}}, G)$ (see Section 8.1).

Theorem 29. *The expected communication complexity of the correlated equilibrium stochastic procedure Π_{BAYES} is $O(n^4)$, i.e.,*

$$\text{E}[\text{CCS}(\Pi_{\text{BAYES}}, \mathcal{G})] = O(n^4),$$

where the expectation E is according to the probability distribution $\Pr$ of Theorem 7.

Proof. Each one of the n sets S_i has $2m = 8n^2$ elements of n bits each, and so the first stage has a communication complexity of $8n^4$ bits. Then each player sends a single bit, which adds n bits of communication. With probability at most $2ne^{-n}$ the resulting ζ_S is not a correlated equilibrium, and then there is an additional communication complexity of $n2^n$ of broadcasting the payoff functions. Altogether, $\text{E}[\text{CCS}(\Pi_{\text{BAYES}}, \mathcal{G})] = 8n^4 + n + 2ne^{-n}n2^n \leq 8n^4 + n + 2n^2 = O(n^4)$.   □

B.3. Correlated approximate equilibria

We will now show that "regret-minimization" procedures that reach correlated approximate equilibria have a low communication complexity. Let $\varepsilon > 0$; a *correlated ε-equilibrium* is defined in the same way as a correlated equilibrium, except that now all the inequalities must hold only within ε.

Theorem 30. *For every $n \geq 2$ and $\mathfrak{u} \geq 1$, let $\mathcal{U}^n_\mathfrak{u} \subset \Gamma^n_2$ be the family of n-person binary-action games with payoffs of magnitude at most $\mathfrak{u}$, i.e., $\max_{1 \leq i \leq n} \mathfrak{mag}(u_i) \leq \mathfrak{u}$. Then for every $\varepsilon > 0$ there exists a correlated ε-equilibrium stochastic procedure $\Pi_{\varepsilon\text{-CORR}}$ whose expected communication complexity over $\mathcal{U}^n_\mathfrak{u}$ is polynomial in n and $\mathfrak{u}$, i.e.,*

$$\mathrm{CCS}(\Pi_{\varepsilon\text{-CORR}}, \mathcal{U}^n_\mathfrak{u}) = \mathrm{O}\left(\frac{n^2}{\varepsilon^2} + n\mathfrak{u}\right).$$

Proof. For every $\varepsilon, \delta > 0$ there exists a regret-minimization procedure that guarantees that, after $T = C \log(n/\delta)/\varepsilon^2$ periods, there is a probability of at least $1 - \delta$ that the time average of the played joint actions constitutes a correlated ε-equilibrium (where C is an appropriate constant); e.g., see Cesa-Bianchi and Lugosi (2006, Section 7.4). Let Π_{RM} be such a procedure for $\delta = 2^{-n}$; its communication complexity is $nT = \mathrm{O}(n^2/\varepsilon^2)$, since in each one of the T periods every player's action corresponds to one bit of communication.

The procedure $\Pi_{\varepsilon\text{-CORR}}$ starts by running Π_{RM}; let $a^t = (a_{1,t}, \ldots, a_{n,t}) \in A$ be the joint action at time t. When Π_{RM} terminates, after T periods, each player i computes his average regrets and checks whether they are both at most ε (i.e., $(1/T) \sum_{t=1}^{T} [u_i(b, a^{-i,t}) - u_i(a^t)] \leq \varepsilon$ for[20] $b = 0, 1$); he then broadcasts the result using one bit. With probability at least $1 - 2^{-n}$ all average regrets are at most ε, and then the uniform distribution on $(a^1, \ldots, a^T)$ is a correlated ε-equilibrium and $\Pi_{\varepsilon\text{-CORR}}$ terminates.

Otherwise, all players broadcast their complete payoff functions, after which they all compute a correlated equilibrium using the same algorithm. The communication at this stage requires $\mathrm{O}(n2^n\mathfrak{u})$ bits, but it only happens with probability at most 2^{-n}. The total expected communication complexity is therefore $\mathrm{O}(n^2/\varepsilon^2) + n + 2^{-n}\mathrm{O}(n2^n\mathfrak{u}) = \mathrm{O}(n^2/\varepsilon^2 + n\mathfrak{u})$. $\qquad\square$

References

Aumann, R. J. (1974), "Subjectivity and Correlation in Randomized Strategies," *Journal of Mathematical Economics*, 1, 67–96.

Blum, A. and Y. Mansour (2007), "From External to Internal Regret," *Journal of Machine Learning Research*, 8, 1307–1324.

---

[20]For binary-action games, conditional ("internal") and unconditional ("external") regrets are the same.

Cahn, A. (2004), "General Procedures Leading to Correlated Equilibria," *International Journal of Game Theory*, 33, 21–40. [Chapter 6]

Cesa-Bianchi, N. and G. Lugosi (2003), "Potential-Based Algorithms in Online Prediction and Game Theory," *Machine Learning*, 51, 239–261.

———— (2006), *Prediction, Learning and Games*. Cambridge, UK: Cambridge University Press.

Conitzer, V. and T. Sandholm (2004), "Communication Complexity as a Lower Bound for Learning in Games," in *Proceedings of the Twenty-First International Conference on Machine Learning*, ed. by C. E. Brody. New York: ACM, pp. 185–192.

Foster, D. and S. M. Kakade (2004), "Deterministic Calibration and Nash Equilibrium," in *Learning Theory, 17th Annual Conference on Learning Theory (COLT 2004)*, ed. by J. S. Taylor and Y. Singer. New York: Springer, pp. 33–18.

Foster, D. and R. Vohra (1997), "Calibrated Learning and Correlated Equilibrium," *Games and Economic Behavior*, 21, 40–55.

Foster, D. and H. P. Young (2003), "Learning, Hypothesis Testing, and Nash Equilibrium," *Games and Economic Behavior*, 45, 73–96.

———— (2006), "Regret Testing: Learning to Play Nash Equilibrium without Knowing You Have an Opponent," *Theoretical Economics*, 1, 341–367.

Fudenberg, D. and D. K. Levine (1998), *The Theory of Learning in Games*. Cambridge, MA: MIT Press.

Germano, F. and G. Lugosi (2007), "Global Nash Convergence of Foster and Young's Regret Testing," *Games and Economic Behavior*, 60, 135–154.

Hart, S. (2005), "Adaptive Heuristics," *Econometrica*, 73, 1401–1430. [Chapter 11]

Hart, S. and A. Mas-Colell (2000), "A Simple Adaptive Procedure Leading to Correlated Equilibrium," *Econometrica*, 68, 1127–1150. [Chapter 2]

———— (2001), "A General Class of Adaptive Strategies," *Journal of Economic Theory*, 98, 26–54. [Chapter 3]

———— (2003), "Uncoupled Dynamics Do Not Lead to Nash Equilibrium," *American Economic Review*, 93, 1830–1836. [Chapter 7]

———— (2006), "Stochastic Uncoupled Dynamics and Nash Equilibrium," *Games and Economic Behavior*, 57, 286–303. [Chapter 8]

Hart, S. and D. Schmeidler (1989), "Existence of Correlated Equilibria," *Mathematics of Operations Research*, 14, 18–25. [Chapter 1]

Hoeffding, W. (1963), "Probability Inequalities for Sums of Bounded Random Variables," *Journal of the American Statistical Association*, 58, 13–30.

Hurwicz, L. (1960), "Optimality and Informational Efficiency in Resource Allocation Processes," in *Mathematical Models in the Social Sciences*, ed. by K. J. Arrow, S. Karlin, P. Suppes. Stanford: Stanford University Press.

Jordan, J. (1991), "Bayesian Learning in Normal Form Games," *Games and Economic Behavior*, 3, 60–81.

———— (1993), "Three Problems in Learning Mixed Equilibria," *Games and Economic Behavior*, 5, 368–386.

———— (2008), "Information Aggregation and Prices," in *The New Palgrave Dictionary of Economics*, 2nd edn., ed. by S. Durlauf and L. Blume. Hampshire: Palgrave Macmillan.

Kalai, E. and E. Lehrer (1993), "Rational Learning Leads to Nash Equilibrium," *Econometrica*, 61, 1019–1045.

Kushilevitz, E. and N. Nisan (1997), *Communication Complexity*. Cambridge, UK: Cambridge University Press.

Monderer, D. and L. S. Shapley (1996), "Potential Games," *Games and Economic Behavior*, 14, 124–143.

Papadimitriou, C. H. and T. Roughgarden (2008), "Computing Correlated Equilibria in Multi-Player Games," *Journal of the ACM*, 55, 14:1–14:29.

Stoltz, G. and G. Lugosi (2005), "Internal Regret in OnLine Portfolio Selection," *Machine Learning*, 59, 125–159.

——— (2007), "Learning Correlated Equilibria in Games with Compact Sets of Strategies," *Games and Economic Behavior*, 59, 187–208.

Young, H.P. (2004), *Strategic Learning and Its Limits*. Oxford, UK: Oxford University Press.

Part IV

Dynamics and Equilibria

Chapter 11

ADAPTIVE HEURISTICS*

Sergiu Hart

We exhibit a large class of simple rules of behavior, which we call *adaptive heuristics*, and show that they generate rational behavior in the long run. These adaptive heuristics are based on natural regret measures, and may be viewed as a bridge between rational and behavioral viewpoints. Taken together, the results presented here establish a solid connection between the dynamic approach of adaptive heuristics and the static approach of correlated equilibria.

1. Introduction

Consider dynamic settings where a number of decision-makers interact repeatedly. We call a rule of behavior in such situations an *adaptive heuristic* if, on the one hand, it is simple, unsophisticated, simplistic, and myopic (a so-called "rule of thumb"), and, on the other, it leads to movement in seemingly "good" directions (like stimulus-response or reinforcement). One example of an adaptive heuristic is to always choose a best reply to the actions of the other players in the previous period—or, for that matter,

Originally published in *Econometrica*, 73 (2005), 1401–1430.

*Walras-Bowley Lecture 2003, delivered at the North American Meeting of the Econometric Society in Evanston, Illinois. A presentation is available at *http://www. ma.huji.ac.il/hart/abs/adaptdyn.html*.

Keywords: dynamics; heuristics; adaptive; correlated equilibrium; regret; regret matching; uncoupled dynamics; joint distribution of play; bounded rationality; behavioral; calibration; fictitious play; approachability.

Research partially supported by the Israel Science Foundation.

It is a great pleasure to acknowledge the joint work with Andreu Mas-Colell over the years, upon which this paper is based. I also thank Ken Arrow, Bob Aumann, Maya Bar-Hillel, Avraham Beja, Elchanan Ben-Porath, Gary Bornstein, Toni Bosch, Ido Erev, Drew Fudenberg, Josef Hofbauer, Danny Kahneman, Yaakov Kareev, Aniol Llorente, Yishay Mansour, Eric Maskin, Abraham Neyman, Bezalel Peleg, Motty Perry, Avi Shmida, Sorin Solomon, Menahem Yaari, and Peyton Young, as well as the editor and the anonymous referees, for useful discussions, suggestions, and comments.

to the frequency of their actions in the past (essentially, the well-known "fictitious play").

Adaptive heuristics are boundedly rational strategies (in fact, highly "bounded away" from full rationality). The main question of interest is whether such simple strategies may in the long run yield behavior that is nevertheless highly sophisticated and rational.

This paper is based mainly on the work of Hart and Mas-Colell (2000, 2001a, 2001b, 2003a, 2003b), which we try to present here in a simple and elementary form (see Section 10 and the pointers there for the more general results). Significantly, when the results are viewed together new insights emerge—in particular, into the relations of adaptive heuristics to rationality on the one hand, and to behavioral approaches on the other. See Section 9, which may well be read immediately.

The paper is organized as follows. In Section 2 we provide a rough classification of dynamic models. The setting and notations are introduced in Section 3, and the leading adaptive heuristic, *regret matching*, is presented and analyzed in Section 4. Behavioral aspects of our adaptive heuristics are discussed in Section 5, and Section 6 deals with the notion of correlated equilibrium, to which play converges in the long run. Section 7 presents the large class of generalized regret matching heuristics. In Section 8 we introduce the notion of "uncoupledness" (which is naturally satisfied by adaptive heuristics) and show that uncoupled dynamics cannot be guaranteed to always lead to Nash equilibria. A summary together with the main insights of our work are provided in Section 9. Section 10 includes a variety of additional results and discussions of related topics, and the Appendix presents Blackwell's approachability theory, a basic technical tool in this area.

2. A Rational Classification of Dynamics

We consider dynamic models where the same game is played repeatedly over time. One can roughly classify dynamic models in game theory and economic theory into three classes: learning dynamics, evolutionary dynamics, and adaptive heuristics.

2.1. *Learning dynamics*

In a *(Bayesian) learning dynamic*, each player starts with a prior belief on the relevant data (the "state of the world"), which usually includes the game being played and the other players' types and (repeated-game) strategies.[1] Every period, after observing the actions taken (or, more generally, some information about these actions), each player updates his beliefs (using Bayes' rule). He then plays optimally given his updated beliefs.

Such dynamics are the subject of much study; see, for example, the books of Fudenberg and Levine (1998, Chapter 8) and Young (2004, Chapter 7). Roughly speaking, conditions like "the priors contain a grain of truth" guarantee that in the long run play is close to the Nash equilibria of the repeated game; see Kalai and Lehrer (1993) and the ensuing literature.[2]

2.2. Evolutionary dynamics

Here every player i is replaced by a *population* of individuals, each playing the given game in the role of player i. Each such individual always plays the same one-shot action (this fixed action is his "genotype"). The relative frequencies of the various actions in population i may be viewed as a *mixed action* of player i in the one-shot game (for instance, one third of the population having the "gene" L and two thirds the "gene" R corresponds to the mixed action $(1/3, 2/3)$ on (L,R)); one may think of the mixed action as the action of a randomly chosen individual.

Evolutionary dynamics are based on two main "forces": selection and mutation. *Selection* is a process whereby better strategies prevail; in contrast, *mutation*, which is rare relative to selection, generates actions at random, whether better or worse. It is the combination of the two that allows for natural adaptation: new mutants undergo selection, and only the better ones survive. Of course, selection includes many possible mechanisms: biological (the payoff determines the number of descendants, and thus the share of better strategies increases), social (imitation, learning), individual (experimentation, stimulus-response), and so on. What matters is that selection is "adaptive" or "improving," in the sense that the proportion of better strategies is likely to increase.

Dynamic evolutionary models have been studied extensively; see, for example, the books of Fudenberg and Levine (1998, Chapters 3 and 5), Hofbauer and Sigmund (1998), Weibull (1995), and Young (1998).

---

[1] To distinguish the choices of the players in the one-shot game and those in the repeated game, we refer to the former as *actions* and to the latter as *strategies*.

[2] Under weaker assumptions, Nyarko (1994) shows convergence to correlated equilibria.

2.3. *Adaptive heuristics*

We use the term *heuristics* for rules of behavior that are simple, unsophisticated, simplistic, and myopic (unlike the "learning" models of Section
2.1 above). These are "rules of thumb" that the players use to make their
decisions. We call them *adaptive* if they induce behavior that reacts to
what happens in the play of the game, in directions that, loosely speaking, seem "better." Thus, always making the same fixed choice, and always
randomizing uniformly over all possible choices, are both heuristics. But
these heuristics are not adaptive, since they are not at all responsive to
the situation (i.e., to the game being played and the behavior of the other
participants). In contrast, *fictitious play* is a prime example of an adaptive heuristic: at each stage one plays an action that is optimal against the
frequency distribution of the past actions of the other players.

Adaptive heuristics commonly appear in behavioral models, such as
reinforcement, feedback, and stimulus-response. There is a large literature,
both experimental and theoretical, on various adaptive heuristics and their
relative performance in different environments; see for example, Fudenberg
and Levine (1998, Chapters 2 and 4) and the literature in psychology (where
this is sometimes called "learning"; also, the term "heuristics and biases" is
used by Kahneman and Tversky—see, e.g., Kahneman, Slovic, and Tversky
1982).

2.4. *Degrees of rationality*

One way to understand the distinctions between the above three classes
of dynamics is in terms of the degree of rationality of the participants;
see Figure 1. *Rationality* is viewed here as a process of optimization in
interactive (multiplayer) environments.

Learning dynamics require high levels of rationality. Indeed, repeated-
game strategies are complex objects; even more so are beliefs (i.e., probability distributions) over such objects; moreover, in every period it is necessary
to update these beliefs, and, finally, to compute best replies to them.

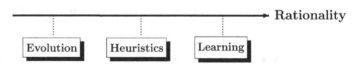

Fig. 1.   A classification of dynamics.

At the other extreme are evolutionary dynamics. Here the individuals in each population do not exhibit any degree of rationality; their behavior ("phenotype") is completely mechanistic, dictated by their "genotype." They do not compute anything—they just "are there" and play their fixed actions. What may be viewed as somewhat rational is the aggregate dynamic of the population (particularly the selection component), which affects the relative proportions of the various actions.

Adaptive heuristics lie in between: on the one hand, the players do perform certain usually simple computations given the environment, and so the behavior is not fixed as in evolutionary dynamics; on the other hand, these computations are far removed from the full rationality and optimization that is carried out in learning models.

3. Preliminaries

The setup is as follows. The basic game Γ is an N-person game in strategic (or normal) form. N is a positive integer, the players are $i = 1, 2, \ldots, N$, and to each player i there corresponds a set of *actions* S^i and a *payoff* (or *utility*) *function* $u^i : S \to \mathbb{R}$, where $S := S^1 \times S^2 \times \cdots \times S^N$ is the set of N-tuples of actions ("action combinations" or "action profiles") and $\mathbb{R}$ denotes the real line. When dealing with a fixed player i, it is at times convenient to take his set of actions to be[3] $S^i = \{1, 2, \ldots, m\}$.

The game Γ is repeatedly played over time. In a *discrete-time* dynamic model, the time periods are $t = 1, 2, \ldots$, and the action played by player i at time t is denoted s^i_t, with $s_t = (s^1_t, s^2_t, \ldots, s^N_t)$ standing for the N-tuple of actions at period t (these are the *actual* realized actions when randomizations are used). In a *continuous-time* dynamic model, the time t becomes a continuous variable. We will refer to Γ as the *stage* (or *one-shot*) *game*.

A standard assumption is that of *perfect monitoring*: at the end of each period t, all players observe s_t, the actions taken by everyone.

Some notations: when $s = (s^1, s^2, \ldots, s^N) \in S$ is an N-tuple of actions, we write s^i for its ith coordinate and s^{-i} for the $(N-1)$-tuple of coordinates of all players except i (so $s^{-i} \in S^{-i} := S^1 \times \cdots \times S^{i-1} \times S^{i+1} \times \cdots \times S^N$ and $s = (s^i, s^{-i})$). When randomized strategies are used, σ^i_t denotes the *mixed action* of player i at time t; thus $\sigma^i_t \equiv (\sigma^i_t(1), \sigma^i_t(2), \ldots, \sigma^i_t(m)) \in \Delta(S^i) := \{x \in \mathbb{R}^m_+ : \sum_{k=1}^m x(k) = 1\}$ is a probability distribution over

---

[3]The number of actions $m \equiv m^i$ may be different for different players.

$S^i = \{1, 2, \ldots, m\}$, with $\sigma_t^i(k)$ denoting, for each k in S^i, the probability that the action s_t^i taken by player i at time t is k.

4. Regret Matching

We start by considering our basic adaptive heuristic, "regret matching," introduced in Hart and Mas-Colell (2000). While it may appear to be quite special, we will see later (in Sections 7 and 10) that the results generalize to a wide class of heuristics—and our comments and interpretations in Sections 4 and 5 apply to all of them.

Regret matching is defined by the following rule:

> Switch next period to a different action
> with a probability that is
> *proportional* to the *regret* for that action,
> where *regret* is defined as the increase in payoff
> had such a change always been made in the past.

That is, consider player i at time $T + 1$. Denote by[4] U the average payoff that i has obtained up to now, i.e.,

$$U := \frac{1}{T} \sum_{t=1}^{T} u^i(s_t), \tag{1}$$

and let $j = s_T^i$ in S^i be the action that i played in the previous period T. For each alternative action $k \neq j$ in S^i, let $V(k)$ be the average payoff i would have obtained had he played k instead of j every time in the past that he actually played j; i.e.,

$$V(k) := \frac{1}{T} \sum_{t=1}^{T} v_t, \tag{2}$$

where

$$v_t := \begin{cases} u^i(k, s_t^{-i}), & \text{if } s_t^i = j, \\ u^i(s_t^i, s_t^{-i}) \equiv u^i(s_t), & \text{if } s_t^i \neq j. \end{cases} \tag{3}$$

---

[4]We drop the indices for readability (thus U_T^i is written U, and similarly for $V_T^i(k)$, $v_t^i(j, k)$, $R_T^i(k)$, and so on).

The *regret* $R(k)$ for action k is defined as the amount, if any, by which $V(k)$ exceeds the actual payoff U:

$$R(k) := [V(k) - U]_+, \tag{4}$$

where $[x]_+ := \max\{x, 0\}$ is the "positive part" of x. (The actual computation of the regrets consists of a simple updating from one period to the next; see Section 10.7.)

Regret matching stipulates that each action k different from the previous period's action j is played with a probability that is proportional to its regret $R(k)$, and, with the remaining probability, the same action j as last period is played again. That is, let $c > 0$ be a fixed constant;[5] then the probability of playing the action k at time $T + 1$ is given by

$$\sigma_{T+1}(k) := \begin{cases} cR(k), & \text{if } k \neq j, \\ 1 - \sum_{k:k\neq j} cR(k), & \text{if } k = j. \end{cases} \tag{5}$$

There are no requirements on the action in the first period: σ_1 is arbitrary.

Thus, player i considers whether to continue to play next period the same action j as in the previous period, or to switch to a different action k. Specifically, he looks at what would have happened to his average payoff had he always replaced j by k in the past:[6] he compares what he got, U, to what he would have gotten, $V(k)$. If the alternative payoff is no higher, i.e., if $V(k) \leq U$, then he has no regret for k (the regret $R(k)$ equals 0) and he does not switch to action k. If the alternative payoff is higher, i.e., if $V(k) > U$, then the regret for k is positive ($R(k)$ equals the increase $V(k) - U$) and i switches to action k with a probability that is proportional to this regret.

In particular, if i has no regret (i.e., all regrets $R(k)$ equal 0), then i plays for sure the same action j as last period. If some regrets are positive, then the higher the regret for some action, the higher the probability of switching to that action next period.

---

[5]For instance, any c less than $1/(2mM)$ will do, where m is the number of actions of i and $M = \max_{s \in S} |u^i(s)|$ bounds the possible payoffs. Such a c guarantees that (5) yields a probability distribution over S^i and, moreover, that the probability of j is *strictly* positive.

[6]Since one is looking at the long-run average payoff, it makes sense to consider replacing j by k not just in the previous period, but also in all other periods in the past when j was played; after all, the effect of one period goes to zero as T increases. (Interestingly, one-period-regret matching yields essentially the evolutionary "replicator dynamic"; see Schlag 1998.)

The main result of Hart and Mas-Colell (2000) is:

Theorem 1. Regret Matching: *Let each player play regret matching. Then the joint distribution of play converges to the set of correlated equilibria of the stage game.*

The "joint distribution of play" (also known as the "empirical distribution" or "sample distribution") measures the relative frequency of each N-tuple of actions being played; i.e., the *joint distribution of play* for the first T periods is a probability distribution z_T on S, where, for each s in S,

$$z_T(s) := \frac{1}{T} |\{1 \le t \le T : s_t = s\}|$$

is the proportion of periods up to T in which the combination of actions s has been played.[7] See Section 4.1 for a discussion of the role of the joint distribution of play.

The concept of *correlated equilibrium* was introduced by Aumann (1974); it is a Nash equilibrium where the players may receive payoff-irrelevant signals before playing the game. This is discussed in Section 6.

The *Regret Matching Theorem* says that, for almost every history of play, the sequence of joint distributions of play $z_1, z_2, \ldots, z_T, \ldots$ converges to the set of correlated equilibria CE of Γ. This means that z_T is close to a correlated equilibrium, or, equivalently, is a correlated approximate equilibrium, for all T large enough. The convergence is to the *set* CE, *not* necessarily to a point in that set. See Section 10.8 for formal statements.

Note that it is the empirical distributions that become essentially correlated equilibria—not the actual play. Our results imply that the long-run *statistics of play* of heuristics-playing players (such as "regret-matchers") and of fully rational players (who play a correlated equilibrium each period) are indistinguishable.

The proof of the Regret Matching Theorem consists in showing, first, that all regrets vanish in the limit (this uses arguments suggested by Blackwell's approachability; see the Appendix) and, second, that such "no regret" situations precisely correspond to correlated equilibria; see Section 10.2 for further details.

It is interesting to note that as the regrets become small, so does the probability of switching (see (5)). Therefore, regret matching leads to longer and longer stretches of time in which the action is constant, and the play

---

[7]For a finite set A, we denote by $|A|$ the number of elements of A.

exhibits much "inertia" and infrequent switches. (This is similar to the behavior of fictitious play in the classic 3 × 3 example of Shapley (1964), where the play cycles among six outcomes with increasingly longer time intervals in each cycle.)

Finally, where does the "correlation" come from? The answer is, of course, from the commonly observed history of play. Indeed, each player's action is determined by his regrets, which are in turn determined by the history.[8]

4.1. *Joint distribution of play*

At each stage the players randomize independently of one another. This however does *not* imply that the joint distribution of play should be independent across players (i.e., the product of its marginal distributions) or that it should become independent in the long run.[9] The reason is that the probabilities the players use may change over time. To take a simple example, assume that in odd periods player 1 chooses T or B with probabilities $(3/4, 1/4)$, and, independently, player 2 chooses L or R with probabilities $(3/4, 1/4)$, whereas in even periods these probabilities become $(1/4, 3/4)$ for each player. The joint distribution of play will then converge almost surely to $(5/16, 3/16, 3/16, 5/16)$ (for TL, TR, BL, and BR, respectively)—which is *not* the product of its marginals, $(1/2, 1/2)$ on (T,B) and $(1/2, 1/2)$ on (L,R).

The joint distribution of play is fully determined by the history of play, which players standardly observe. So having players determine their actions based on the joint distribution of play (rather than just the marginal distributions) does not go beyond the "standard monitoring" assumption that is commonly used. It is information that the players possess anyway.

Finally—and this is a behavioral observation—people do react to the joint distribution. Think of a two-person Matching Pennies game, where, say, half the time they play HH, and half the time TT. The players will very quickly notice this, and at least one of the players (the "mismatching" player in this case) will change his behavior; but, if he were to look only at

---

[8]At this point one may be tempted to conclude that, since the signal is common (all players observe the history of play), the convergence is in fact to *publicly* correlated equilibria (cf. Section 6). That is not so, however: our players are *not* fully rational; they apply heuristics, whereby each one uses the history to determine only his *own* regrets. Since all regrets are based on the common history, they are correlated among the players—but except for special cases they are far from being fully correlated.

[9]This point is, of course, not new; see, for instance, Fudenberg and Kreps (1988, 1993).

the marginal distributions of play, he would see $(1/2, 1/2)$ for each player, and have no reason to change. In general, people are very much aware of coincidences, signals, communications, and so on (even to the point of interpreting random phenomena as meaningful)—which just goes to show that they look at the joint distribution, and not only at the marginals.

To summarize: reasonable models of play can—and should—take into account the joint distribution of play.

5. Behavioral Aspects

Regret matching, as well as its generalizations below, embodies commonly used rules of behavior. For instance, if all the regrets are zero ("there is no regret"), then a regret matching player will continue to play the same action of the previous period. This is similar to common behavior, as expressed in the saying "Never change a winning team."

When some regrets are positive, actions may change—with probabilities that are proportional to the regrets: the higher the payoff would have been from switching to another action in the past, the higher the tendency is to switch to that action now. Again, this seems to fit standard behavior; we have all seen ads of the sort "Had you invested in A rather than B, you would have gained X more by now. So switch to A now!" (and, the larger X is, the larger the size of the ad).

It has been observed that people tend to have too much "inertia" in their decisions: they stick to their current state for a disproportionately long time (as in the "status quo bias"; see Samuelson and Zeckhauser 1988 and, recently, Moshinsky 2003). Regret matching has "built-in" inertia: the probability of not switching (i.e., repeating the previous period's action) is always strictly positive (see footnote 6). Moreover, as we saw in Section 4, regret matching leads to behavior where the same action is played over and over again for long time intervals.

Regret matching is not very sophisticated; players neither develop beliefs nor reply optimally (as in the learning dynamics of Section 2.1). Rather, their rule of behavior is simple and defined directly on actions; "propensities" of play are adjusted over time. In the learning, experimental, and behavioral literature there are various models that bear a likeness to regret matching (see Bush and Mosteller 1955; Roth and Erev 1995; Erev and Roth 1998; Camerer and Ho 1998, 1999; and others; probably the closest are the models of Erev–Roth). Also, incorporating regret measures into the utility function has been used to provide alternative theories of decision-making under uncertainty; see Bell (1982) and Loomes and Sugden (1982).

Recently, the study of Camille *et al.* (2004) has shown that certain measures of regret influence choices, and that the orbitofrontal cortex is involved in experiencing regret.

In summary, while we have arrived at regret matching and the other heuristics of this paper from purely theoretical considerations, it turns out that they have much in common with actual rules of behavior that are frequently used in real decisions.

6. Correlated Equilibria

In this section we leave the dynamic framework and discuss the notion of correlated equilibrium. Its presentation (which may well be skipped by the expert reader) is followed by a number of comments showing that this concept, belonging to the realm of full rationality, is particularly natural and useful.

We thus consider the one-shot game Γ. Assume that, before playing the game Γ, each player i receives a signal θ^i. These signals may be correlated: the combination of signals $\theta = (\theta^1, \theta^2, \ldots, \theta^N)$ occurs according to a joint probability distribution F, commonly known to all players. Moreover, the signals do not affect the payoffs of the game. Can this affect the outcome?

Indeed, it can: the players may use these signals to correlate their choices. For a simple example, take the Battle of the Sexes game (see Figure 2). Consider a public coin toss (i.e., the common signal $\theta^1 = \theta^2$ is either H or T, with probabilities $(1/2, 1/2)$), after which both go to the hockey game if H and to the theater if T; this constitutes a Nash equilibrium of the extended game (with the signals)—which cannot be achieved in the original game.

Formally, a *correlated equilibrium* (introduced by Aumann 1974) of the game Γ is a Nash equilibrium of a "pregame signals extension" of Γ. Clearly,

	HOCKEY	THEATER
HOCKEY	2, 1	0, 0
THEATER	0, 0	1, 2

	HOCKEY	THEATER
HOCKEY	1/2	0
THEATER	0	1/2

Fig. 2. The Battle of the Sexes game (left) and a (publicly) correlated equilibrium (right).

only the probability distribution of the signals matters. Let thus ψ be the induced probability distribution on the N-tuples of actions; i.e., for each action combination s in S, let $\psi(s)$ be the probability of all those signal combinations after which the players choose[10] s. The conditions for ψ to be a correlated equilibrium are that

$$\sum_{s^{-i} \in S^{-i}} \psi(j, s^{-i}) u^i(j, s^{-i}) \geq \sum_{s^{-i} \in S^{-i}} \psi(j, s^{-i}) u^i(k, s^{-i}) \qquad (6)$$

for all players i and all actions j, k in S^i. Indeed, if there are no deviations, then the expected payoff of player i when he chooses action j is the expression on the left-hand side; if i were the only one to deviate and choose instead of j some other action k, then his expected payoff would be the expression on the right-hand side. As a *canonical* setup, think of a "referee" who chooses, according to the distribution ψ, an N-tuple of actions $s = (s^1, s^2, \ldots, s^N)$ for all players, and then sends to each player i the message "s^i" (a "recommendation to play s^i"); a correlated equilibrium ensues if for each player it is always best to follow the recommendation (i.e., to play the recommended s^i), assuming that all other players also do so.

We denote by CE the set of correlated equilibria; it is a subset of $\Delta(S)$, the set of probability distributions on S. If the inequalities (6) hold only within some $\varepsilon > 0$, we will say that ψ is a *correlated approximate equilibrium*; more precisely, a *correlated ε-equilibrium*.

In the special case where the signals are *independent* across players (i.e., when the joint distribution ψ satisfies $\psi(s) = \psi^1(s^1) \cdot \psi^2(s^2) \cdot \ldots \cdot \psi^N(s^N)$ for all s, with ψ^i denoting the ith marginal of ψ), a correlated equilibrium is just a Nash equilibrium of Γ. At the other extreme, when the signals are *fully correlated* (i.e., common, or public—like "sunspots"), each signal must necessarily be followed by a Nash equilibrium play; hence such correlated equilibria—called *publicly correlated equilibria*—correspond to weighted averages (convex combinations) of Nash equilibria of Γ, as in the Battle of the Sexes example of Figure 2.

In general, when the signals are neither independent nor fully correlated, new equilibria arise. For example, in the Chicken game, there is a correlated equilibrium that yields equal probabilities of $1/3$ to each action combination except (STAY, STAY) (see Figure 3). Indeed, let the signal to

---

[10]In general, since players may randomize their choices, $\psi(s)$ is the corresponding total probability of s.

	LEAVE	STAY
LEAVE	5, 5	3, 6
STAY	6, 3	0, 0

	LEAVE	STAY
LEAVE	1/3	1/3
STAY	1/3	0

Fig. 3. A correlated equilibrium in the Chicken game.

each player be L or S; think of this as a recommendation to play LEAVE or STAY, respectively. When the row player gets the signal L, he assigns a (conditional) probability of 1/2 to each one of the two pairs of signals (L, L) and (L, S); so, if the column player follows his recommendation, then the row player gets an expected payoff of $4 = (1/2)5 + (1/2)3$ from playing LEAVE, and only of $3 = (1/2)6 + (1/2)0$ from deviating to STAY. When the row player gets the signal S, he deduces that the pair of signals is necessarily (S, L), so if the column player indeed plays LEAVE then the row player is better off choosing STAY. Similarly for the column player.

For examples of correlated equilibria in biology, see Hammerstein and Selten (1994, Section 8.2 and the references there) and Shmida and Peleg (1997) (speckled wood butterflies, studied by Davies 1978, "play" the Chicken game). Other examples can be found in team sports, like basketball and football. Teams that are successful due to their so-called "team play" develop signals that allow correlation among their members but yield no information to their opponents.[11] For a stylized example, consider a signal that tells the team members whether to attack on the left or on the right—but is unknown (or unobservable) to the opposing team.

In fact, signals are all around us—whether public, private, or mixed. These signals are mostly irrelevant to the payoffs of the game that is being played. Nevertheless, it is hard to exclude the possibility that they may find their way into the equilibrium—so the notion of correlated equilibrium becomes all the more relevant.

Finally, Aumann (1987) shows that all players being "Bayesian rational" is equivalent to their playing a correlated equilibrium (under the "common prior" or "consistency" assumption of Harsanyi (1967–1968)).[12] Correlated equilibrium is thus a concept that embodies full rationality.

---

[11] These signals may well be largely inexplicit and unconscious; they are recognized due to the many repeated plays of the team.

[12] In this light, our result may appear even more surprising: non-Bayesian and far-from-rational behavior leads in the long run to outcomes that embody full Bayesian rationality and a common prior.

7. Generalized Regret Matching

The regret matching strategy of Section 4 appears to be very specific: the play probabilities are directly proportional to the regrets. It is natural to enquire whether this is necessary for our result of Theorem 1. What would happen were the probabilities proportional to, say, the square of the regrets? Another issue is the connection to other dynamics leading to correlated equilibria, particularly variants of conditional smooth fictitious play (Fudenberg and Levine 1999a; see Section 10.4 below).

This leads us to consider a large class of adaptive heuristics that are based on regrets. Specifically, instead of $cR(k)$ in (5), we now allow functions $f(R(k))$ of the regret $R(k)$ that are *sign-preserving*, i.e., $f(x) > 0$ for $x > 0$ and $f(0) = 0$.

A strategy σ of player i is called a *generalized regret matching* strategy if the action at time $T + 1$ is chosen according to probabilities

$$\sigma_{T+1}(k) := \begin{cases} f(R(k)), & \text{if } k \neq j, \\ 1 - \sum_{k:k\neq j} f(R(k)), & \text{if } k = j, \end{cases} \tag{7}$$

where f is a Lipschitz continuous sign-preserving real function,[13] $j = s_T^i$ is the previous period's action, and $R(k)$ is the regret for action k (as given in Section 4); again, the play in the first period is arbitrary. The following result is based on Hart and Mas-Colell (2001a, Sections 3.2 and 5.1) and proved in Cahn (2004, Theorem 4.1):

Theorem 2. Generalized Regret Matching: *Let each player play a generalized regret matching strategy.*[14] *Then the joint distribution of play converges to the set of correlated equilibria of the stage game.*

In fact, the full class of generalized regret matching strategies (for which Theorem 2 holds) is even larger; see Section 10.3. In particular, one may use a different $f_{k,j}$ for each pair $k \neq j$, or allow $f_{k,j}$ to depend on the whole vector of regrets and not just on the kth regret.

As a special case, consider the family of functions $f(x) = cx^r$, where $r \geq 1$ and $c > 0$ is an appropriate constant.[15] At one extreme, when $r = 1$, this is regret matching. At the other extreme, the limit as $r \to \infty$ is such that the switching probability $1 - \sigma(j)$ is equally divided among those

---

[13]There is L such that $|f(x) - f(y)| \leq L|x - y|$ for all x, y, and also $\sum_k f(R(k)) < 1$.
[14]Different players may use different such strategies.
[15]The condition $r \geq 1$ is needed for Lipschitz continuity.

actions $k \neq j$ with maximal regret (i.e., with $R(k) = \max_{\ell \neq j} R(\ell)$). This yields a variant of fictitious play, which however no longer satisfies the continuity requirement; therefore this strategy does not belong to our class and, indeed, the result of Theorem 2 does not hold for it (see Section 10.4). To regain continuity one needs to smooth it out, which leads to *smooth conditional fictitious play*; see Cahn (2004, Section 5).

8. Uncoupled Dynamics

At this point it is natural to ask whether there are adaptive heuristics that lead to *Nash equilibria* (the set of Nash equilibria being, in general, a strict subset of the set of correlated equilibria).

The answer is positive for *special* classes of games. For instance, two-person zero-sum games, two-person potential games, dominance-solvable games, and supermodular games are classes of games where fictitious play or general regret-based strategies make the marginal distributions of play converge to the set of Nash equilibria of the game (see Hofbauer and Sandholm 2002 and Hart and Mas-Colell 2003a for some recent work). But what about general games? Short of variants of exhaustive search (deterministic or stochastic),[16] there are no general results in the literature. Why is that so?

A natural requirement for adaptive heuristics (and adaptive dynamics in general) is that each player's strategy not depend on the payoff functions of the other players; this condition was introduced in Hart and Mas-Colell (2003b) and called *uncoupledness*. Thus, the strategy may depend on the actions of the other players—what they *do*—but not on their preferences—*why* they do it. This is an *informational* requirement: actions are observable, utilities are not. Almost all dynamics in the literature are indeed uncoupled: best-reply, better-reply, payoff-improving, monotonic, fictitious play, regret-based, replicator dynamics, and so on.[17] They all use the history of actions and determine the play as some sort of "good" reply to it, using only the player's own utility function.

---

[16] See Foster and Young (2003a, 2003b), Kakade and Foster (2004), Young (2004), Hart and Mas-Colell (2006), and Germano and Lugosi (2004).

[17] One example of a "non-uncoupled" dynamic is to compute a Nash equilibrium $\bar{x} = (\bar{x}^1, \bar{x}^2, \ldots, \bar{x}^N)$ and then to let each player i converge to $\bar{x}^i$; of course, the determination of $\bar{x}$ generally requires knowing *all* payoff functions.

Formally, we consider here dynamic systems in continuous time,[18] of the general form

$$\dot{x}(t) = F(x(t); \Gamma), \qquad (8)$$

where Γ is the game and the state variable is $x(t) = (x^1(t), x^2(t), \ldots, x^N(t))$, an N-tuple of (mixed) actions in $\Delta(S^1) \times \Delta(S^2) \times \cdots \times \Delta(S^N)$; equivalently, this may be written as

$$\dot{x}^i(t) = F^i(x(t); \Gamma) \quad \text{for each } i \in N, \qquad (9)$$

where $F = (F^1, F^2, \ldots, F^N)$. Various dynamics can be represented in this way; the variable $x^i(t)$ may be, for instance, the choice of i at time t, or the long-run average of his choices up to time t (see Hart and Mas-Colell 2003b, Footnote 3).

To state the condition of "uncoupledness," fix the set of players N and the action spaces $S^1, S^2, \ldots, S^N$; a game Γ is thus given by its payoff functions $u^1, u^2, \ldots, u^N$. We consider a family of games $\mathcal{U}$ (formally, a family of N-tuples of payoff functions $(u^1, u^2, \ldots, u^N)$). A general dynamic (9) is thus

$$\dot{x}^i(t) = F^i(x(t); u^1, u^2, \ldots, u^N) \quad \text{for each } i \in N.$$

We will call the dynamic $F = (F^1, F^2, \ldots, F^N)$ *uncoupled* on $\mathcal{U}$ if each F^i depends on the game Γ only through the payoff function u^i of player i, i.e.,

$$\dot{x}^i(t) = F^i(x(t); u^i) \quad \text{for each } i \in N.$$

That is, let Γ and Γ' be two games in the family $\mathcal{U}$ for which the payoff function of player i is the same (i.e., $u^i(\Gamma) = u^i(\Gamma')$); uncoupledness requires that, if the current state of play of all players, $x(t)$, is the same, then player i will adapt his action in the same way in the two games Γ and Γ'.

To study the impact of uncoupledness, we will deal with games that have *unique* Nash equilibria; this eliminates difficulties of coordination (different players may converge to different Nash equilibria). If there are dynamics that always converge to the set of Nash equilibria, we can apply them in particular when there is a unique Nash equilibrium.

---

[18]The results up to now on regret matching and generalized regret matching carry over to the continuous-time setup—see Section 10.6. For a discrete-time treatment of "uncoupledness," see Hart and Mas-Colell (2006).

We thus consider families of games $\mathcal{U}$ such that each game Γ in $\mathcal{U}$ possesses a unique Nash equilibrium, which we denote $\bar{x}(\Gamma)$. A dynamic F is *Nash-convergent* on $\mathcal{U}$ if, for each game Γ in $\mathcal{U}$, the unique Nash equilibrium $\bar{x}(\Gamma)$ is a rest-point of the dynamic (i.e., $F(\bar{x}(\Gamma); \Gamma) = 0$), which is moreover stable for the dynamic (i.e., $\lim_{t \to \infty} x(t) = \bar{x}(\Gamma)$ for any solution $x(t)$ of (8); some regularity assumptions are used here to facilitate the analysis).

The result of Hart and Mas-Colell (2003b) is:

Theorem 3. Uncoupled Dynamics: *There exist no uncoupled dynamics that guarantee Nash convergence.*

It is shown that there are simple families of games $\mathcal{U}$ (in fact, arbitrarily small neighborhoods of a single game), such that every uncoupled dynamic on $\mathcal{U}$ is *not* Nash-convergent; i.e., the unique Nash equilibrium is unstable for every uncoupled dynamic. The properties of uncoupledness and Nash-convergence are thus incompatible, even on simple families of games (and thus, a fortiori, on any larger families).

It follows that there can be no uncoupled dynamics that always converge to the set of Nash equilibria, or, for that matter, to the convex hull of Nash equilibria (which is the set of publicly correlated equilibria), since, in our games, both sets consist of the single Nash equilibrium.

The result of Theorem 3 indicates why dynamics that are to some extent "adaptive" or "rational" *cannot* always lead to Nash equilibria (see the references in Hart and Mas-Colell 2003b, Section IV(d)). In contrast, correlated equilibria may be obtained by uncoupled dynamics, such as regret matching and the other adaptive heuristics of this paper.[19]

9. Summary

Our results may be summarized as follows:

1. *There* are *simple adaptive heuristics that always lead to correlated equilibria* (the Regret Matching Theorem in Section 4).

---

[19]This suggests a "Coordination Conservation Law": some form of coordination must be present, either in the equilibrium concept (such as correlated equilibrium) or, if not (as in the case of Nash equilibrium), then in the dynamics leading to it (see Hart and Mas-Colell 2003b). As a further illustration, consider the learning dynamics of Section 2.1; while they are usually uncoupled, the convergence to Nash equilibria is obtained there only under certain initial conditions—such as "beliefs that contain a grain of truth"—which are in fact a form of coordination.

2. *There is a large class of adaptive heuristics that always lead to correlated equilibria* (the Generalized Regret Matching Theorem in Section 7).

3. *There can be no adaptive heuristics that always lead to Nash equilibria, or to the convex hull of Nash equilibria* (the Uncoupled Dynamics Theorem in Section 8).

Taken together, these results establish a solid connection between the dynamic approach of adaptive heuristics and the static approach of correlated equilibria.

From a more general viewpoint, the results show how simple and far-from-rational behavior in the short run may well lead to fully rational outcomes in the long run. Adaptive heuristics are closely related to behavioral models of what people do, whereas correlated equilibria embody fully rational considerations (see Sections 5 and 6, respectively). Our results show that rational behavior, which has been at times quite elusive and difficult to exhibit in single acts, may nevertheless be obtained in the long run.[20]

In short, adaptive heuristics may serve as a natural bridge connecting "behavioral" and "rational" approaches.

9.1. *Directions of research*

There are many interesting questions that arise in connection with this research. We will mention a few.

First, we need to further understand the relations between dynamics and equilibria. Which equilibria are obtained from adaptive heuristics, and which are not? At this point, we only know that the joint distribution of play converges to the *set* of correlated equilibria, and that it converges to a *point* only when it is a pure Nash equilibrium.[21] We do not know if *all* correlated equilibria are obtained from adaptive heuristics, or if only a strict subset of them are; recall that the Uncoupled Dynamics Theorem implies that the set of limit points can be neither the set of Nash equilibria nor

---

[20] Aumann, in various lectures since the late nineties (e.g., Aumann 1997; see also his interview in Hart 2005), has argued that rationality should be examined in the context of *rules* rather than *acts*; "rational rules" (i.e., rules of behavior that are best when compared to other rules) may well lead to single acts that are not rational. Here, we argue that rationality should be examined also *in the long run*; single acts that are not rational may nevertheless generate long-run behavior that is rational.

[21] See Hart and Mas-Colell (2000, p. 1132, comment (4)).

its convex hull. A more refined question is to characterize which dynamics lead to which equilibria. Finally, one needs to understand the behavior of these dynamics not just in the limit, but also along the way.

Second, alternative notions of regret—in particular, those that are obtained by different ways of time-averaging, like discounting, and finite recall or memory—should be analyzed. For such an analysis of approachability, see Lehrer and Solan (2003).

Third, one needs to strengthen the ties between the behavioral, experimental, and empirical approaches on the one hand, and the theoretical and rational approaches on the other. Adaptive heuristics that arise from theoretical work may be tested in practice, and theoretical work may be based on the empirical findings.

Fourth, some of the focus needs to be shifted from Nash equilibria to the more general class of correlated equilibria—in both static and dynamic setups. Problems of coordination, correlation, and communication have to be studied extensively.

Finally, we emphasize again (see Section 4.1) that looking at what each player does separately—i.e., considering the mixed actions independently— misses much relevant information; one needs to look at the *joint* distribution of play.[22]

10. Additional Results

This final section is devoted to a number of additional results and discussions of related issues.

10.1. *Hannan consistency and the Hannan set*

The regret for action k has been defined relative to the previous period's action j. One may consider a rougher measure instead: the increase in average payoff, if any, were one to replace *all* past plays, and not just the j-plays, by k. We thus define the *unconditional regret* for action k as

$$\widetilde{R}(k) := [\widetilde{V}(k) - U]_+, \tag{10}$$

---

[22]Unfortunately, almost all the experimental and behavioral literature deals only with the marginal distributions. (Two books where the joint distribution appears are Suppes and Atkinson 1960 and Rapoport, Guyer, and Gordon 1976.)

where

$$\widetilde{V}(k) := \frac{1}{T} \sum_{t=1}^{T} u^i(k, s_t^{-i}) \tag{11}$$

and U is the average payoff (see (1)). *Unconditional regret matching* prescribes play probabilities at each period that are directly proportional to the vector of unconditional regrets; i.e.,

$$\sigma_{T+1}(k) := \frac{\widetilde{R}(k)}{\sum_{\ell=1}^{m} \widetilde{R}(\ell)} \quad \text{for each } k = 1, 2, \ldots, m$$

(of course, this applies only when there is some positive unconditional regret, i.e., $\widetilde{R}(k) > 0$ for some k; σ_{T+1} is arbitrary otherwise). Unlike with regret matching, here we do not use a constant proportionality factor c, but rather normalize the vector of unconditional regrets to get a probability vector.

A strategy of player i is said to be *Hannan-consistent* (following Hannan 1957[23]) if it guarantees, for any strategies of the other players, that all the unconditional regrets of i become nonnegative in the limit, i.e., $\widetilde{R}(k) \to 0$ (almost surely) as $T \to \infty$ for all $k = 1, 2, \ldots, m$. We have:

PROPOSITION 4: *Unconditional regret matching is Hannan-consistent. Moreover, if all players play unconditional regret matching, then the joint distribution of play converges to the Hannan set of the stage game.*

Proposition 4 is Theorem B in Hart and Mas-Colell (2000). The proof applies Blackwell's Approachability Theorem 5 (see the Appendix) to the m-dimensional vector of unconditional regrets $(\widetilde{R}(1), \widetilde{R}(2), \ldots, \widetilde{R}(m))$: the negative orthant is shown to be approachable, and unconditional regret matching is the corresponding Blackwell strategy.

The *Hannan set* (see Hart and Mas-Colell 2003a and Moulin and Vial 1978), like the set of correlated equilibria, consists of joint distributions of play (i.e., it is a subset of $\Delta(S)$).[24] In contrast to correlated equilibria, the requirement now is that no player can gain unilaterally by playing a *constant* action (regardless of his signal). The set of correlated equilibria is contained in the Hannan set (and the two sets coincide when every player

---

[23]Fudenberg and Levine (1995) call this "universal consistency."
[24]Hannan-consistency is a property of strategies in the *repeated* game, whereas the Hannan set is a concept defined for the *one-shot* game.

has at most two strategies); moreover, the Hannan distributions that are independent across players are precisely the Nash equilibria of the game.

Hannan-consistent strategies have been constructed by Hannan (1957), Blackwell (1956b) (see also Luce and Raiffa 1957, pp. 482–483), Foster and Vohra (1993, 1998), Fudenberg and Levine (1995), and Freund and Schapire (1999).[25] Many of these strategies are smoothed-out variants of fictitious play, which, by itself, is not Hannan-consistent; see Section 10.4.[26] For a general class of Hannan-consistent strategies, which includes the unconditional regret matching of Proposition 4—apparently the simplest Hannan-consistent strategy—as well as smooth fictitious play, see Section 10.3.

10.2. *Regret eigenvector strategies*

Returning to our ("conditional") setup where regrets are defined relative to the previous period's action, Blackwell's approachability (see the Appendix) leads to the following construction (see Hart and Mas-Colell 2000, Section 3).[27] Start by defining the *regret* $R(j, k)$ *from* j *to* k using the same formulas (1)–(4) of Section 4 for *every* pair $j \neq k$ (i.e., j need no longer be the previous period's action). Take as payoff vector the $m(m-1)$-dimensional vector of *signed regrets* (i.e., *before* taking the positive part $[\cdot]_+$ in (4)). The negative orthant turns out to be approachable, and the Blackwell strategy translates into playing at each stage a randomized action σ that satisfies

$$\sum_{k:k\neq j} \sigma(j)R(j,k) = \sum_{k:k\neq j} \sigma(k)R(k,j) \quad \text{for all } j = 1, 2, \ldots, m. \quad (12)$$

That is, σ is a "regret-invariant vector": for all j, the average regret *from* j equals the average regret *to* j. Equivalently, put $q(j,k) := cR(j,k)$ for $j \neq k$ and $q(j,j) := 1 - \sum_{k:k\neq j} q(j,k)$, where $c > 0$ is large enough to guarantee that $q(j,j) > 0$ for all j; then $Q = (q(j,k))_{j,k=1,2,\ldots,m}$ is a stochastic (or Markov) matrix and (12) is easily seen to be equivalent to $\sigma = \sigma Q$. Thus σ is a left eigenvector of Q (corresponding to the eigenvalue 1); or, regarding

---

[25] See also the references in Hart and Mas-Colell (2001a, footnote 6) for related work.
[26] The original strategy of Hannan (1957) essentially uses at each stage an average of the best replies to a small neighborhood of the distribution of the opponents' past play.
[27] Interestingly, similar "regrets" (on probabilistic forecasts) as well as formula (12) appear in the earlier work—not based on approachability—of Foster and Vohra (1998); see Section 10.9.

Q as the one-step transition probability matrix of a Markov chain, σ is an invariant vector of Q.

We will call a strategy satisfying (12) a *regret eigenvector* strategy. By comparison, regret matching is defined by $\sigma(k) = q(j, k)$, which amounts to using Q as a Markov one-step transition probability matrix (from the previous period's action j to the current period's action k).[28]

Hart and Mas-Colell (2000, Corollary to Theorem A) show that if every player plays a regret eigenvector strategy, then, again, the joint distribution of play converges almost surely to the set of correlated equilibria. While the proof of this result is much simpler than that of Theorem 1, we do not regard the regret eigenvector strategies as heuristics (since they require computing each period an eigenvector of a matrix Q that changes over time).

10.3. *Generalized regret matching strategies*

It is convenient to consider first the "unconditional" Hannan setup of Section 10.1. Hart and Mas-Colell (2001a) characterize the class of *generalized unconditional regret* strategies, which are Hannan-consistent, as follows. For each $k = 1, 2, \ldots, m$ there is a function f_k defined on the m-dimensional vector of signed regrets $x = (\widetilde{V}(1) - U, \widetilde{V}(2) - U, \ldots, \widetilde{V}(m) - U)$, such that f_k is continuous, $\sum_{k=1}^{m} x_k f_k(x) > 0$ for all $x \not\leq 0$, and the vector of functions $f = (f_1, f_2, \ldots, f_m)$ is integrable (i.e., there exists a continuously differentiable function $P : \mathbb{R}^m \to \mathbb{R}$, a "potential function," such that $f_k = \partial P / \partial x_k$ for all k, or f is the gradient ∇P of P). Finally, the strategy is given by $\sigma(k) = f_k(x) / \sum_{\ell=1}^{m} f_\ell(x)$ for each $k = 1, 2, \ldots, m$. (Unconditional regret matching is obtained when $P(x) = \sum_{k=1}^{m} ([x_k]_+)^2$.)[29] The proof is based on characterizing *universal* approachability strategies.

Next, in the "conditional" setup, the generalization proceeds in two steps. The first step yields *generalized regret eigenvector* strategies, which are obtained by replacing each $R(j, k)$ in (12) with a function $f_{j,k}(R)$ defined on the $m(m-1)$-dimensional vector of (signed) regrets R, such that the vector of $m(m-1)$ functions $(f_{j,k})_{j \neq k}$ is the gradient ∇P of a continuously differentiable potential function P with $x \cdot \nabla P(x) > 0$ for all $x \not\leq 0$ (again,

---

[28] If Q were constant over time, the Ergodic Theorem for Markov chains would imply that regret matching and regret eigenvector strategies lead to essentially the same long-run distribution of play. The proof of the Regret Matching Theorem in Hart and Mas-Colell (2000) shows that this also holds when Q changes over time as a function of the regrets.
[29] See Sandholm (2004) for related work in an evolutionary setup.

these strategies are too complex to be viewed as heuristics). In the second step, we dispose of the computation of eigenvectors in (12) and get the full class of *generalized regret matching* strategies: $\sigma(k) = cf_{j,k}(R)$ for $k \neq j$ and $\sigma(j) = 1 - \sum_{k:k \neq j} \sigma(k)$, where j is the previous period's action. (The strategies given by (7) in Section 7 correspond to the "separable" special case where $P(x) = \sum \sum_{j \neq k} F(x_{j,k})$ and $F = \int f$.) If every player uses such a strategy, convergence to the set of correlated equilibria is obtained. For precise statements and proofs, see Hart and Mas-Colell (2001a, Section 5.1) and Cahn (2004, Theorem 4.1).

10.4. *Fictitious play and variants*

Fictitious play is an extensively studied adaptive heuristic; it prescribes playing at each period a best reply to the distribution of the past play of the opponents. Now the action k is such a best reply if and only if k maximizes $\widetilde{V}(k)$ or, equivalently, the signed unconditional regret $\widetilde{V}(k) - U$ (see (11) and (10)). Thus fictitious play turns out to be a function of the regrets too; however, since choosing a maximizer does not yield a continuous function, fictitious play does not belong to the class of unconditional regret-based strategies of Section 10.3—and it is indeed not Hannan-consistent. Therefore some smoothing out is needed, as in the strategies mentioned at the end of Section 10.1.

Similarly, *conditional fictitious play* consists of playing at each period a best reply to the distribution of the play of the opponents in those periods where i played the same action j as in the previous period. Smoothing this out yields *smooth fictitious play eigenvector* strategies (see Fudenberg and Levine 1998, 1999a) and *smooth conditional fictitious play* (see Cahn 2004, Section 4.5), which lead to the set of correlated approximate equilibria; for a discussion of the reason that one gets only approximate equilibria, see Hart and Mas-Colell (2001a, Section 4.1).

10.5. *The case of the unknown game*

Consider now the case where player i knows initially only his set of actions S^i, and is informed, after each period of play, of his realized payoff.[30] He does not know what game he is playing: how many players there are and

---

[30]Following a suggestion of Dean Foster; see Foster and Vohra (1993).

what their actions and payoffs are. In particular, he does not know his own payoff function—but only the payoffs he did actually receive every period. Thus at time $T + 1$ he knows the T numbers $u^i(s_1), u^i(s_2), \ldots, u^i(s_T)$; in addition, he recalls what he did in the past (i.e., his actual actions, $s_1^i, s_2^i, \ldots, s_T^i$ in S^i, and the probabilities that he used, $\sigma_1^i, \sigma_2^i, \ldots, \sigma_T^i$ in $\Delta(S^i)$). This is essentially a standard stimulus-response setup.

At each period the player can compute his realized average payoff U, but he cannot compute his regrets $R(k)$ (see (2) and (4)): he knows neither what the other players did (i.e., s_t^{-i}) nor what his payoff would have been had he played k instead (i.e., $u^i(k, s_t^{-i})$). We therefore define the *proxy regret* $\widehat{R}(k)$ for action k by using the payoffs he got when he did actually play k:

$$\widehat{R}(k) := \left[\frac{1}{T} \sum_{t \leq T : s_t^i = k} \frac{\sigma_t^i(j)}{\sigma_t^i(k)} u^i(s_t) - \frac{1}{T} \sum_{t \leq T : s_t^i = j} u^i(s_t) \right]_+$$

(the normalizing factor $\sigma_t^i(j)/\sigma_t^i(k)$ is needed, roughly speaking, to offset the possibly unequal frequencies of j and k being played in the past).

In Hart and Mas-Colell (2001b) it is shown that convergence to correlated approximate equilibria is obtained also for *proxy regret matching* strategies.

10.6. *Continuous time*

The regret-based dynamics up to this point have been *discrete-time* dynamics: the time periods were $t = 1, 2, \ldots$. It is natural to study also *continuous-time* models, where the time t is a continuous variable and the change in the players' actions is governed by appropriate differential equations. It turns out that the results carry over to this framework (in fact, some of the proofs become simpler). See Hart and Mas-Colell (2003a) for details.

10.7. *Computing regrets*

The regrets, despite depending on the whole history, are easy to compute. A player needs to keep record only of his $m(m - 1)$ signed regrets $D_T(j, k)$ (one for each $j \neq k$) and the "calendar" time T. The updating is simply $D_T(j, k) = (1 - 1/T) D_{T-1}(j, k) + (1/T)(u^i(k, s_T^{-i}) - u^i(s_T))$ for $j = s_T^i$ and $D_T(j, k) = (1 - 1/T) D_{T-1}(j, k)$ for $j \neq s_T^i$, and the regrets at time T are $R_T(k) = [D_T(s_T^i, k)]_+$.

10.8. *Convergence*

The convergence of the joint distributions play z_T to the set of correlated equilibria CE, i.e., $z_T \to$ CE (a.s.) as $T \to \infty$, means that[31]

$$\text{dist}(z_T, \text{CE}) \xrightarrow[T \to \infty]{} 0 \quad (\text{a.s.}).$$

That is, the sequence z_T eventually enters any neighborhood of the set CE, and stays there forever: for every $\varepsilon > 0$ there is a time $T_0 \equiv T_0(\varepsilon)$ such that for each $T > T_0$ there is a correlated equilibrium within ε of z_T; i.e., there is $\psi_T \in$ CE with $\|z_T - \psi_T\| < \varepsilon$. Since the players randomize, all of the above are random variables, and all statements hold with probability 1 (i.e., for almost every history); in particular, T_0 and ψ_T depend on the history.

An equivalent way of stating this is as follows. Given $\varepsilon > 0$, there is a time $T_1 \equiv T_1(\varepsilon)$ after which the joint distribution of play is always a correlated ε-equilibrium; i.e., z_T satisfies the correlated equilibrium constraints (see (6)) within ε, for all $T > T_1$.

As for the rate of convergence, it is essentially of the order of $1/\sqrt{T}$; see the Proof of the Approachability Theorem in the Appendix and, for more precise recent bounds, Cesa-Bianchi and Lugosi (2003) and Blum and Mansour (2005).

10.9. *A summary of strategies*

At this point the reader may well be confused by the plethora of regret-based strategies that have been presented above. We therefore provide a summary, with precise references, in Table 1.

An important additional dynamic leading to correlated equilibria is the *calibrated learning* of Foster and Vohra (1997). Here each player computes "calibrated forecasts" on the behavior of the other players, and then plays a best reply to these forecasts. Forecasts are *calibrated* if, roughly speaking, the probabilistic forecasts and the long-run frequencies are close: for example, it must have rained on approximately 75% of all days for which the forecast was "a 75% chance of rain" (and the same holds when replacing 75% with any other percentage). There are various ways to generate

---

[31]The distance between a point x and a set A is $\text{dist}(x, A) := \inf_{a \in A} \|x - a\|$.

Table 1. Regret-based strategies.

Correlated Equilibria (conditional setup)	Hannan Consistency (unconditional setup)
Regret matching[a] Regret eigenvector[b]	Unconditional regret matching[c]
Generalized regret matching[d] Generalized regret eigenvector[e]	Generalized unconditional regret matching[f]
Conditional fictitious play[g]	Fictitious play[h]
Smooth conditional fictitious play[i] Smooth fictitious play eigenvector[j]	Smooth fictitious play[k]
Proxy regret matching[l]	Proxy unconditional regret matching[m]
Continuous-time regret matching[n]	Continuous-time unconditional regret matching[o]

[a]Section 4; Hart and Mas-Colell (2000, Main Theorem).
[b]Section 10.2; Hart and Mas-Colell (2000, Theorem A).
[c]Section 10.1; Hart and Mas-Colell (2000, Theorem B).
[d]Sections 7 and 10.3; Hart and Mas-Colell (2001a, Section 5.1), Cahn (2004, Theorem 4.1).
[e]Section 4; Hart and Mas-Colell (2001a, Section 5.1).
[f]Section 10.3; Hart and Mas-Colell (2001b, Theorem 3.3).
[g]Section 10.4; it does *not* converge to the set of correlated equilibria.
[h]Section 10.4; it is *not* Hannan-consistent.
[i]Section 10.4; Cahn (2004, Proposition 4.3).
[j]Section 10.4; Fudenberg and Levine (1998, 1999a).
[k]Section 10.4; Fudenberg and Levine (1995).
[l]Section 10.5; Hart and Mas-Colell (2001b).
[m]Section 10.5; Hart and Mas-Colell (2000, Section 4(j); 2001a, Section 5.3).
[n]Section 10.6; Hart and Mas-Colell (2003a).
[o]Section 10.6; Hart and Mas-Colell (2003a).

calibrated forecasts; see Foster and Vohra (1997, 1998, 1999), Foster (1999), Fudenberg and Levine (1999b), and Kakade and Foster (2004).[32]

There is also a significant body of work in the computer science literature (where conditional regrets are called "internal regrets," and unconditional ones, "external"), with connections to machine learning, on-line prediction, experts, classification, perceptrons, and so on. For a recent study (and earlier references), see Cesa-Bianchi and Lugosi (2003).

---

[32]One construction, due to Hart and Mas-Colell, uses approachability to prove "no regret"; see Foster and Vohra (1999, Section 2). Interestingly, calibration is also closely related to the "merging of opinions" that arises in the study of the learning dynamics of Section 2.1; see Kalai, Lehrer, and Smorodinsky (1999).

10.10. *Variable game*

Even if the one-shot game changes every period (as in stochastic games), our results—that the regrets converge to zero—continue to hold, provided that the payoffs are uniformly bounded and the players are told at the end of each period which game has been played. This follows easily from our proofs (replace u^i by u^i_t throughout) and is related to the "universality" of the regret-based strategies; see Fudenberg and Levine (1998, Chapter 4, footnote 19) and Hart and Mas-Colell (2001a, Section 5.2).

10.11. *The set of correlated equilibria*

Correlated equilibria always exist in finite games. This follows from the existence of Nash equilibria (which requires fixed-point arguments), or directly (by linear duality arguments; see Hart and Schmeidler 1989[33] and Nau and McCardle 1990).

A natural question is, how large (or small) is the set of correlated equilibria? An interesting result in this direction is provided by Keiding and Peleg (2000). Fix the number of players N, the action sets $S^1, S^2, \ldots, S^N$, and a bound on the possible payoffs M. If one chooses at random (uniformly) a game Γ and a joint distribution of play $z \in \Delta(S)$, then the probability that z is a correlated equilibrium of Γ is at most $1/2^N$ (which goes to zero as N increases).

There is also work on the structure of the set of correlated equilibria; see Evangelista and Raghavan (1996), Myerson (1997), Nau, Gomes Canovas, and Hansen (2004), Calvó-Armengol (2004), and Nitzan (2005). For some recent results on the computation of correlated equilibria, see Kakade, Kearns, Langford, and Ortiz (2003) and Papadimitriou (2005).

Appendix. Approachability

A most useful technical tool in this area is the approachability theory originally introduced by Blackwell (1956a). The setup is that of games where the payoffs are *vectors* (rather than, as in standard games, scalar real numbers). For instance, the coordinates may represent different commodities; or contingent payoffs in different states of the world (when there is incomplete information—see Aumann and Maschler (1995, Section I.6 and postscript

---

[33]The starting point for the research presented here was the application of fictitious play to the auxiliary two-person zero-sum game of Hart and Schmeidler (1989); see Hart and Mas-Colell (2000, Section 4(i)).

I.d)); or, as in the current setup, regrets for the various actions in a standard game.

Let thus $A : S \equiv S^i \times S^{-i} \to \mathbb{R}^m$ be the payoff function of player i, where $\mathbb{R}^m$ denotes the m-dimensional Euclidean space (thus $A(s^i, s^{-i}) \in \mathbb{R}^m$ is the payoff vector when player i chooses s^i and the other players, $-i$, choose s^{-i}), which is extended bilinearly to mixed actions, i.e., $A : \Delta(S^i) \times \Delta(S^{-i}) \to \mathbb{R}^m$. The time is discrete, $t = 1, 2, \ldots$, and let $s_t = (s_t^i, s_t^{-i}) \in S^i \times S^{-i}$ be the actions chosen by i and $-i$, respectively, at time t, with payoff vector $a_t := A(s_t)$; put $\bar{a}_T := (1/T) \sum_{t=1}^{T} a_t$ for the average payoff vector up to T.

Let $C \subset \mathbb{R}^m$ be a convex and closed set.[34] We define:

- The set C is *approachable* by player i (cf. Blackwell 1956a)[35] if there exists a strategy of i such that, no matter what the opponents $-i$ do, $\text{dist}(\bar{a}_T, C) \to 0$ almost surely as $T \to \infty$.

- The set C is *enforceable* by player i if there exists a mixed action σ^i in $\Delta(S^i)$ such that, no matter what the opponents $-i$ do, the one-shot vector payoff is guaranteed to lie in C; i.e., $A(\sigma^i, s^{-i}) \in C$ for all s^{-i} in S^{-i} (and so also $A(\sigma^i, \sigma^{-i}) \in C$ for all σ^{-i} in $\Delta(S^{-i})$).

Approachability is a notion in the long-run repeated game, whereas enforceability is a notion in the one-shot game.

We restate the result of Blackwell (1956a) as follows:

Theorem 5. Approachability:

(i) *A half-space H is approachable if and only if it is enforceable.*

(ii) *A convex set C is approachable if and only if every half-space H containing C is approachable.*

The statement of Theorem 5 seems like a standard convexity result (based on the fact that a closed convex set is the intersection of the half-spaces containing it). This is however not so, since the intersection of approachable sets need not be approachable. For a simple example, consider the game of Figure 4, where player i has two actions: T yields the payoff

---

[34] For nonconvex sets, see Blackwell (1956a), Vieille (1992), and Spinat (2002).

[35] Blackwell's definition requires in addition that the approachability be *uniform* over the strategies of the opponents; namely, for every $\varepsilon > 0$ there is $T_0 \equiv T_0(\varepsilon)$ such that $E[\text{dist}(\bar{a}_T, C)] < \varepsilon$ for all $T > T_0$ and all strategies of $-i$ (i.e., T_0 is independent of the strategy of $-i$). It turns out that for convex sets C this strengthening is always satisfied.

T	$(1,0)$
B	$(0,1)$

Fig. 4. A game with vector payoffs.

vector $(1,0)$, and B yields $(0,1)$ (the opponent $-i$ has only one action). The half-space $\{x = (x_1, x_2) \in \mathbb{R}^2 : x_1 \geq 1\}$ is approachable (by playing T), and so is the half-space $\{x : x_2 \geq 1\}$ (by playing B)—whereas their intersection $\{x : x \geq (1,1)\}$ is clearly not approachable. What the Approachability Theorem says is that if *all* half-spaces containing the convex set C are approachable, then, and only then, their intersection C is approachable.

Let $H = \{x \in \mathbb{R}^m : \lambda \cdot x \geq \rho\}$ be a half-space, where $\lambda \neq 0$ is a vector in $\mathbb{R}^m$ and ρ is a real number. Consider the game $\lambda \cdot A$ with scalar payoffs given by $\lambda \cdot A(s^i, s^{-i})$ (i.e., the linear combination of the m coordinates with coefficients λ). Then H is enforceable if and only if the minimax value of this game, $\mathrm{val}(\lambda \cdot A) = \max\min \lambda \cdot A(\sigma^i, \sigma^{-i}) = \min\max \lambda \cdot A(\sigma^i, \sigma^{-i})$, where the max is over $\sigma^i \in \Delta(S^i)$ and the min over $\sigma^{-i} \in \Delta(S^{-i})$, satisfies $\mathrm{val}(\lambda \cdot A) \geq \rho$. Given a convex set C, let φ be the "support function" of C, namely, $\varphi(\lambda) := \inf\{\lambda \cdot c : c \in C\}$ for all $\lambda \in \mathbb{R}^m$. Since, for every direction λ, only the minimal half-space containing C, i.e., $\{x : \lambda \cdot x \geq \varphi(\lambda)\}$, matters in (ii),[36] the Approachability Theorem may be restated as follows: C is approachable if and only if

$$\mathrm{val}(\lambda \cdot A) \geq \varphi(\lambda) \quad \text{for all } \lambda \in \mathbb{R}^m.$$

Proof of Theorem 5. (i) If the half-space $H = \{x \in \mathbb{R}^m : \lambda \cdot x \geq \rho\}$ is enforceable then there exists a mixed action $\sigma^i \in \Delta(S^i)$ such that $\lambda \cdot A(\sigma^i, \sigma^{-i}) \geq \rho$ for all $\sigma^{-i} \in \Delta(S^{-i})$. Player i, by playing σ^i every period, guarantees that $b_t := E[a_t | h_{t-1}]$, the expected vector payoff conditional on the history h_{t-1} of the previous periods, satisfies $\lambda \cdot b_t \geq \rho$. Put $\bar{b}_T := (1/T) \sum_{t=1}^{T} b_t$; then $\lambda \cdot \bar{b}_T \geq \rho$, or $\bar{b}_T \in H$; the Strong Law of Large Numbers[37] implies that almost surely $\bar{a}_T - \bar{b}_T \to 0$, and therefore $\bar{a}_T \to H$, as $T \to \infty$. Conversely, if H is not enforceable then $\mathrm{val}(\lambda \cdot A) < \rho$.

---

[36] Trivially, a superset of an approachable set is also approachable.

[37] The version used in this proof is: $(1/T) \sum_{t=1}^{T} (X_t - E[X_t | h_{t-1}]) \to 0$ almost surely as $T \to \infty$, where the X_t are uniformly bounded random variables; see Loève (1978, Theorem 32.1.E). While player i's actions constitute an independent and identically distributed sequence, those of the other players may well depend on histories—and so

Therefore, given a strategy of player i, for every history h_{t-1} the other players can respond to player i's mixed action at t so that[38] $\lambda \cdot b_t \leq v$; hence $\lambda \cdot \bar{b}_T \leq v < \rho$ and $\bar{a}_T$ cannot converge to H.

(ii) The condition that every half-space containing C be approachable is clearly necessary. To see that it is also sufficient, for every history h_{t-1} such that $\bar{a}_{t-1} \notin C$, let $c \in C$ be the closest point to $\bar{a}_{t-1}$, and so $\delta_{t-1} := [\text{dist}(\bar{a}_{t-1}, C)]^2 = \|\bar{a}_{t-1} - c\|^2$. Put $\lambda := \bar{a}_{t-1} - c$ and $H := \{x : \lambda \cdot x \leq \lambda \cdot c\}$; then $C \subset H$ (since C is a convex set and λ is orthogonal to its boundary at c; see Figure 5). By assumption the half-space H is approachable, and thus enforceable; the *Blackwell strategy* prescribes to player i to play a mixed action $\sigma^i \in \Delta(S^i)$ corresponding to this[39] H—and so the expected payoff vector $b := E[a_t | h_{t-1}]$ satisfies $b \in H$, or[40] $\lambda \cdot b \leq \lambda \cdot c$. Now

$$\delta_t = [\text{dist}(\bar{a}_t, C)]^2 \leq \|\bar{a}_t - c\|^2 = \left\| \left(\frac{t-1}{t} \bar{a}_{t-1} + \frac{1}{t} a_t \right) - c \right\|^2$$

$$= \frac{(t-1)^2}{t^2} \|\bar{a}_{t-1} - c\|^2 + \frac{2(t-1)}{t^2} (\bar{a}_{t-1} - c) \cdot (a_t - c) + \frac{1}{t^2} \|a_t - c\|^2.$$

Taking expectation conditional on h_{t-1} yields in the middle term $\lambda \cdot (b - c)$, which is ≤ 0 by our choice of σ^i, and so

$$E[t^2 \delta_t | h_{t-1}] \leq (t-1)^2 \delta_{t-1} + M^2, \tag{13}$$

for some bound[41] M. Taking overall expectation and then using induction implies that $E[t^2 \delta_t] \leq M^2 t$; hence $E[\text{dist}(\bar{a}_t, C)] = E[\sqrt{\delta_t}] \leq \sqrt{E[\delta_t]} \leq M/\sqrt{t}$ and so $\bar{a}_t \to C$ in probability.

To get almost sure convergence,[42] put $\zeta_t := t\delta_t - (t-1)\delta_{t-1}$. Then $E[\zeta_t | h_{t-1}] \leq -(1 - 1/t)\delta_{t-1} + M^2/t \leq M^2/t \to 0$ (by (13)) and $|\zeta_t| \leq M$ (since $|\delta_t - \delta_{t-1}| \leq \|\bar{a}_t - \bar{a}_{t-1}\| = (1/t)\|a_t - \bar{a}_{t-1}\|$), from which it follows

---

we need a Strong Law of Large Numbers for *dependent* random variables (essentially, a Martingale Convergence Theorem).

[38] These responses may be chosen pure—which shows that whether or not the opponents $-i$ can correlate their actions is irrelevant for approachability; see Hart and Mas-Colell (2001a, footnote 12).

[39] When $\bar{a}_{t-1} \in C$ take $\lambda = 0$ and an arbitrary σ^i.

[40] As can be seen in Figure 5, it follows that (the conditional expectation of) $\bar{a}_t$ is closer to C than $\bar{a}_{t-1}$ is. The computation below will show that the distance to C not only decreases but, in fact, converges to zero.

[41] Take $M := \max_{b,b' \in B} |b - b'| + \max_{b \in B} \text{dist}(b, C)$, where B is the compact set $B := \text{conv}\{A(s) : s \in S\}$.

[42] See also Mertens, Sorin, and Zamir (1995, Proof of Theorem 4.3) or Hart and Mas-Colell (2001a, Proof of Lemma 2.2).

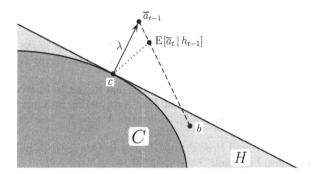

Fig. 5. The Blackwell strategy for the set C.

that $\delta_T = (1/T)\sum_{t=1}^{T} \zeta_t \to 0$ almost surely (by the Strong Law of Large Numbers; see Footnote 38). □

Historically, Blackwell (1956b) used the Approachability Theorem to provide an alternative proof of the Hannan (1957) result (see Rustichini 1999 for an extension); the proof was indirect and nonconstructive. The direct application of approachability to regrets, with the vector of regrets as payoff vector and the negative orthant as the approachable set, was introduced in the 1996 preprint of Hart and Mas-Colell (2000) (see Section 3 there). This eventually led to the simple regret matching strategy of Section 4, to the universal approachability strategies and the generalized regret matching of Section 7, and to the other results presented here—as well as to various related uses of approachability (for example, Foster 1999; Foster and Vohra 1999; Lehrer 2001, 2002, 2003; Sandroni, Smorodinsky, and Vohra 2003; Cesa-Bianchi and Lugosi 2003; Greenwald and Jafari 2003; and Lehrer and Solan 2003).

References

Aumann, R. J. (1974), "Subjectivity and Correlation in Randomized Strategies," *Journal of Mathematical Economics*, 1, 67–96.

———(1987), "Correlated Equilibrium as an Expression of Bayesian Rationality," *Econometrica*, 55, 1–18.

———(1997), "Rationality and Bounded Rationality," in *Cooperation: Game Theoretic Approaches*, ed. by S. Hart and A. Mas-Colell. Berlin: Springer, pp. 219–232.

Aumann R. J. and M. Maschler (1995), *Repeated Games of Incomplete Information*. Cambridge, MA: MIT Press.

Bell, D. E. (1982), "Regret in Decision Making under Uncertainty," *Operations Research*, 30, 961–981.

Blackwell, D. (1956a), "An Analog of the Minmax Theorem for Vector Payoffs," *Pacific Journal of Mathematics*, 6, 1–8.

——— (1956b), "Controlled Random Walks," in *Proceedings of the International Congress of Mathematicians 1954*, Vol. III. Amsterdam: North-Holland, pp. 335–338.

Blum, A. and Y. Mansour (2005), "From External Regret to Internal Regret," in *Learning Theory*, ed. by P. Auer and R. Meir. Berlin: Springer, pp. 621–636.

Bush, R. R. and F. Mosteller (1955), *Stochastic Models for Learning*. New York: Wiley.

Cahn, A. (2004), "General Procedures Leading to Correlated Equilibria," *International Journal of Game Theory*, 33, 21–40. [Chapter 6]

Calvó-Armengol, A. (2004), "The Set of Correlated Equilibria of 2 × 2 Games," Universitàt Autonoma de Barcelona (mimeo).

Camerer, C. and T.-H. Ho (1998), "Experience-Weighted Attraction Learning in Coordination Games: Probability Rules, Heterogeneity, and Time-Variation," *Journal of Mathematical Psychology*, 42, 305–326.

——— (1999), "Experience-Weighted Attraction Learning in Normal Form Games," *Econometrica*, 67, 827–874.

Camille, N., G. Coricelli, J. Sallet, P. Pradat-Diehl, J.-R. Duhamel, and A. Sirigu (2004), "The Involvement of the Orbitofrontal Cortex in the Experience of Regret," *Science*, 304, 1167–1170.

Cesa-Bianchi, N. and G. Lugosi (2003), "Potential-Based Algorithms in On-Line Prediction and Game Theory," *Machine Learning*, 51, 239–261.

Davies, N. B. (1978), "Territorial Defense in the Speckled Wood Butterfly *Pararge aegeria*: The Resident Always Wins," *Animal Behavior*, 26, 138–147.

Erev, I. and A. E. Roth (1998), "Predicting How People Play Games: Reinforcement Learning in Experimental Games with Unique, Mixed Strategy Equilibria," *American Economic Review*, 88, 848–881.

Evangelista, F. and T. E. S. Raghavan (1996), "A Note on Correlated Equilibrium," *International Journal of Game Theory*, 25, 35–41.

Foster, D. P. (1999), "A Proof of Calibration via Blackwell's Approachability Theorem," *Games and Economic Behavior*, 29, 73–78.

Foster, D. P. and R. V. Vohra (1993), "A Randomization Rule for Selecting Forecasts," *Operations Research*, 41, 704–709.

——— (1997), "Calibrated Learning and Correlated Equilibrium," *Games and Economic Behavior*, 21, 40–55.

——— (1998), "Asymptotic Calibration," *Biometrika*, 85, 379–390.

——— (1999), Regret in the On-Line Decision Problem, *Games and Economic Behavior*, 29, 7–35.

Foster, D. P. and H. P. Young (2003a), "Learning, Hypothesis Testing, and Nash Equilibrium," *Games and Economic Behavior*, 45, 73–96.

——— (2003b), "Regret Testing: A Simple Payoff-Based Procedure for Learning Nash Equilibrium," University of Pennsylvania and Johns Hopkins University (mimeo).

Freund, Y. and R. E. Shapire (1999), "Adaptive Game Playing Using Multiplicative Weights," *Games and Economic Behavior*, 29, 79–103.

Fudenberg, D. and D. Kreps (1988), A Theory of Learning, Experimentation, and Equilibrium in Games, Stanford University (mimeo).

―――― (1993), "Learning Mixed Equilibria," *Games and Economic Behavior*, 5, 320–367.

Fudenberg, D. and D. Levine (1995), "Consistency and Cautious Fictitious Play," *Journal of Economic Dynamics and Control*, 19, 1065–1090.

―――― (1998): *Theory of Learning in Games*, Cambridge, MA: MIT Press.

―――― (1999a), "Conditional Universal Consistency," *Games and Economic Behavior*, 29, 104–130.

―――― (1999b), "An Easier Way to Calibrate," *Games and Economic Behavior*, 29, 131–137.

Greenwald, A. and A. Jafari (2003), "A General Class of No-Regret Algorithms and Game-Theoretic Equilibria," in *Learning Theory and Kernel Machines* (*Lecture Notes on Artificial Intelligence* 2777), ed. by J. G. Carbonell and J. Siekmann. Berlin: Springer, pp. 2–12.

Hammerstein, P. and R. Selten (1994), "Game Theory and Evolutionary Biology," in *Handbook of Game Theory, with Economic Applications*, Vol. 2, ed. by R. J. Aumann and S. Hart. Amsterdam: Elsevier, pp. 929–993.

Hannan, J. (1957), "Approximation to Bayes Risk in Repeated Play," in *Contributions to the Theory of Games*, Vol. III (*Annals of Mathematics Studies* 39), ed. by M. Dresher, A. W. Tucker, and P. Wolfe. Princeton: Princeton University Press, pp. 97–139.

Harsanyi, J. C. (1967–1968), "Games with Incomplete Information Played by Bayesian Players," Parts I, II, III, *Management Science*, 14, 159–182, 320–334, 486–502.

Hart, S. (2005), "An Interview With Robert Aumann," *Macroeconomic Dynamics*, 5, 683–740.

Hart, S. and A. Mas-Colell (2000), "A Simple Adaptive Procedure Leading to Correlated Equilibrium," *Econometrica*, 68, 1127–1150. [Chapter 2]

―――― (2001a), "A General Class of Adaptive Strategies," *Journal of Economic Theory*, 98, 26–54. [Chapter 3]

―――― (2001b), "A Reinforcement Procedure Leading to Correlated Equilibrium," in *Economic Essays: A Festschrift for Werner Hildenbrand*, eds. by G. Debreu, W. Neuefeind and W. Trockel. Berlin: Springer, pp. 181–200. [Chapter 4]

―――― (2003a), "Regret-Based Continuous-Time Dynamics," *Games and Economic Behavior*, 45, 375–394. [Chapter 5]

―――― (2003b), "Uncoupled Dynamics Do Not Lead to Nash Equilibrium," *American Economic Review*, 93, 1830–1836. [Chapter 7]

―――― (2006), "Stochastic Uncoupled Dynamics and Nash Equilibirum," *Games and Economic Behavior*, 57, 286–303. [Chapter 8]

Hart, S. and D. Schmeidler (1989), "Existence of Correlated Equilibria," *Mathematics of Operations Research*, 14, 18–25. [Chapter 1]

Hofbauer, J. and W. H. Sandholm (2002), "On the Global Convergence of Stochastic Fictitious Play," *Econometrica*, 70, 2265–2294.

Hofbauer, J. and K. Sigmund (1998), *Evolutionary Games and Population Dynamics.* Cambridge, UK: Cambridge University Press.

Kahneman, D., P. Slovic, and A. Tversky (eds.) (1982), *Judgement under Uncertainty: Heuristics and Biases.* Cambridge, UK: Cambridge University Press.

Kakade, S. and D. P. Foster (2004), "Deterministic Calibration and Nash Equilibrium," *Lecture Notes in Computer Science,* 3120, 33–48.

Kakade, S., M. Kearns, J. Langford, and L. Ortiz (2003), "Correlated Equilibria in Graphical Games," *Proceedings of the Fourth ACM Conference on Electronic Commerce.* New York: ACM Press, pp. 42–47.

Kalai, E. and E. Lehrer (1993), "Rational Learning Leads to Nash Equilibrium," *Econometrica,* 61, 1019–1045.

Kalai, E., E. Lehrer, and R. Smorodinsky (1999), "Calibrated Forecasting and Merging," *Games and Economic Behavior,* 29, 151–169.

Keiding, H. and B. Peleg (2000), "Correlated Equilibria of Games with Many Players," *International Journal of Game Theory,* 29, 375–389.

Lehrer, E. (2001), "Any Inspection is Manipulable," *Econometrica,* 69, 1333–1347.

———— (2002), "Approachability in Infinitely Dimensional Spaces," *International Journal of Game Theory,* 31, 255–270.

———— (2003), "A Wide Range No-Regret Theorem," *Games and Economic Behavior,* 42, 101–115.

Lehrer, E. and E. Solan (2003), No-Regret with Bounded Computational Capacity, Tel Aviv University (mimeo).

Loève, M. (1978), *Probability Theory,* Vol. II, 4th Edn. Berlin: Springer.

Loomes, G. and R. Sugden (1982), "Regret Theory: An Alternative Theory of Rational Choice under Uncertainty," *Economic Journal,* 92, 805–824.

Luce, R. D. and H. Raiffa (1957), *Games and Decisions.* New York: Wiley.

Mertens, J.-F., S. Sorin, and S. Zamir (1995), "Repeated Games," Part A, CORE DP-9420, Université Catholique de Louvain (mimeo).

Moshinsky, A. (2003), "The Status-Quo Bias in Policy Judgements," Ph.D. Thesis, The Hebrew University of Jerusalem (mimeo).

Moulin, H. and J. P. Vial (1978), "Strategically Zero-Sum Games: The Class of Games Whose Completely Mixed Equilibria Cannot Be Improved upon," *International Journal of Game Theory,* 7, 201–221.

Myerson, R. B. (1997), "Dual Reduction and Elementary Games," *Games and Economic Behavior,* 21, 183–202.

Nau R., S. Gomez Canovas, and P. Hansen (2004), "On the Geometry of Nash Equilibria and Correlated Equilibria," *International Journal of Game Theory,* 32, 443–453.

Nau R. and K. F. McCardle (1990), "Coherent Behavior in Noncooperative Games," *Journal of Economic Theory,* 50, 424–444.

Nitzan, N. (2005), Tight Correlated Equilibrium, DP-394, Center for the Study of Rationality, The Hebrew University of Jerusalem (mimeo).

Nyarko, Y. (1994), "Bayesian Learning Leads to Correlated Equilibria in Normal Form Games," *Economic Theory,* 4, 821–841.

Papadimitriou, C. H. (2005), "Computing Correlated Equilibria in Multi-Player Games," University of California at Berkeley (mimeo).

Rapoport, A., M. J. Guyer, and D. J. Gordon (1976), *The 2 × 2 Game*. Ann Arbor: University of Michigan Press.

Roth, A. E. and I. Erev (1995), "Learning in Extensive-Form Games: Experimental Data and Simple Dynamic Models in the Intermediate Term," *Games and Economic Behavior*, 8, 164–212.

Rustichini, A. (1999), "Optimal Properties of Stimulus-Response Learning Models," *Games and Economic Behavior*, 29, 244–273.

Samuelson, W. and R. Zeckhauser (1988), "Status Quo Bias in Decision Making," *Journal of Risk and Uncertainty*, 1, 7–59.

Sandholm, W. H. (2004), "Excess Payoff Dynamics, Potential Dynamics, and Stable Games," University of Wisconsin (mimeo).

Sandroni, A., R. Smorodinsky, and R. V. Vohra (2003), "Calibration with Many Checking Rules," *Mathematics of Operations Research*, 28, 141–153.

Schlag, K. H. (1998), "Why Imitate, and If So, How? A Bounded Rational Approach to Multi-Armed Bandits," *Journal of Economic Theory*, 78, 130–156.

Shapley, L. S. (1964), "Some Topics in Two-Person Games," in *Advances in Game Theory*, Annals of Mathematics Studies 52, ed. by M. Dresher, L. S. Shapley, and A. W. Tucker. Princeton: Princeton University Press, pp. 1–28.

Shmida, A. and B. Peleg (1997), "Strict and Symmetric Correlated Equilibria are the Distributions of the ESS's of Biological Conflicts with Asymmetric Roles," in *Understanding Strategic Interaction*, ed. by W. Albers, W. Güth, P. Hammerstein, B. Moldovanu, and E. van Damme. Berlin: Springer, pp. 149–170.

Spinat, X. (2002), "A Necessary and Sufficient Condition for Approachability," *Mathematics of Operations Research*, 27, 31–44.

Suppes, P. and R. C. Atkinson (1960), *Markov Learning Models for Multiperson Interactions*. Palo Alto: Stanford University Press.

Vieille, N. (1992), "Weak Approachability," *Mathematics of Operations Research*, 17, 781–791.

Weibull, J. W. (1995), *Evolutionary Game Theory*. Cambridge, UK: Cambridge University Press.

Young, H. P. (1998), *Individual Strategy and Social Structure*. Princeton: Princeton University Press.

—— (2004), *Strategic Learning and Its Limits*. Oxford, UK: Oxford University Press.

Chapter 12

NASH EQUILIBRIUM AND DYNAMICS*

Sergiu Hart

John F. Nash, Jr., submitted his Ph.D. dissertation entitled *Non-Cooperative Games* to Princeton University in 1950. Read it 58 years later, and you will find the germs of various later developments in game theory. Some of these are presented below, followed by a discussion of dynamic aspects of equilibrium.

1. Nash Equilibrium

What is a *Nash equilibrium*?[1] John Nash defines an *equilibrium point* as

"an n-tuple s such that each player's mixed strategy maximizes his pay-off if the strategies of the others are held fixed. Thus each player's strategy is optimal against those of the others." (page 3)

2. Non-cooperative Games

Nash emphasizes the *non-cooperative* aspect of his theory, in contrast to the "cooperative" approach of von Neumann and Morgenstern (1944) that was prevalent at the time:

"Our theory, in contradistinction, is based on the *absence* of coalitions in that it is assumed that each participant acts independently,

Originally published in *Games and Economic Behavior*, 71 (2011), 6–8.

*Presented at the Opening Panel of the Conference in Honor of John Nash's 80th Birthday at Princeton University in June 2008. Updated: October 2010.

JEL classification: C70; C72; C73

Keywords: Nash equilibrium; dynamics; non-cooperative games; uncoupled dynamics.
[1]I have seen "nash equilibrium" in print, as if "nash" were an English word. Having one's name spelled with a lower-case initial letter is surely a sign of lasting fame!

without collaboration or communication with any of the others. [...] The basic requirement for a non-cooperative game is that there should be no pre-play communication among the players [...]. Thus, by implication, there are no coalitions and no side-payments." (pages 1, 21)

3. The "Nash Program"

In the last section of his thesis, "Applications," Nash introduces what has come to be known as the *Nash program*, namely:

"a 'dynamical' approach to the study of cooperative games based upon reduction to non-cooperative form. One proceeds by constructing a model of the pre-play negotiation so that the steps of negotiation become moves in a larger non-cooperative game [...] describing the total situation. This larger game is then treated in terms of the theory of this paper [...]. Thus the problem of analyzing a cooperative game becomes the problem of obtaining a suitable, and convincing, non-cooperative model for the negotiation." (pp. 25–26)

Nash himself used this approach for the two-person bargaining problem in his 1953 paper. Since then, the Nash program has been applied to a large number of models (see, e.g., the surveys of Mas-Colell 1997 and Reny 1997). The whole area of *implementation*—discussed in this panel by Eric Maskin—may also be regarded as a successful offshoot of the Nash program.

4. The "Mass-Action" Interpretation

The next-to-last section of the dissertation is entitled "Motivation and Interpretation." Two interpretations of the concept of Nash equilibrium are provided. The first, the "mass-action" interpretation, assumes that

"there is a population [...] of participants for each position in the game." (p. 21)

This framework brings us to a significant area of research known as *evolutionary game theory*, with important connections to biology; it is discussed in this panel by Peyton Young.

5. The "Rational" Interpretation

The second interpretation of Nash equilibrium provided in the dissertation refers to

"a 'rational' prediction of the behavior to be expected of rational playing of the game [...]. In this interpretation we need to assume the players know the full structure of the game [...]. It is quite strongly a rationalistic and idealizing interpretation." (page 23)

With the development of the area known as *interactive epistemology*, which deals formally with the issues of knowledge and rationality in multi-player situations, one can now state precise conditions for Nash equilibria. For instance, to obtain Nash equilibria in pure strategies, it suffices for each player to be rational and to know his own payoff function as well as the choices of the pure strategies of the other players (see Aumann and Brandenburger 1995; for mixed strategies the conditions are more complicated).

6. Dynamics

Nash equilibrium is by definition a *static* concept. So what happens in dynamic setups where the players adjust their play over time? Since Nash equilibria are essentially "rest points," it is reasonable to expect that appropriate dynamical systems should lead to them.

However, after more than half a century of research, it turns out that

There are no general, natural dynamics leading to Nash equilibria. (∗)

Let us clarify the statement (∗). First, "general" means "for all games," rather than "for specific classes of games only." [2] Second, "leading to Nash equilibria" means that the process reaches Nash equilibria (or is close to them) from some time on. [3] Finally, what is a "natural dynamic"? While it is not easy to define the term precisely, there are certain clear requirements for a dynamic to be called "natural": it should be in some sense adaptive—i.e., the players should react to what happens, and move in generally improving

---

[2]Indeed, there are classes of games—in fact, no more than a handful—for which such dynamics have been found, e.g., two-player games, and potential games.

[3]Dynamics that are close to Nash equilibria *most of the time* have been constructed in particular by Foster and Young (2003) and recently by Young (2009). These dynamics will reach a neighborhood of equilibria and stay there for a long time, but always leave this neighborhood eventually and the process then restarts.

directions (this rules out deterministic and stochastic variants of "exhaustive search" where the players blindly search through all the possibilities); and it should be simple and efficient—in terms of how much information the players possess, how complex their computations are at each step, and how long it takes to reach equilibrium. What (∗) says is that, despite much effort, no such dynamics have yet been found.

A natural informational restriction is for each player to know initially only his own payoff (or utility) function, but not those of the other players; this is called *uncoupledness* (Hart and Mas-Colell 2003) and is a usual condition in many setups. However, it leads to various impossibility results on the existence of "natural" uncoupled dynamics leading to Nash equilibria (Hart and Mas-Colell 2003, 2006; Hart and Mansour 2010). Thus (∗) turns out to be not just a statement about the current state of research in game theory, but in fact a result:

There cannot be *general, natural dynamics leading to Nash equilibria.*

7. Correlated Equilibrium

A generalization of the concept of Nash equilibrium is the notion of *correlated equilibrium* (Aumann 1974): it is a Nash equilibrium of the extended game where each player may receive certain information (call it a "signal") before playing the game; these signals do not affect the game directly. Nevertheless, the players may well take these signals into account when deciding which strategy to play. When the players' signals are independent, this reduces to nothing more than Nash equilibria; when the signals are public, this yields a weighted average of Nash equilibria; and when the signals are correlated (but not fully), new equilibria obtain.

Interestingly, there *are* general, natural dynamics leading to correlated equilibria (such as "regret matching": Hart and Mas-Colell 2000, Hart 2005).

8. Summary

The dissertation that John Nash submitted in 1950 is relatively short (certainly by today's standards). But not in content! Besides the definition of the concept of non-cooperative equilibrium and a proof of its existence, one can find there intriguing and stimulating ideas that predate a lot of modern game theory.

References

Aumann, R. J. (1974), "Subjectivity and Correlation in Randomized Strategies," *Journal of Mathematical Economics*, 1, 67–96.

Aumann, R. J. and A. Brandenburger (1995), "Epistemic Conditions for Nash Equilibrium," *Econometrica*, 63, 116–180.

Foster, D. and H. P. Young (2003), "Learning, Hypothesis Testing, and Nash Equilibrium," *Games and Economic Behavior*, 45, 73–96.

Hart, S. (2005), "Adaptive Heuristics," *Econometrica*, 73, 1401–1430. [Chapter 11]

Hart, S. and Y. Mansour (2010), "How Long to Equilibrium? The Communication Complexity of Uncoupled Equilibrium Procedures," *Games and Economic Behavior*, 69, 107–126. [Chapter 10]

Hart, S. and A. Mas-Colell (2000), "A Simple Adaptive Procedure Leading to Correlated Equilibrium," *Econometrica*, 68, 1127–1150. [Chapter 2]

Hart, S. and A. Mas-Colell (2003), "Uncoupled Dynamics Do Not Lead to Nash Equilibrium," *American Economic Review*, 93, 1830–1836. [Chapter 7]

Hart, S. and A. Mas-Colell (2006), "Stochastic Uncoupled Dynamics and Nash Equilibrium," *Games and Economic Behavior*, 57, 286–303. [Chapter 8]

Mas-Colell, A. (1997), "Bargaining Games," in *Cooperation: Game Theoretic Approaches*, ed. by S. Hart and A. Mas-Colell. New York: Springer, pp. 69–90.

Nash, J. F. (1950), "Non-cooperative Games," Ph.D. Dissertation, Princeton University.

Nash, J. F. (1953), "Two-person Cooperative Games," *Econometrica*, 21, 128–140.

Reny, P. H. (1997), "Two Lectures on Implementation under Complete Information: General Results and the Core," in *Cooperation: Game Theoretic Approaches*, ed. by S. Hart and A. Mas-Colell. New York: Springer, pp. 91–113.

Von Neumann, J. and O. Morgenstern (1944), *Theory of Games and Economic Behavior*. Princeton: Princeton University Press.

Young, H. P. (2009), "Learning by Trial and Error," *Games and Economic Behavior*, 65, 626–643.

INDEX